Advances in

ENVIRONMENTAL SCIENCE AND TECHNOLOGY

Volume 2

Advances in ENVIRONMENTAL SCIENCE AND TECHNOLOGY

Edited by
JAMES N. PITTS, JR.
University of California
Riverside, California

and

ROBERT L. METCALF
University of Illinois
Urbana, Illinois

Volume 2

WILEY-INTERSCIENCE
A DIVISION OF JOHN WILEY & SONS, INC.
NEW YORK / LONDON / SYDNEY / TORONTO

Library of Congress Catalog Card Number: 69–18013

ISBN 0-471-69085-6

Printed in the United States of America.

10 9 8 7 6 5 4 3 2 1

INTRODUCTION TO THE SERIES

Advances in Environmental Science and Technology is a series of multiauthored books devoted to the study of the quality of the environment and to the technology of its conservation. Environmental sciences relate, therefore, to the chemical, physical, and biological changes in the environment through contamination or modification; to the physical nature and biological behavior of air, water, soil, food, and waste as they are affected by man's agricultural, industrial, and social activities; and to the application of science and technology to the control and improvement of environmental quality.

The deterioration of environmental quality, which began when man first assembled into villages and utilized fire, has existed as a serious problem since the industrial revolution. In the second half of the twentieth century, under the ever-increasing impacts of exponentially growing population and of industrializing society, environmental contamination of air, water, soil, and food has become a threat to the continued existence of many plant and animal communities of the ecosytem and may ultimately threaten the very survival of the human race.

It seems clear that if we are to preserve for future generations some semblance of the existing biological order and if we hope to improve on the deteriorating standards of urban public health, environmental sciences and technology must quickly come to play a dominant role in designing our social and industrial structure for tomorrow. Scientifically rigorous criteria of environmental quality must be developed and, based in part on these, realistic standards must be established, so that our technological progress can be tailored to meet such standards. Civilization will continue to require increasing amounts of fuels, transportation, industrial chemicals, fertilizers, pesticides, and countless other products, as well as to produce waste products of all descriptions. What is urgently needed is a total systems approach to modern civilization through which the pooled talents of scientists and engineers, in cooperation with social scientists and the medical profession, can be focused on the development of order and equilibrium among the presently disparate segments of the human environment. Most of the skills and tools that are needed already exist. Surely a technology that has created manifold environmental problems is also capable of solving them. It is our hope that the series in Environmental Sciences and Technology will not only serve to make this challenge more explicit to the established professional but will also help to stimulate the student toward the career opportunities in this vital area.

Finally, the chapters in this series of Advances are written by experts in their respective disciplines, who also are involved with the broad scope of environmental science. As editors, we asked the authors to give their "points of view" on key questions; we were not concerned simply with literature surveys. They have responded in a gratifying manner with thoughtful and challenging statements on critical environmental problems.

James N. Pitts, Jr.
Robert L. Metcalf

Contents

Advances in

ENVIRONMENTAL SCIENCE AND TECHNOLOGY

Volume 2

Air Pollution: Present and Future Threat to Man and His Environment

DAVID L. COFFIN, V.M.D.
Health Effects Research Division,
Air Pollution Control Office, EPA, and
School of Medicine,
University of Cincinnati,
Cincinnati, Ohio

Man has been subject to air pollution since his primordial ancestor lit the first fire. It was not, however, until people became crowded together in cities that pollution was more than a family problem associated with smoke from the hearth. With the coming of the use of coal for heating and the Industrial Revolution, the problem became intensified. Today, with the phenomenal growth of both the population and the use of power in the technologically advanced countries, pollution has reached such magnitude that it not only threatens the health and well-being of the population in a particular locality, but also produces effects on a global scale. It can now truly be said that man not only pollutes his own house but also has the ability to foul the atmosphere of the entire world.

The root cause of air pollution and, in fact, water pollution, land pollution, and other forms of deterioration of the environment is "more." In terms of air quality deterioration, there are three "mores" of particular significance: (*1*) more people, (*2*) more urbanization, and (*3*) more technology.

It has been predicted that by the year 2000 the United States will contain 320 million people, 85% of whom will be in cities concentrated on 10% of the total land area. As we all know, the increase of the human population has been almost incredible. On a global basis, the population doubles each 30 years. Furthermore, worldwide, people tend to concentrate in cities. Obviously the greater the concentration of people in a given area, the greater the problems created locally by the by-products of their living.

Such a concentration of people through urbanization (see Fig. 1) is also proceeding at an accelerated rate even in the less technologically advanced countries where sociological trends and agricultural practices appear to make it inevitable. In our own country the development of huge industrial-urban complexes encompassing several states is so well known that it scarcely needs mentioning. From the standpoint of air pollution, such complexes compound the situation because in these areas numerous local pollutant sources contribute to a continuum of

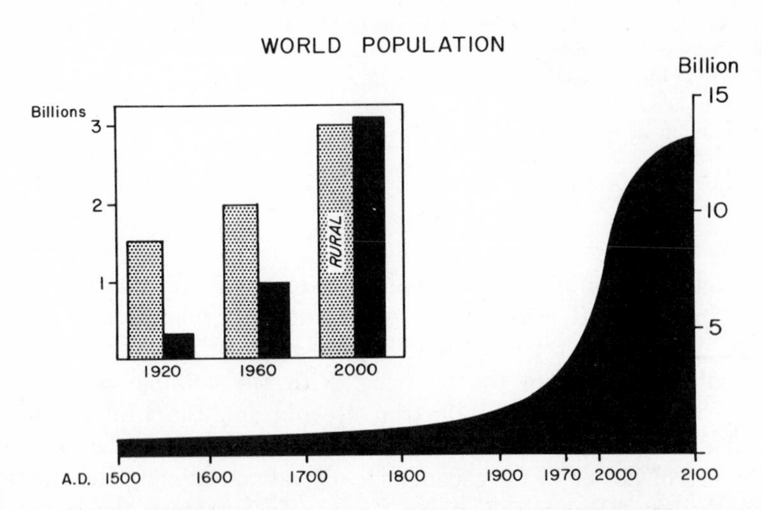

Fig. 1. Population growth curve for world human population with concomitant trend for urbanization indicated. Courtesy of *Natural History Mag.*, American Museum of Natural History, New York, January 1970, p. 62.

discharge of noxious material into the atmosphere. Thus pollutants accumulate despite considerable movement by wind and by diffusion into the upper air.

By and large, the major pollutants are products of combustion, including domestic fires and combustion for various technological processes, such as smelting, roasting, desalting, and ignition of fuels for the production of power. Classically, the coal-fired domestic heating unit has been very important and still contributes much to air pollution in coal-producing areas of the world. However, the change to central heating and

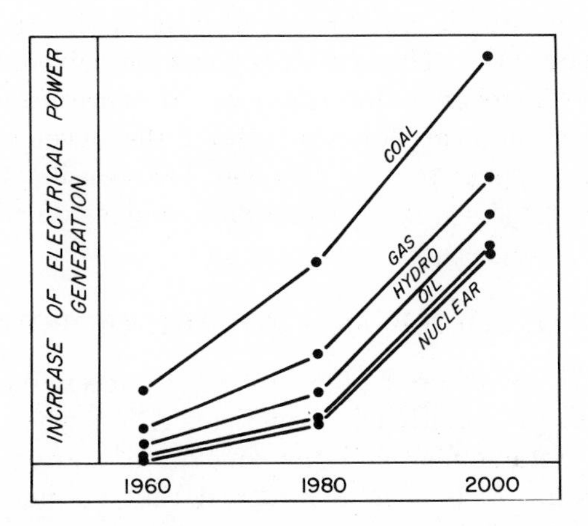

Fig. 2. The amount of electricity used in the United States is higher than in any other country, and is increasing each year. By 1980 estimations are that three times as much electricity will be used as in 1960; by the end of the century, five and six times as much. In 1960 nuclear and water energy, neither of which contributes significantly to air pollution, supplied a small fraction of our needs; nuclear energy undoubtedly will be used increasingly in the future. However, the greatest source of energy is now and will continue to be the burning of coal and oil. The combustion of these fuels is already a major contributor to the pollution of our atmosphere. From *U.S. Public Health Serv. Publ.* No. 1555.

the use of other fuels has reduced to some extent the relative importance of this contribution to air pollution. Strangely enough, the switch to electric power (Fig. 2) has not eliminated the consumption of fuel for power since, despite the large output of hydroelectric power, the majority of electricity is still generated by the combustion of fuel; coal and oil presently generate 95%. Although nuclear power will be used increasingly in the future, fossil fuel generation of power will continue to rise for several decades.

The development and application of atomic reactors for the generation of electric power offer hope that ultimately much of our power will be produced by this source, which is low in conventional air pollution emissions. Even atomic power, however, is not innocent in the production of environmental pollution. Large amounts of coal are presently used in the preparation of reagents for reaction. Of much greater significance in the long run is the pollution potential of the reactors themselves in the form of radioactive wastes and of heat discharged into the water of seas, streams, and lakes. Since an enormous proliferation of such reactors would be necessary to replace fossil fuels for power production to any significant extent, the possible ecological implications of these two by-products of atomic reactors are now the subject of intense private and public debate (1). This problem must be solved before any real consensus can be reached. However, even if concern over pollution by radioactive wastes or heat does not inhibit the development of atomic reactors, it would appear that by the year 2000 considerably more power will still be generated by the conventional sources of coal and oil than today.

THE AUTOMOBILE AS A SOURCE OF POLLUTION

The internal combustion motor has rightly been incriminated as a major source of air pollution in American cities, where the growth of the number of automobiles and the amount of automobile travel has been astronomical. It is generally agreed that automobiles proliferate

Source	Carbon monoxide	Particulates	Hydrocarbons	Nitrogen oxides	Sulfur oxides	Total
Transportation	64.5	1.2	17.6	7.6	0.4	91.3
Fuel Combustion in Stationary Sources	1.9	9.2	0.7	6.7	22.9	41.4
Industrial Processes	10.7	7.6	3.5	0.2	7.2	29.2
Solid Waste Disposal	7.6	1.0	1.5	0.5	0.1	10.7
Miscellaneous	9.7	2.9	6.0	0.5	0.6	19.7
Total	94.4	21.9	29.3	15.5	31.2	192.3
Forest Fires	7.2	6.7	2.2	1.2	N	17.3
Total	101.6	28.6	31.5	16.7	31.2	209.6

Fig. 3. There is no "worse source" of air pollution. In this country automobiles are the major contributor to carbon monoxide and hydrocarbon pollution. Stationary sources account for most of the sulfur oxides and formed particles, whereas the nitrogen oxides are evenly divided. From the Third Report of the Secretary of Health, Education, and Welfare to Congress in compliance with Public Law 90-148, Department of Health, Education, and Welfare, Air Pollution Control Office, March 1970.

approximately at the rate of twice the population growth. In the face of such burgeoning pollution sources, it seems rather hopeless to expect any real results from partially effective pollution controls. While certain cities have become notorious for auto smog, an examination of air pollution data shows that this form of pollution is probably of some importance in most American cities. From the standpoint of air pollution, the internal combustion engine can be considered a pollution machine for the production of pollutants and a pump to put them into the atmosphere. (See Fig. 3)

II. FACTORS AFFECTING LOCAL CONCENTRATION OF POLLUTANTS

If factors affecting pollution from stacks and tail pipes are assumed to be equal, a number of meteorologic and geographic conditions alter the pollution in a local area: concentrations of emission sources, topography, wind velocity, wind direction, and the frequency of temperature inversions. The concentration of emission sources is a result of the urbanization mentioned earlier. Pollutants tend to accumulate more when wind speed is low and in bowls in the hills in the face of the prevailing wind and in valleys or other depressions (Figs. 4 and

Fig. 4. Typical smog in New York City with only the top of the Empire State Building emerging above the temperature inversion. Courtesy of Fairchild Aerial Surveys.

5). Temperature inversions (when cool air next to the earth is overlaid by warmer air) are especially important since they serve as a "lid" or ceiling (Fig. 6) that prevents the dissipation of pollutants into the upper air by means of convection currents.

Inversions occur nightly when land cools as a result of heat irradiation to the sky and air cools next to the earth. This phenomenon is more prevalent and long lasting in valleys or other depressions. Inversions may also occur, however, over large areas or plains when a warm front overrides cooler air. Such inversions may persist for a considerable time and, invariably, are associated with the extreme buildup and persistence of pollutants in the atmosphere. Another meteorological factor of great importance to air pollution is the formation of anticyclonic

Fig. 5. Midmorning smog in St. Louis obscuring the base of the Memorial Arch. Courtesy of World-Wide Photo.

Fig. 6. Typical smoggy day in Los Angeles, showing the smog accumulating below the temperature inversion that serves as a lid above which the auto exhaust constituents do not disperse. Courtesy of Los Angeles County Air Pollution Control District.

centers associated with large areas of atmospheric stagnation and intense accumulations of air pollutants (2). Most of the so-called episodes occur during such stagnation periods. Although these factors are especially important in the production of local air pollution problems, they do not aggravate the global problem (see ecological effects in Section III).

III. SPECIFIC POLLUTANTS

Air pollution atmospheres tend to fall into two main categories: (*1*) those derived from the combustion of fossil fuels for heat and stationary power sources, which we might term the classical form of air pollution and which are characteristic of the great urban and industrial complexes (Fig. 7) of Europe and Eastern and Central North America; and (*2*) those derived from automobile traffic which are most characteristic of Western North American cities. The latter are termed photochemical pollution or auto smog. Auto smog tends to be more concentrated and to occur more frequently in the presence of an irradiative climate such as that of the Southwestern United States. Almost no American city,

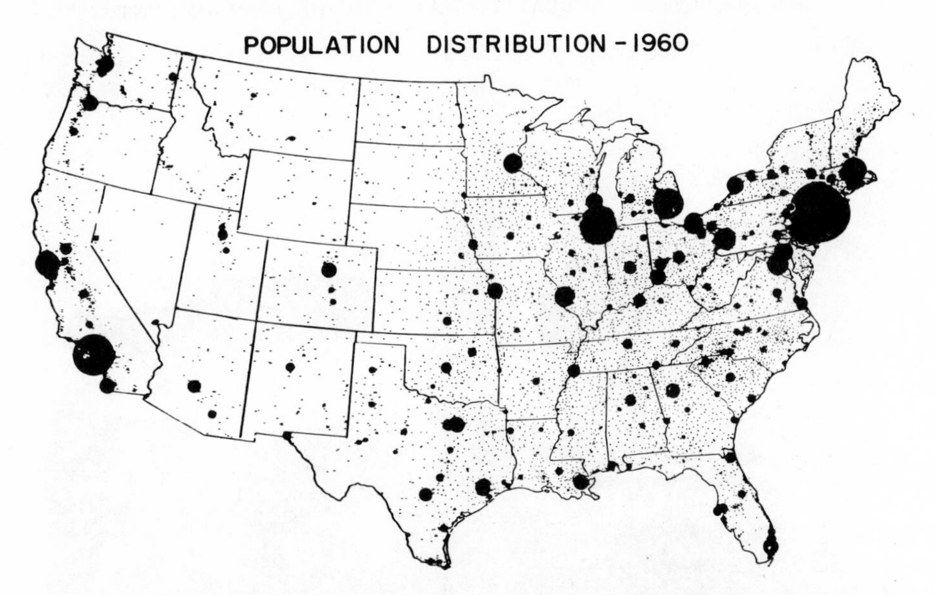

Fig. 7. In 1960 population distribution of the United States shows that major air pollution problems coincide with the urban industrial complexes. From the Department of Commerce, Bureau of the Census, *U.S. Public Health Serv. Publ.* No. 975.

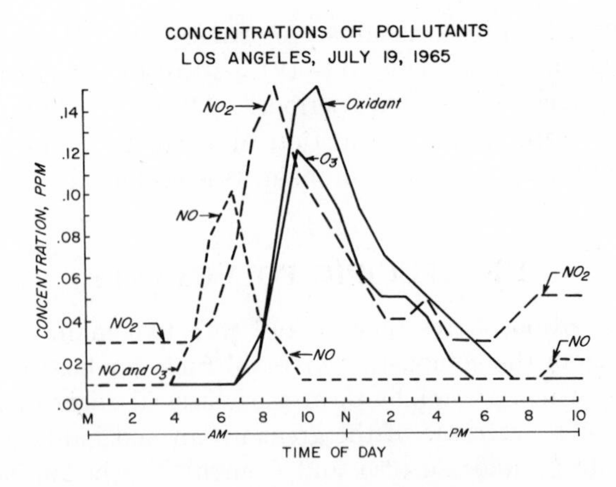

Fig. 8. Primary exhaust pollutants are usually bimodel with morning and evening peaks reflecting traffic patterns. The secondary pollutants shown here, however, peak at midday, since they result from the interaction of sunlight on emissions produced in the morning. From J. N. Pitts, "Environmental Appraisal: Oxidants, Hydrocarbons, and Oxides of Nitrogen," *J. Air Pollution Control Assoc.*, **19,** 658–667 (1969).

however, is free of days in which strong sunlight reacts with auto exhaust to produce significant photochemical smog. As is mentioned later, problems with other constituents of auto exhaust, such as carbon monoxide, lead, and particulates, are not dependent on photochemical reactions. They may occur anywhere and at any time that there is a sufficient amount of automobile traffic, since their development is not dependent on radiant energy derived from sunlight (Fig. 8). Furthermore, constituents of auto exhaust (both irradiated smog and nonirradiated exhaust) are added to the list of classical pollutants, intensifying the noxious mixture in the air of many cities.

IV. CLASSICAL POLLUTION

Many gases and solid particles are the product of the combustion of fossil fuels (coal and oil) in stationary units for the production of heat mainly for warming space and the production of power. The proportions of the various constituents vary according to the quality of the fuel, size of the firebox, temperature of the fire, and other factors. For instance, it has been the experience of certain cities that their atmospheric sulfur dioxide and nitrogen oxide levels rose even though more efficient firing practices reduced their smoke. In addition, the presence of particles from other sources such as fog and dust appears to accentuate the effects on health of these atmospheres. The major constituents of this class of pollution generally are gases—sulfur oxides, nitrogen oxides—and particulates—sulfur trioxide, sulfates, sulfuric acid mists, hydrocarbons, and constituents of fly ash (silica, alumina, iron oxides, carbon).

V. AUTOMOBILE EXHAUST

As emitted from the tail pipe of a car, exhaust has the following major constituents: gases—carbon monoxide, nitric oxides, unburned hydrocarbons, partial oxidation products (formaldehyde, acetaldehyde, acrolein, etc.); and particulates—including lead compounds, carbon, and various other inorganic and organic materials in particulate form. Furthermore, additional hydrocarbons and fuel additives are contributed by the evaporation of gasoline from automobile fuel tanks and carburetors. In the absence of bright sunlight no great alteration of automobile effluent occurs, which is thus mixed with whatever other pollutants are present. Therefore, oxides of nitrogen, carbon monoxide, and lead oxide particles are thrown into atmospheres already polluted by sulfur oxides,

oxides of nitrogen, hydrocarbons, and other particulates derived from smokestacks.

In the presence of bright sunlight, however, a complex chemical interaction occurs between various constituents in the exhaust and the oxygen of the air. This results in photochemical smog, which produces intense eye irritation in human beings and damage to plants and forms haze, oxidant, and a characteristic odor. The haze is light, slightly brownish, and foglike. This is not to be confused with heavy smog, such as that occurring typically in Eastern American and European cities, caused by a physical and chemical interaction between smoke, sulfur oxides, and fog. A number of highly irritating compounds are formed as secondary pollutants in photochemical auto smog. Among these compounds are ozone, peroxyacetyl nitrates, formaldehyde, and acrolein, which develop in a chain reaction involving the constituents of auto exhaust and oxygen as reactants and sunlight as the energy source. Hydrocarbons are essential in this reaction, serving as multipliers of photoenergy. In the process, nitrogen dioxide absorbs ultraviolet light from the sun and is broken down to nitric oxide and atomic oxygen. The atomic oxygen unites with molecular oxygen to form ozone, which then reacts with nitric oxide to form nitrogen dioxide and molecular oxygen. Atomic oxygen also is thought to react with hydrocarbons to form special complexes called free radicals which undergo complex interactions that give rise to secondary pollutants. The whole process is a complex phenomenon still not completely understood. An important component of this atmosphere is termed oxidant, which is actually a chemical activity rather than a chemical entity. Its presence in smog is usually measured by its propensity for oxidizing potassium iodide. The bulk of this oxidizing property during peak daylight hours can be identified as ozone. Several thousand tons of ozone are formed on a smoggy day in Los Angeles, and as much as 500 tons are present over the city at any given moment (3). Because of the importance of ozone in photochemical smog and its extreme toxicity for plants and animals, it has been the subject of intensive research.

VI. SPECIAL POLLUTANTS

Effluents from various classes of industries contain specific pollutants which may be peculiar to the industry or the particular process of manufacture. Raw acids may be emitted by chemical plants, aliphatic compounds are derived from soap factories, iron oxide from steel mills, and so on. Although these contribute to the overall pollution problem, they usually are more important locally. An effluent of great veterinary interest is, of course, fluoride in dust and gases from rock crushing, fertilizer manufacture, and smelting processes (see Section XV).

VII. SPECIFIC EFFECTS OF AIR POLLUTANTS ON MAN AND HIS ENVIRONMENT

Pollution of the air affects man by aesthetically degrading his environment, interfering with his visibility, soiling and corroding his property, blighting his crops, affecting his animals, and, finally, interfering with his health and well-being. Aesthetic degradation of the environment is conspicuous in many of our communities. Foul odors from specific sources bring remarks of distaste generally whenever they are experienced. Often the odor of various sulfur compounds may pervade large communities where fossil fuels are consumed in large amounts in stationary sources. What traveler in a forested area is not shocked by the stench emanating from a paper mill or by the pall of smoke from a waste burner at a sawmill in areas of otherwise great air purity and scenic beauty?

The significant reduction of visibility by the various kinds of air pollution has become aesthetically objectionable in many communities on many days of the year. Often the populace of a community have become so inured to air pollution, industrial haze, and smog that they are scarcely aware of it until they have had an opportunity to view it in perspective from a distant high point or from an airplane. Citizens of communities in a mountainous environment seem more aware of decreased visibility since it ruins their view. Reduction of visibility has more than an aesthetic effect, however, since days of smogged-in airports and hazardous freeway driving are markedly increased by air pollution's contribution to reduced visibility. The haze, smoke, and smog from community air pollution are highly visible from the air where one can observe the haze over a city or industrial area sharply contrasting with the comparative clarity of the air of the surrounding countryside. It is also believed that air pollution haze may have regional and even global implications (see Section XVII).

The corrosive effects of acid products of pollution on metals and their destructive action on paint have long been recognized, since pollution by sulfur oxides is an old phenomenon present wherever fossil fuels are burned. Furthermore, the oxidants derived from auto smog and other sources are responsible for the cracking of rubber in tires and the deterioration of insulation on electric wires. Sulfur oxides cause weakening of natural and synthetic fiber in clothing. It has been estimated that the cost of air pollution damage to materials and the various attempts to prevent it, may amount from $2 to 12 billion yearly in the United States alone (4).

VIII. AIR POLLUTION DAMAGE TO PLANTS

The deleterious effects of various air pollutants on plants are extremely well documented. Whole communities have in the past been completely devastated by fumes emanating from smelters (Fig. 9), for example.

Plants are generally more susceptible to damage from air pollutants than are mammals, since they have no effective means of protecting their respiratory surface from the ingress of pollutants and no adequate excretory apparatus for eliminating absorbed materials from their tissues. Sulfur oxides, oxidants, ozone, fluoride, peroxyacetyl nitrates, and many other pollutants in very low concentrations produce specific recognizable lesions on plants. Economic loss occurs to field crops, ornamental plants, and forest trees (Fig. 10). Citrus fruits, leafy row

Fig. 9. Devastation of a once-forested countryside by sulfur dioxide from a smelter, Copper Hill, Tenn. While such ravages from localized emissions have occurred in many places, more subtle damage results from auto smog that may selectively kill certain species over widespread areas, often by increasing their susceptibility to intercurrent natural disease. Courtesy of U.S. Forest Service.

Fig. 10. Damage to crops, ornamental plants, and forest trees shows correlation with expansion of ozone around our cities. From *U.S. Public Health Serv. Publ.* No. 977.

vegetables, and Christmas tree plantations and forest evergreens are particularly affected.

Of greater significance is the evidence that auto smog is causing the deterioration of forest quality over large areas, as reported in California. In Sweden, the deposition of acid products of fossil-fuel pollution may result in ecological alterations hundreds of miles from the primary sources (5–7). Auto smog also has far-reaching effects on food crops, often rendering the cultivation of particularly susceptible plants impossible over wide areas in polluted environments—for example, romaine lettuce in the Los Angeles area (8). A specific example of the widespread effects of auto smog on forest trees is the destruction of ponderosa pine in the San Bernardino forest more than 80 miles from Los Angeles, requiring extensive salvage lumbering operations (9,10).

The total loss attributed to air pollution to cultivated food crops, forage, ornamental plants, and forest trees is incalculable, being estimated at $500 million in agricultural losses alone in the United States. A vast literature is available on this subject.

IX. HUMAN HEALTH EFFECTS OF AIR POLLUTION

Even though there had been occasional speculation and complaints about the health effects of air pollution since the days of Edward I, it was not until the occurrence of certain air pollution episodes (in the Valley of the Meuse, in London, and in Donora, Pennsylvania) that people finally came to realize that it was a threat to human health. These episodes made it obvious that under certain conditions air pollution could kill.

Since the classic episodes, a number of others have been noted in which pollutants were excessively high over a number of consecutive days. In fact, these occur each year in both Eastern and Western North America. Excess deaths in extreme episodes are reported from New York City (11).

Despite the drama of the episodes, the greatest impact of air pollutants on human health results from day-to-day exposure under unexceptional conditions. It is believed that these effects occur through continued irritation by pollutants interacting with other environmental or biologic factors to initiate disease or exacerbate previously existing disease.

Perhaps the most common complaint in connection with air pollution in American cities is acute pain and irritation of the eyes. Although no association between this curable irritation and lasting disease of the eye is apparent at this time, it is responsible for an enormous amount of discomfort and mental aggravation. Eye irritation is usually associated with photochemical smog and is apparently related to the presence of ozone, aldehydes, and hydrocarbons. Its presence is thus a source of either embarrassment or perverse pride in Southern California cities. Photochemical smog also occurs to a lesser degree in many other areas. The exact etiologic chemical or chemicals in this complex mixture is not completely known at this time, but peroxyacetyl nitrate, formaldehyde, and acrolein are believed to be important (12). Research has been inhibited by the lack of suitable animal models, since the symptoms are largely subjective.

X. EXACERBATION OF CHRONIC DISEASE

It is the experience of physicians that chronic cardiopulmonary and asthmatic patients feel poorly at times of high air pollution. Chronic bronchitis is very prevalent in the heavily polluted cities of Great Britain and other highly polluted areas, where it is an important contributor to the mortality rate. In areas of high pollution, evidenced by sulfur oxides and particulates in the atmosphere, the so-called cough-and-spit syndrome is aggravated when these pollutants are high. Cigarette smoking is an important contributory factor in this disease (13–15). Emerging evidence indicates that the combined results of smoking and air pollution contribute to the production or aggravation of chronic bronchitis in man (16,19). It is believed that similar problems occur to a lesser degree in the United States in like atmospheres. Auto smog also is responsible for the aggravation of certain chronic diseases (20). Results of experiments conducted at the Los Angeles County Hospital confirm the clinical impressions in the latter instance. A double-blind study was conducted in which neither the patients nor the attending physicians knew whether the patients were breathing smoggy air or air cleansed by filtration. Patients breathing the filtered air claimed to feel better and, furthermore, their pulmonary physiological parameters showed improvement (21).

Bronchial asthma is a fairly well-defined state in which acute attacks are more or less related to season and to pollen count. In certain areas, however, episodic asthma occurs that does not appear to be associated with those cycles. In New Orleans, for instance, hospital admissions for asthma increase for brief periods during normally nonallergenic seasons (22,23). Although it is believed that some air pollutants may be responsible for such episodic asthma, no epidemiological association appears firm at this time. A study conducted at Nashville, Tennessee, indicated a statistical association, however, between hospital admissions for asthma and pollution, using rate of sulfation as an index (24).

Possible mechanisms for the deleterious effects of various pollutants on chronic disease includes their influence on airway resistance and mucociliary clearance (25–27). Both sulfur oxides and irradiated auto smog have been shown in animal experiments to cause a reflex increase in expiratory resistance. In animals the toxicity of sulfur oxides is enhanced by the addition of particulates (28). Furthermore, higher levels of sulfur dioxide are remarkably mucogenic in the rat (29). Although the complete mechanism for these reactions in man is not yet known, future clarification can be expected from the development of animal models.

XI. INFLUENCE OF AIR POLLUTION IN THE ETIOLOGY OF ACUTE PRIMARY LUNG DISEASE

Pollution as indicated by smoke and sulfur dioxide has been shown to influence the morbidity from acute lower respiratory tract disease in schoolchildren (30,31). Several studies have shown that adults in urban communities, presumably more subject to pollution, have excessive lower pulmonary tract disease (32). Animal studies with this type of atmosphere have not yet confirmed the relationship beyond demonstrating differences in mucociliary clearance rates and bacterial persistence in the upper airway (33–35). Numerous animal studies, however, have indicated clearly that the exposure of mice and other animals to auto smog or its individual oxidant constituents predisposes to bacterial and viral infections of the deep lung (36–40). Exposure to artificial auto smog and to ozone for as little as 3 hours at levels below those occurring in polluted air have resulted in increased susceptibility to infection with aerosolized bacteria (36,37). Similar results, but at higher levels, have been produced by brief exposures to nitrogen dioxide (38). Experiments in which longer exposure to nitrogen dioxide (3 to 12 months) were made and have resulted in similar enhancement at levels below those frequently seen in the atmosphere (38). Squirrel monkeys exposed to aerosolized influenza virus, followed by nitrogen dioxide exposure, showed elevated virus titer and increased mortality (39,40).

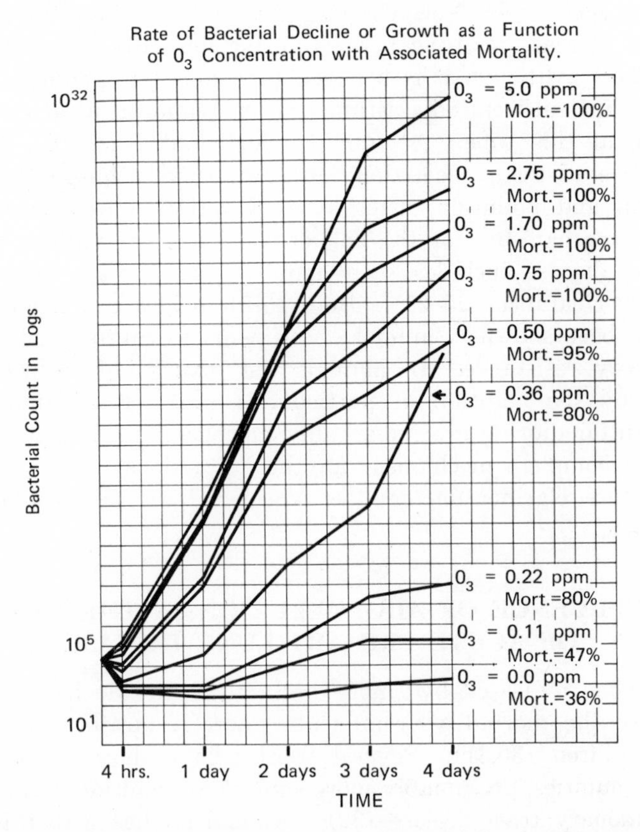

Fig. 11. Exposure of mice to aerosolized streptococci following exposure to various concentrations of ozone results in a dose-related slowing of the rate of bacterial decline, shortening of the lag phase, and acceleration of growth. Mortality is proportional (41).

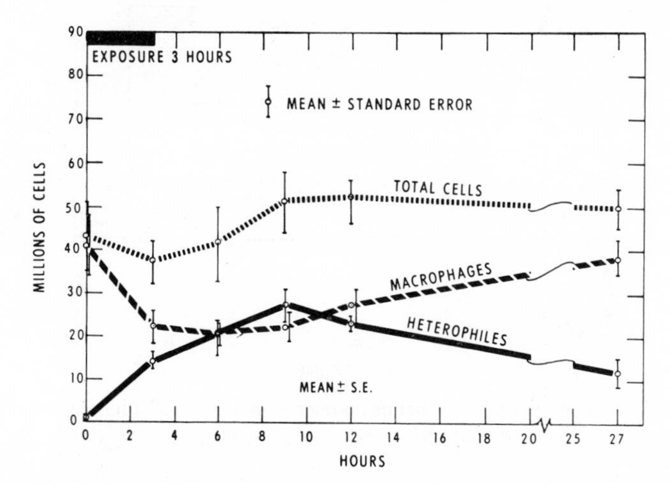

Fig. 12. Exposure to ozone in the rabbit elicits an increase in heterophiles (polymorphonuclear leukocytes) and a decrease in the number of macrophages and the total number of cells obtained by pulmonary lavage. There is a gradual return to an approximately normal number after more than 24 hours (42).

The mechanisms that might be responsible for increased susceptibility to bacterial infection have been the subject of considerable investigation. When mice are exposed to ozone (Fig. 11), followed by a bacterial aerosol of pathogenic organisms, the following sequences occur: slowing of the rate of bacterial kill in the lung (delayed bacterial clearance rate), shortening of the lag phase, and acceleration of bacterial growth (41). Furthermore, the number of positive blood cultures is augmented, and the time required for septicemia and for death is shortened. Experiments conducted in rabbits indicate that when exposure to ozone (Fig. 12) is followed by pulmonary lavage the pulmonary macrophages obtained are reduced in number, and the phagocytic competence (Fig. 13) of the remaining cells is impaired (42). The activity of acid hydrolase enzymes (lysozyme, acid phosphatase, and beta-glucuronidase) is also lessened in the macrophages (43), and lysozyme activity unassociated with cells in the washout fluid also lessens following ozone exposure (44). Studies carried out *in vitro* suggest that the toxic effect of ozone may be mediated through its action on noncellular components in the lung airway associated with, but not identical to the so-called surface-active material (45). Experiments have shown that ozone degrades the cell-protecting properties of this material both *in vivo* and

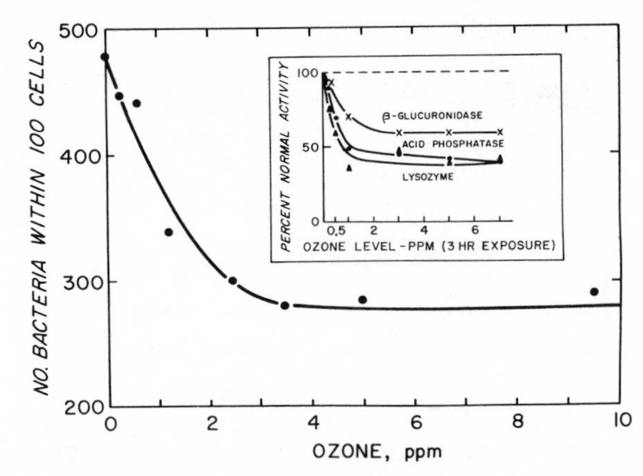

Fig. 13. Exposure of rabbits to ozone causes depression of phagocytosis and bacteriolytic tysosomal enzymes.

in vitro (46). Macrophages do not survive well in the absence of this material in a chemically intact state.

Since the alveolar macrophage is highly active against introduced bacteria and other particles, it is likely that impairment of the macrophage system by oxidant air pollutants is responsible for the increased susceptibility resulting through exposure to these gases (47).

Impairment of interferon formation by nitrogen dioxide has been noted in interesting combined *in vivo* and *in vitro* studies (48). Rabbits were infected with parainfluenza virus and exposed to nitrogen dioxide. Then the rabbits were subjected to pulmonary lavage, and tissue cultures of surviving pulmonary macrophages were prepared. Cells cultured from the rabbits exposed to nitrogen dioxide were more susceptible to subsequent infection with rabbit pox virus. These experiments and other confirmatory work led to the conclusion that the increased susceptibility of the cells in tissue culture is due to lowered interferon production. It is probable that the interferon-inhibiting effect of nitrogen dioxide may explain the increased susceptibility of intact animals to influenza virus. While the biological studies yield clear-cut evidence that bacterial and viral pulmonary defenses are reduced by oxidant air pollutant gases, little confirmatory support has been derived thus far from epidemiological studies in human beings. It seems unlikely, however, that such basic mechanisms would not be subject to similar alterations in man. Under appropriate circumstances, therefore, man would be increasingly susceptible to the appropriate invader or virus. Some confirmatory evidence has been derived recently from studies in which elevated morbidity, presumably caused by influenza, appeared associated with nitrogen dioxide exposure (49).

XII. ALTERATIONS OF PULMONARY STRUCTURE BY AIR POLLUTANTS

Pulmonary emphysema, a disease in which enlargement of the pulmonary air spaces is accompanied by loss of tissue, distortion, and scarring, is becoming increasingly prevalent. It is thought to be more common in urbanized areas affected by air pollution (50,51). A statistical association between cigarette smoking and hospital admissions or deaths from this disease exists (52). Exposure to nitrogen dioxide over a considerable period in a variety of animal species regularly produces enlarged rigid lungs having enlarged air spaces (53–59).

The pathological sequences in the rat appear to be epithelial metaplasia in the terminal bronchioles, accumulation of mucus and macrophages, hyperplasia of the adjacent alveolar epithelium, and enlargement of the alveolar ducts and alveoli, with scarring in the septae. It has been postulated that obstruction at the level of the terminal bronchiole leads to the dilation of the distal air spaces (55).

Compensatory changes occur in the rat in response to the lesions induced by nitrogen dioxide. These consist of enlargement of the red cell mass (compensatory polycythemia) and physical alteration of the thoracic cavity by development of kyphosis and a lateral flaring of the ribs (58). Such compensatory changes are, of course, well-known in advanced emphysema in man. Another interesting facet of the toxicity of nitrogen dioxide for the lung has been the claim that very brief exposures to very low levels cause the denaturation of pulmonary collagen (6).

If the findings above are confirmed, it would indicate that not only is obstructive disease induced by nitrogen dioxide but also weakened alveolar walls increase the susceptibility of the septae to dilate and break under back pressure from the obstruction.

Support for the epidemiological evidence linking smoking to emphysema was gained from an experimental study in which emphysema-like lesions were produced in dogs by exposing them to tobacco smoke (61). The facts may be stated somewhat simplistically as follows: an association exists betweeen human emphysema and smoking and air pollution (at least in urban areas). Exposure to nitrogen dioxide, a common air pollutant, produces a state in animals resembling human emphysema. A similar state may be produced in animals by exposing them to tobacco smoke. Nitric oxide, a precursor to nitrogen dioxide, is an important component of tobacco smoke.

Might it not be reasonable to expect that nitrogen oxides are the common emphysematous influence in smoking and air pollution exposures?

XIII. ROLE OF AIR POLLUTANTS IN THE ETIOLOGY OF LUNG CANCER

Pulmonary cancer has become epidemic in man, ranking fourth as a cause of death (62). Although it is evident that smoking must assume the major blame, certain facts suggest strongly that air pollution may be responsible for an appreciable number of the cases. Several basic facts link air pollution to lung cancer (63). Surveys invariably show a greater incidence of human lung cancer in cities (Fig. 14). In most

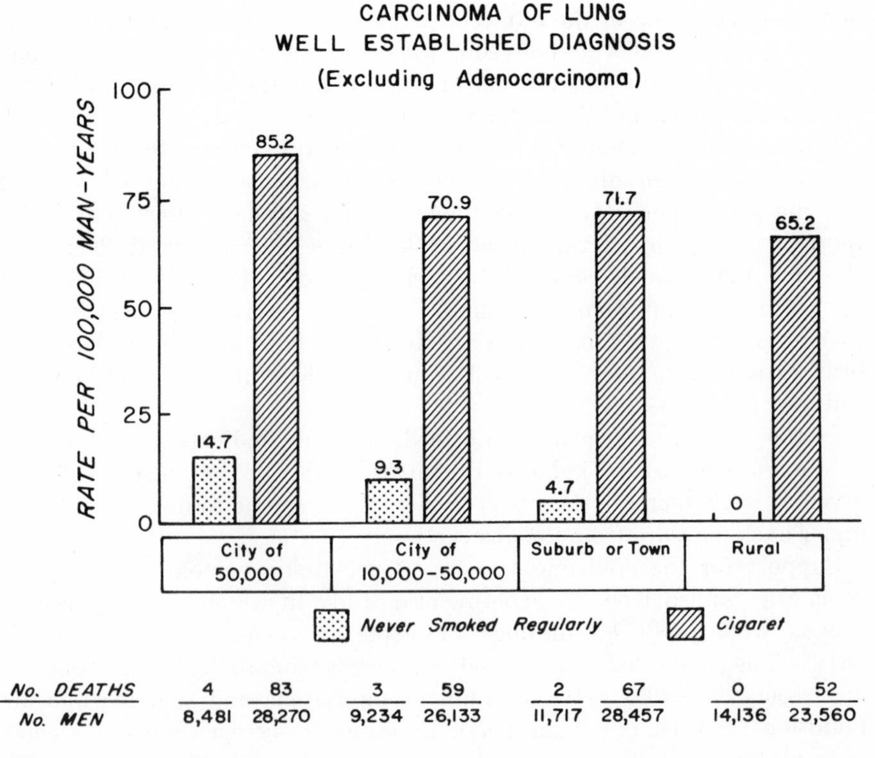

Fig. 14. Association of lung cancer with smoking and correlation of the "urban factor" is shown here. From *U.S. Public Health Serv. Publ.* No. 1022 (63).

State	Urban		Rural	
	Crude organic	Benzo(a) pyrene	Crude organic	Benzo(a) pyrene
California	15.2	2.06	2.6	0.35
Colorado	10.0	2.30	1.2	0.07
Indiana	12.7	10.37	2.5	0.47
Missouri	7.8	5.32	1.4	0.70
North Carolina	11.2	5.71	1.5	0.22
New York	10.3	4.10	1.3	0.22
Pennsylvania	8.1	3.79	2.5	1.45
Wyoming	2.0	0.52	0.7	0.06
Nevada	7.6	1.26	0.6	0.05

Fig. 15. Relative amounts of organic compounds (crude organic extract) and benzo(a)pyrene (an aromatic polycyclic hydrocarbon carcinogen) at urban and rural sites in nine states. The crude organic extract is expressed in micrograms per cubic meter of air and the benzo(a)pyrene in nanograms per cubic meter. The sites polluted chiefly by automobiles have relatively less of this carcinogen—Los Angeles, Denver, and Las Vegas. From "Air Quality Data from the National Air Suveillance Networks and Contributing State and Local Networks," 1966.

surveys the incidence roughly equates with the population: large cities having the highest and rural areas the lowest (64). Concentrations of carcinogens derived from incomplete combustion of carbonaceous material are also higher in urban areas (Fig. 15). Benzo(a)pyrene and associated hydrocarbon carcinogens are 10- to 1000-fold higher in urban-industrial complexes than in their rural counterparts (65). Epidemiological studies by some investigators have suggested that the urban effect and smoking are synergistic, the greater effect occurring in smokers (66). Other epidemiological studies have suggested that urban-rural differences could be explained by the additive effects of chemical carcinogens from the air plus those derived from smoking. These investigators were able to show a fairly convincing relationship between the contribution of carcinogens expected from both sources and lung tumor incidence for locations of different pollution concentrations and for different smoking habits (68). Other epidemiological studies suggest the opposite: the urban increment in lung cancer may be best associated with factors other than air pollution, including population density (14,69). Despite the variability of epidemiologic results, cancer can regularly be produced in mice by applying extracts of airborne hydro-

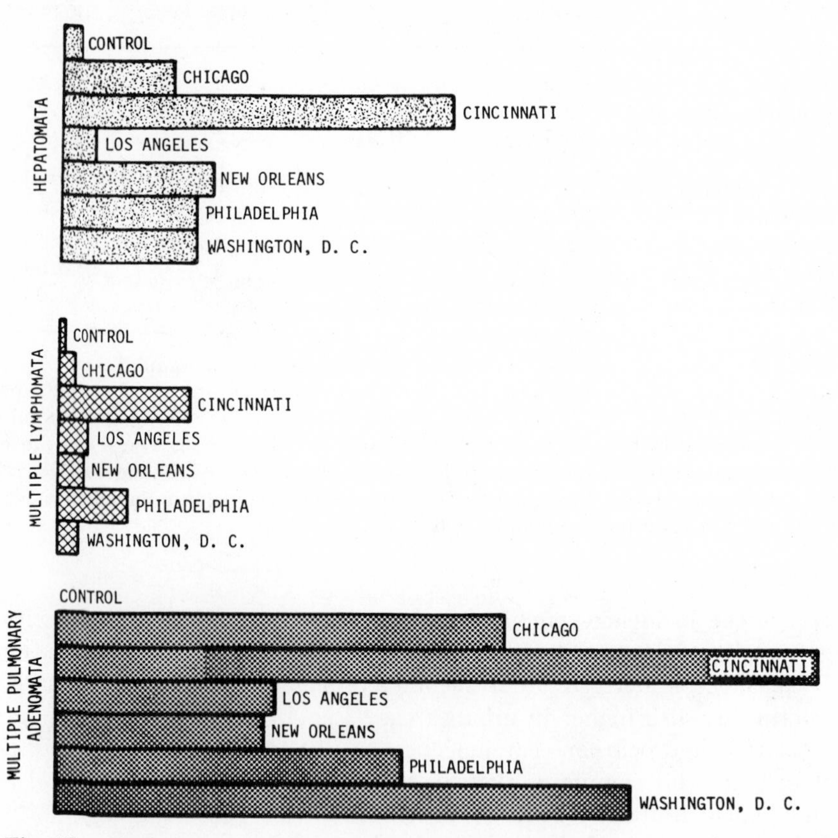

Fig. 16. Newborn mice react at distal sites from injections of carcinogens. The quantitative differences in this study are not understood but agree in certain sites with a former study in adult mice. From Epstein *et al*, *Nature*, **212**, 1305–1307 (1966).

carbons to the skin (Fig. 16) or by injecting the animal subcutaneously (70). More important, the instillation of benzo(a)pyrene, constantly present in polluted air, together with iron oxide particles into the tracheae of hamsters regularly induces lung tumors resembling those seen in man (71).

Besides the hydrocarbon carcinogens in air, other substances serve as cofactors in the formation of cancers experimentally. For example, both hematite and carbon, which are common contaminants of polluted air, have been shown to interact with carcinogens to produce lung tumors in animals. Asbestos exposure produces an enhancement of lung tumors only in smokers (72). It thus appears that particulates have a marked enhancing effect in the production of lung cancer. It has been postulated that particles, even though they are inert, act by sequestering the chemical carcinogen, thereby permitting its slow release into the target tissue for a maximum carcinogenic effect (73,74). It is probable that such action is proportional to the degree to which the particle can absorb the carcinogen (75). The enhancing effects of irritants when used in conjunction with chemical carcinogens are well known. Recent experiments have indicated that the common pollutant, sulfur dioxide, has this propensity (76). Finally, the subtle effects of trace minerals contained in atmospheric pollution cannot be overlooked. Certain of these elements (nickel, copper, and iron) have been shown to inhibit the induction of anticarcinogenic enzymes when tissues are brought in contact with the carcinogen (77). It would appear at this time that although evidence for a causal relationship between air pollution and lung cancer is inconclusive, the preponderance of the data argues that a causal relationship may be implicated. Certainly the problem is of such great interest that intensive research should be undertaken for its final determination.

XIV. CARBON MONOXIDE

From the point of view of the total amount in the air, carbon monoxide is the most important pollutant, amounting to over 70 million tons a year in the United States. It is discussed here as a single substance since, so far as is known at this time, it differs from most other pollutants in that it produces no organic diseases or even residual functional disturbances, even though, as generally known, it is an extremely toxic substance. Anyone who drives a car or has a source of illuminating gas or even a coal furnace in his home is in some danger from poisoning with this highly toxic substance. From the standpoint of community air pollution, however, it is not the accidental sporadic high-concentration exposures that concern us but the continued daily concentrations over the city streets and in metropolitan areas generally (usually below 50 ppm). Therefore, the possible interference of carbon monoxide with problem solving, complex operations requiring integration of mental activity, and physical response (such as in auto driving) are the subject of considerable research (78). Equally important is the effect of carbon monoxide on those individuals in a precarious oxygen balance, such as cardiopulmonary patients. It is known that continuous exposure to rather low levels of carbon dioxide is detrimental to such persons. It is feared also that chronic hypoxia imposed by carbon monoxide exposure may injure tissues with great oxygen demand, such as the heart (79,80). The possibility exists that areas where oxygen partial pressure is low may suffer oxygen deprivation because of competing carboxyhemoglobin. Some precedent for this view may be had in the supposed effect of cigarette smoking on the fetus, an effect which may presumably be the result of oxygen deprivation by carbon monoxide exposure (81). The effects of carbon monoxide are augmented by high altitude, anemia, cardiovascular disease, and smoking (82).

Since carbon monoxide acts by combining with hemoglobin at a greater rate than oxygen and dissociates less readily than oxygen, the end result of exposure is oxygen deprivation. The rate of formation of carboxyhemoglobin is governed by the concentration of carbon monoxide, and, for any given concentration, a maximum level is reached after a certain period of exposure. Equilibrium is maintained at this level, unless death intervenes. It is thus possible to compute the expected carboxyhemoglobin level resulting from any given CO concentration and period of exposure. It is also possible to quantitate carboxyhemoglobin and relate it to the symptoms seen in man and the signs in experimental animals. Because of the reputed reversibility of its acute effects on healthy individuals, carbon monoxide is one of the few toxic gases whose effects are investigated through experiments in human beings. Thus it might be expected that there would be agreement on the threshold of toxicity for man. Unfortunately, this is not the case. For instance, in a recent study in which subjects were deprived of normal attention-getting stimuli such as sight and sound, decrements occurred in this ability to detect differences in the duration of a tone after exposure to as little as 50 ppm of carbon monoxide over a 2-hour period (83). Effects were noted from an exposure which might be predicted to produce a carboxyhemoglobin level of 2% above background, a level well below that which would result from prolonged exposure to downtown atmospheres or from moderate cigarette smoking. In contrast, another recent study in which the subjects were in a competitive situation indicated that decrements in performance tests occurred only at levels of exposure yielding approximately 20% carboxyhemoglobin (84). Such concentrations regularly produce headache and other subjective symptoms and are well above those usually occurring in natural auto smog, smoking, and the like.

Differences in test procedures may explain some of the divergence in results. For instance, the deprivation of sensory stimuli in the first study might well have increased the sensitivity of the subjects in their response to the tests. The divergence of the results of these and other studies argues strongly, however, for further work under standardized conditions before a solution of this vitally important problem can be achieved.

XV. AIR POLLUTION AND DOMESTIC ANIMALS

Air pollution effects in domestic animals have been of much interest because of concern with animals *per se* as well as their possible employment as monitors or sentinels in the interest of human health. Undoubtedly, because of their location in rural areas, few animals raised for economic reasons are at maximum risk from community air pollution. A considerable number of dairy cattle and chickens, however, are kept close enough to industrial pollution sources or to cities to come under some threat from pollution. Pet animals, on the other hand, undergo the same risks as their masters in cities and elsewhere. Animals have suffered in air pollution episodes, including dogs, cats, sheep, and pigs in Donora, and various house pets in Poza Rica (85,86). It is of interest to note that beef cattle at a livestock show suffered considerable illness and mortality during the London episode of 1953 (85). Illness and mortality associated with these episodes were attributed to respiratory deficiency. A systematic survey in dogs conducted in the city of Columbus, Ohio, showed a correlation between dustfall (a common gross method of measuring air pollution) and pulmonary anthracosis and fibrosis (87). Numerous studies have been conducted on animal populations in the Bay area of California in relation to air pollution (88).

Present knowledge indicates that two air pollutants have been most responsible for economic loss in livestock in the United States: fluorine and arsenic (89). Fluorine, as fluoride, is emitted by various heavy-chemical industries such as those concerned with the production of phosphate fertilizers, aluminum metals, fluorinated hydrocarbons, fluorinated plastics, and others. Fluoride toxicity is mediated through deposition of fluoride on forage by the settling of dust or the precipitation from a sublimed form. Thus grazing animals are subjected to much greater hazards through ingestion than would occur from the inhalation of air via the pulmonary tract. It is a local fallout problem rather than one of community air pollution (90). Acute toxicity is rarely seen as forage that is heavily contaminated is unpalatable. The lag period for the development of poisoning may be as long as several years after animals graze on contaminated pastures. The rate of absorption of the various compounds and postabsorption fate and toxicity are well known (91).

Regardless of the fluoride compound, symptoms in animals are the same: hyperostosis, fluorosis of the teeth, reduction of milk production, loss of weight, stiff posture, lameness, and poor hair coat.

Arsenic poisoning is also a fallout forage contamination problem. While it is of more historical than current interest, poisoning of pastures still occurs (92). A classic case of arsenic poisoning with great mortality in sheep and other animals was reported in detail in 1908 (93).

XVI. ANIMALS AS SENTINELS

Colonies of various laboratory animals have been employed by the Air Pollution Control Office to detect and study the biological effects of air pollutants. In these experiments, animals were housed at polluted sites, with one group exposed to polluted air and a corresponding group protected by air filtration. The sites established in Los Angeles, where photochemical smog occurs regularly, produced considerable biological alterations. The Los Angeles animals evidenced increased respiratory resistance on smoggy days and alteration of their pulmonary alveolar cells (94,95). It is interesting to note that the cellular changes were most severe following initial exposures, suggesting a later adaptative response to the atmosphere (96).

Somewhat similar studies conducted in an area of Detroit that is heavily polluted by auto exhaust were not productive of significant biological effects beyond the elevation of tissue lead levels. The paucity of oxidants in the Detroit atmosphere may account for a part of these differences (97).

Some interest has been expressed in studying various free-living wild animals for their potential in monitoring the effects of air pollution. Pigeons and starlings, for instance, are probably subjected to intense air pollution because of their habit of frequenting heavily industrialized areas. Their continued presence in such areas would seem to indicate that some adaptative response must be operating. Conversely, house martins, formerly residents in London, are said to have returned as nesting birds to the city only after the recent smoke control was achieved (98). Certain animals—Norway rats, ground squirrels, and others—live within very limited boundaries and thus might be of value in determining the effects of air pollution and other environmental agents in microecological situations.

Dairy cattle might offer a suitable model system for the study of air pollutants. Unlike most livestock, they are frequently kept in proximity to large cities and thus share with man the dubious advantages of such a locale. The detailed records that are kept might, when applied to matched herds in polluted and nonpolluted areas, yield interesting information about milk production and other parameters. Symptoms of air pollution effects and deaths in cattle during the Smithfield Show of 1953 were concomitant with a recognized episode and greater than normal human deaths (85,86); however, an earlier episode at the same show, although it occasioned greater sickness and mortality in cattle, appears not to have been correlated with human health at the time (99). One cannot help speculating that human health reporting was insufficiently developed to permit detection of effects in 1873 and that the cattle deaths were the only recognized evidence of an air pollution episode of equal or greater magnitude than the 1953 episode, until retrospective studies showed excess human deaths.

XVII. ECOLOGICAL IMPLICATIONS OF AIR POLLUTION

Since the earliest days of agriculture, man has probably been slowly groping toward a synthetic environment. Whenever sufficient population and technological skill have been available, alteration of the environment from agriculture and the cutting of the forests has occurred. Despite the many gains from irrigation, fertilization, and the like, much of this alteration has been detrimental to man's own existence. For instance, the gradual destruction of the hill forests of Italy since Etruscan times leaves today only a mere skeleton of a once verdant, fertile land. Much change is inevitable. The point must be made, however, that man's folk wisdom has never been able to predict and control environmental alteration consistently for his own betterment. Short-term agricultural gains have been followed by the creation of deserts, eroded hillsides, and intractable floods, leaving much of the classical world in a poorer physical state today than it was in Roman times.

Environmental changes now are taking place at an exponential rate because of the growth of the human population and the advancement of technology. The greatest danger does not lie in the destruction of physical aspects of the earth's surface but in the subtle chemical poisoning of the soil, water, and air through the by-products of man's technological society. To prevent environmental catastrophe, man must look to the future with all the wisdom he can muster to guide his development into nondestructive channels.

In addition to the immediate and local effects previously discussed, pollution of the air has regional and even global ramifications. These concern widespread deterioration of plant growth, interference with radiant energy from sunlight, deposition of acid products in the soil, and alteration of the gaseous components of the thin envelop of air surrounding the globe, the troposphere. Although these effects differ in their immediacy, there can be no doubt that if present trends of population growth and power uses continue, they must receive serious attention to forestall a future crisis. The deterioration of plants as the direct result of air pollution has been discussed previously. Obviously, serious interference with plant growth has ecological implications because of loss of human and animal food and lessened carbon dioxide removal by photosynthesis.

Acid washout results from sulfur dioxide pollution, which may be transported great distances by way of air currents and deposited as sulfates on the surface of the earth by means of rain and snow. In soil derived from acid rock with low buffering action and in water from surface runoff and melting snows, such acid contamination can result in deterioration of the ecological balance in both the soil and water by virtue of change in the hydrogen ion concentration. Concern is being expressed in Sweden about this alteration in the delicate ecological balance characteristic of arctic regions. Such alterations include changes in the biota of streams, slowing of the growth of lichens, and the loss of essential trace elements in the soil (5–7).

Haze from smoke, dirt, smog, fog, and other sources is more prominent in urban than in nonurban areas. It has been estimated that solar irradiation has been lessened by as much as one-half of the visible and two-thirds of the ultraviolet radiation in heavily polluted cities because of absorption and scattering of the sun's rays by haze particles (100). As our urban industrial complexes grow in intensity and size, the urban haze spreads ever more widely over the countryside. For example, a study in the central valley of California showed that air pollution haze geneally increased in intensity and area of coverage over a period of 18 years (101). Although visibility in some urban areas has increased over some years past because of smoke control, there is presently a generally increasing haze over many areas as the result of photochemical smog. Future interference by air pollution with the effects of the sun's energy on global heat and plant growth cannot be gauged accurately at this time, but any significant increase in pollution poses a threat in this regard.

Carbon dioxide is a normal and necessary part of the air, where it is theoretically in equilibrium between evolution and consumption. It is evolved by decomposition of organic matter, respiration of animals, and dark-phase respiration of plants and is consumed during photosynthesis by green plants during the daylight hours and by absorption into the oceans. Because of these and other factors, carbon dioxide fluctuates in the air at various times and places and ranges from approximately 300 to 450 ppm from day to night over areas covered by vegetation. Normally an equilibrium between natural production sources and sinks from carbon dioxide would be expected. However, there appears to be fairly consistent consensus among investigators that a gradual but appreciable increase in the amount of carbon dioxide in the air has occurred during the last part of the nineteenth century and in the present century, resulting in an estimated change from 285 ppm in 1870 to approximately 330 ppm in 1956 (102). Furthermore, the increased carbon dioxide increment is directly related to an increased consumption of fossil fuels during this period. Although some of these estimates are based on data derived prior to 1900 and are of questionable accuracy, recent data based on detailed studies over shorter periods substantiate them (103,104). Observations over a 4-year period in Hawaii indicate an average increase in atmospheric CO_2 levels of 0.70 ppm/year; observations over a 2-year period in the Antarctic revealed an increase of 1.3 ppm/year (105,106). Similar increments have been reported in the city of Paris, where CO_2 levels rose from 320 ppm (1891 to 1900) to 354 ppm (1959 to 1964) respectively, an increment of 34 ppm in 64 years (107). It is clear that an accumulation of carbon dioxide is occurring in the troposphere. Since an increase in carbon dioxide results in an increased absorption of infrared radiation, it has been speculated that the gradual accumulation of carbon dioxide will eventually cause heating of the entire earth because of increased retention of solar heat. This "greenhouse effect" has received much attention in the popular press where the attendant dangers, such as alterations in climate and melting of the polar ice caps, have been prominently mentioned. A calculation of the impact of changes in carbon dioxide levels on world heat indicates that a 300 ppm increment would cause an elevation of temperature of 1.5°C (108). At the rate of change now predicted, an increment of this magnitude would be achieved in 300 to 500 years. Even though we are probably in no immediate danger from severe climatic changes due to the accumulation of carbon dioxide, the steady increase of this gas illustrates that significant and potentially dangerous alterations of the troposphere are occurring as the result of human activities.

XVIII. CONTROL OF AIR POLLUTION

Man's pollution of the air was not merely a frivolous act, but one that was necessary for his domestication. The primitive industry associated with the dawn of civilization could not have been accomplished without the use of fire and the passage of smoke into the air. Thus it has been both natural and necessary from Paleolithic times to the present for burning and related industrial activities to proceed with little regard to their effect on the quality of the atmosphere. The air was free and limitless; it was as natural to pollute it as it was to respire it. The atmosphere miraculously took care of such matters just as the limitless land and water disposed of their respective burdens, the by-products of human living. Some attempts were made to conserve water quality in the early days, not by controlling pollution, but by assuring special sources of potable water for domestic use. Meanwhile, in heavily populated areas the streams gradually deteriorated into open sewers in which fish could no longer survive. When it became apparent that the concentration of air pollutants was becoming detrimental to man, the early solutions that had been used for water were not practical since air pollution to human beings and other terrestrial animals is equivalent to water pollution for fish. The air breather must dwell in it and utilize it freely; he cannot be supplied with a purified sample through a pipe. Early attempts to control pollution consisted mainly of constructing high smokestacks to convey the pollutants from the area. Thinking of this type still pervades air pollution control philosophy, and high stacks are still being constructed in lieu of better methods. Unfortunately, because of the proliferation of pollution sources, their aggregation in vast urban complexes, the rise of the internal combustion engine, and an attendant worldwide deterioration of air quality, such simplistic approaches are no longer valid. It can be said that *dilution is no solution for pollution.*

Control of air pollution or, more accurately, the preservation of the quality of the atmosphere, calls first and foremost for a change in traditional attitudes. We can no longer lightheartedly throw trash into the air. There are too many of us, and we demand too much space, heat, and power to tolerate practices suitable for a less populous agrarian society. Unfortunately, there are so many technologically developed, uncontrolled pollution sources that we must all share the tremendous monetary, social, physical, and political efforts before significant progress can be made. This is a price we must pay for our population density and for our technological society. Furthermore, the cost of control is much smaller than the cost of pollution.

It would be unfair to state that no progress in pollution control has been accomplished. Considerable effort on the part of both private and public sectors has been expended. We have local and state air pollution control boards that study and attempt to control pollution. However, because of the technical, economic, and political problems imposed by local conditions and conflicts between private and public interests, sources of pollution seem to be proliferating faster than effective control measures. Pollution sources today and in the future cannot be coped with adequately by smoke abatement acts, public nuisance laws, and private suits. To supplement local control, federal activity has been focused on air pollution. Consequently, the Clean Air Act was passed in 1963 and was amended numerous times (109). The latest and most vital effort toward effective control of air pollution is vested in the Clean Air Amendments of 1970 (110). By virtue of these amendments, a national research and development program for investigation into the causes, effects, and control of air pollution was instituted. This act also provides for grants-in-aid for local, state, or interstate control agencies. Perhaps one of the most important features of the Clean Air Act is the provision for the development of air quality criteria for the identification of the effects of air pollution on health and welfare and the promulgation of national ambient air quality standards. Air quality criteria for an air pollutant shall accurately reflect the latest scientific knowledge useful in indicating the kind and extent of all identifiable effects of such pollutant in the ambient air, in varying quantities. Also in the Clean Air Amendment of 1970 is the provision for the development of national emission standards for hazardous air pollutants. A hazardous air pollutant means an air pollutant to which no ambient air quality standard is applicable and which in the judgement of the Environmental Protection Agency Administrator may cause or contribute to an increase in mortality or an increase in serious irreversible or incapacitating reversible illness.

The prevention and control of air pollution at its source is the primary responsibility of state and local governments. The federal government sets the ambient air quality standards. The state, through federally approved implementation plans, develops and carries out the control strategy leading to effective control of air pollution. The federal government also sets performance standards for new industrial sources of air pollutants and emission standards for new automobiles. Emission standards for hazardous pollutants are promulgated and enforced by the federal agency.

The technical aspects of air pollution control are very complex, and in general, expensive. In some instances adequate means of control have not yet been adequately demonstrated. Various methods have been developed to remove solids and gases from effluents of stationary sources, such as the precipitators, water scrubbers, and chemical means. In methods employing water, there is an attendant danger of water pollution since the waste-containing effluent must eventually be discharged into natural bodies of water. Thus the control technology must always be evaluated against the environmental impact.

XIX. CONTROL OF AUTOMOBILE EMISSIONS

The control of pollutants from automobiles has been discussed widely in the popular press and among individuals. Concern has been expressed about pollutants from automobiles, not only because of their direct toxic effects, but also because of the role some of them play in the development of secondary pollutants in photochemical smog. Control of automotive pollution has been considered chiefly from three aspects:

1. Control of effluents from present-day cars by slight modifications in the operation of internal combustion engines, such as blow-by recycling, exhaust recycling, air injection, and improved mixture control. Equipping new automobiles with such control systems is already obligatory. Slightly more complex devices such as catalytic mufflers and afterburners are under discussion, but have not been utilized because of technical problems. The modification of gasoline by removing lead and other ingredients may also be required.

2. Substitution of a low-pollution power source for the conventional internal combustion engine have been advocated. Gas turbines, steam engines, and electric motors have come under consideration for use in private automobiles. Such alternatives are not finding ready acceptance for many reasons, one of which undoubtedly is the inertia of established thinking in favor of the internal combustion engine.

3. Development of adequate and efficient mass transit systems in densely populated areas is thought by many to be the solution to motor vehicle pollution, despite the best engineering efforts being expended on control measures. Otherwise the proliferation of automobiles that must occur to keep pace with our growing population eventually will make life very difficult. However, in addition to the political inertia and economic problems besetting the development of such transit systems, we face the public attitude toward the automobile; in no other age has so much power and freedom of movement been in the hands of the ordinary citizen. The automobile makes of every man a *caballero*. Thus it seems likely that a significant amount of discomfort must ensue before we are willing to give up such a remarkable steed as the automobile.

Acknowledgment

The author thanks Dr. Ronald Engel for his assistance with the section pertaining to the Public Laws.

References

1. L. J. Carter, "Thermal Pollution: A Threat to Cayuga's Waters?" *Science*, **162**, 649 (1968).

2. J. Korshover, "Synoptic Climatology of Stagnating Anticyclones East of the Rocky Mountains in the United States for the Period 1936–1956," *Report SEC TR-A60-7,* Robert A. Taft Sanitary Engineering Center, Cincinnati, Ohio, 1960.
3. A. J. Haagen-Smit and L. G. Wayne, "Atmospheric Reactions and Scavenging Processes," in *Air Pollution,* Vol. I, 2nd ed., A. C. Stern, Ed., Academic Press, New York, 1968.
4. J. E. Yocom and R. O. McCaldin, "Effect of Air Pollution on Materials and Economy," in *Air Pollution,* Vol. I, 2nd ed., A. C. Stern, Ed., Academic Press, New York, 1968.
5. E. Gorham, "The Influence and Importance of Daily Weather Conditions in the Supply of Chloride, Sulfate and other Ions to Fresh Waters from Atmospheric Precipitation," *Phil. Trans. Roy. Soc. London, Ser. B* **241,** 147 (1958).
6. E. Barrett and G. Brodin, "The Acidity of Scandinavian Precipitation, *Tellus,* **7,** 251 (1955).
7. American Chemical Society, "Cleaning Our Environment; a Chemical Basis for Action," a Report by the Subcommittee on Environmental Improvement, Committee on Chemistry and Public Affairs, Washington, D.C., American Chemical Society, 1969, pp. 31–32.
8. J. T. Middleton, J. B. Kendrick, and H. W. Schwalm, "Injury to Herbaceous Plants by Smog or Air Pollution," *Plant Disease Reptr.,* **34,** 245 (1950).
9. P. R. Miller, "Air Pollution and the Forests of California," *California Environ.,* **1,** 1 (1969), University of California, Riverside, Statewide Air Pollution Center.
10. P. R. Miller, J. R. Parameter, Jr., H. F. Brigitta, *et al.,* "Ozone Dosage Response on Ponderosa Pine Seedlings," *J. Air Pollution Control Assoc.,* **19,** 435 (1969).
11. L. Greenburg, C. L. Erhardt, F. Field, *et al.,* "Air Pollution Incidents and Morbidity Studies," *Arch. Environ. Health,* **10,** 351 (1965).
12. Air Pollution Foundation, Report 30, Sixth Technical Progress Report by W. L. Faith, Los Angeles, 1960, pp. 14–18.
13. P. J. Lawther, "Climate Air Pollution and Chronic Bronchitis," *Proc. Roy. Soc. Med.,* **51,** 262 (1958).
14. S. F. Buck and D. A. Brown, "Mortality from Lung Cancer and Bronchitis in Relation to Smoke and Sulphur Dioxide Concentration, Population Density and Social Index," Tobacco Research Council, Research Paper 7, 1964.
15. P. M. Lambert and D. D. Reid, "Smoking, Air Pollution, and Bronchitis in Britain," *Lancet,* **1,** 853 (1970).
16. C. M. Fletcher, "Chronic Bronchitis: Its Prevalence, Nature, and Pathogenesis," *Amer. Rev. Respirat. Diseases,* **80,** 483 (1959).
17. I. I. Higgins and J. B. Cochran, "Respiratory Symptoms, Bronchitis and Disability in a Random Sample of an Agricultural Community in Dumfrieshire," *Tubercle,* **39,** 296 (1964).
18. T. Mork, "A Comparative Study of Respiratory Disease in England and Wales and Norway," *Acta Med. Scand. Suppl. 384,* **172** 1 (1962).
19. Y. Oshima, T. Ishizaki, T. Miyamoto, *et al.,* "Air Pollution and Respiratory Diseases in the Tokyo–Yokohama Area," *Amer. Rev. Respirat. Diseases,* **90,** 572 (1964).
20. H. L. Motley *et al.,* "Effect of Polluted Los Angeles Air (Smog) on Lung Volume Measurements," *J. Amer. Med. Assoc.,* **171,** 1469 (1959).

21. J. E. Remmers and O. J. Balchum, "Effects of Los Angeles Urban Air Pollution upon Respiratory Function of Emphysematous Patients: the Effect of the Micro Environment on Patients with Chronic Respiratory Disease," presented at the Air Pollution Control Association Annual Meeting, Toronto, June 1965.
22. H. Weill, M. M. Ziskind, V. J. Derbes, *et al.*, "Recent Developments in New Orleans Asthma," *Arch. Environ. Health,* **10,** 148 (1965).
23. H. Weill, M. M. Ziskind, R. C. Dickerson, *et al.*, "Allergenic Air Pollutants in New Orleans," *J. Air Pollution Control Assoc.,* **15,** 467 (1965).
24. L. D. Zeidberg, R. A. Prindle, and E. Landau, "The Nashville Pollution Study, III. Morbidity in Relation to Air Pollution," *Amer. J. Public Health,* **54,** 85 (1964).
25. N. R. Frank, M. O. Amdur, and J. L. Whittenberger, "A Comparison of the Acute Effects of SO_2 Administered Alone or in Combination with NaCl Particles on the Respiratory Mechanisms of Healthy Adults," *Intern. J. Air Water Pollution,* **8,** 125 (1964).
26. T. Dalham and L. Strandberg, "Synergism Between Sulfur Dioxide and Carbon Particles; Studies on Adsorption and on Ciliary Movements in the Rabbit Trachea *in vivo*," *Intern. J. Air Water Pollution,* **7,** 517 (1963).
27. S. D. Murphy, J. K. Leng, C. E. Ulrich, *et al.*, "Effects on Animals of Exposure to Auto Exhaust," *Arch. Environ. Health,* **7,** 303 (1963).
28. M. O. Amdur, "The Effect of Aerosols on the Response to Irritant Gases," in *Inhaled Particles and Vapors,* C. N. Davies, Ed., Pergamon Press, Oxford, 1961, pp. 281–292.
29. L. Reid, "An Experimental Study of Hypersecretion of Mucus in the Bronchial Tree," *Brit. J. Exp. Pathol.,* **44,** 437 (1963).
30. J. W. Douglas and R. E. Waller, "Air Pollution and Respiratory Infection in Children," *Brit. J. Prevent. Social Med.,* **20,** 1 (1966).
31. J. E. Lunn, J. Knowelden, and A. J. Handyside, "Patterns of Respiratory Illness in Sheffield Infant School Children," *Brit. J. Prevent. Social Med.,* **21,** 7 (1967).
32. W. W. Holland and D. D. Reid, "The Urban Factor in Chronic Bronchitis," *Lancet,* **1,** 445 (1965).
33. R. Rylander, "Plumonary Defense Mechanisms to Airborne Bacteria," *Acta Physiol. Scand. Suppl.,* **306,** 55–57, (1968).
34. T. Dalham and J. Rhodin, "Mucous Flow and Ciliary Activity in the Tracheae of Rats Exposed to Pulmonary Irritant Gas," *Brit. J. Ind. Med.,* **13,** 110 (1956).
35. K. H. Kilburn, "Cilia and Mucus Transport as Determinants of the Response of Lung to Air Pollutants," *Arch. Environ. Health,* **14,** 77 (1967).
36. D. L. Coffin and E. J. Blommer, "Acute Toxicity of Irradiated Auto Exhaust," *Arch. Environ. Health,* **15,** 36 (1967).
37. D. L. Coffin, E. J. Blommer, D. E. Gardner, *et al.*, "Effect of Air Pollution on Alteration of Susceptibility to Pulmonary Infection," *Proc. 3rd Ann. Conf. Atmospheric Contamination Confined Spaces,* Dayton, Ohio, 1965.
38. R. Ehrlich, "Effect of Nitrogen Dioxide on Resistance to Respiratory Infection," *Bacteriol. Rev.,* **30,** 604 (1966).
39. M. C. Henry, J. Findlay, J. Spangler, *et al.*, "Chronic Toxicity of NO_2 in Squirrel Monkeys; III. Effect on Resistance to Bacterial and Viral Infection," *Arch. Environ. Health,* **20,** 566 (1970).

40. R. Ehrlich, unpublished data. Contract No. PHS 86-67-30, under the auspices of the National Air Pollution Control Administration, 1969.
41. D. L. Coffin and E. J. Blommer, "Alteration of the Pathogenic Role of Streptococci Group C in Mice Conferred by Previous Exposure to Ozone," *Proc. 3rd Intern. Symp. Aerobiology,* University of Sussex, England, 1969.
42. D. L. Coffin, D. E. Gardner, R. S. Holzman, *et al.,* "Influence of Ozone on Pulmonary Cells," *Arch. Environ. Health,* **16,** 633 (1968).
43. D. J. Hurst, D. E. Gardner, and D. L. Coffin, "Effect of Ozone on Lysosomes," *Reticuloendothelial Soc. J.,* **8,** 288 (1970).
44. R. S. Holzman, D. E. Gardner, and D. L. Coffin, "*In vivo* Inactivation of Lysozyme by Ozone," *J. Bacteriol.,* **96,** 1562 (1968).
45. D. E. Gardner, E. A. Pfitzer, R. T. Christian, *et al.,* "Loss of Protective Factor for Alveolar Macrophages when Exposed to Ozone," presented at the 10th Annual Hanford Biology Symposium on Biochemical Effects of Air Pollutants, Pasco, Wash., June 1970.
46. D. J. Hurst and D. L. Coffin, "Effect of Ozone on Lysosomal Hydrolases of Alveolar Macrophage *in vitro,*" presented at the 10th Annual Hanford Biology Symposium on Biochemical Effects of Air Pollutants, Pasco, Wash., June 1970.
47. D. L. Coffin, "Study of the Mechanisms of Alteration of Susceptibility to Infection Conferred by Oxidant Air Pollutants," *Inhalation Carcinogenesis* (AEC Symposium Series 18), U.S. AEC, Oak Ridge, Tenn., 1970, pp. 259–269.
48. S. B. Valand, J. D. Acton, and Q. N. Myrvik, "Nitrogen Dioxide Inhibition of Viral-Induced Resistance in Alveolar Macrophages," *Arch. Environ. Health,* **20,** 303 (1970).
49. C. M. Shy, J. P. Creason, M. Pearlman, *et al.,* "The Chattanooga School Children Study: Effects of Atmospheric Nitrogen Dioxide Exposure on the Incidence of Acute Respiratory Illness," to be published.
50. D. Anderson and B. G. Ferris, "Air Pollution Levels and Chronic Respiratory Disease," *Arch. Environ. Health,* **10,** 307 (1965).
51. S. Ishikawa *et al.,* "The 'Emphysema Profile' in Two Mideastern Cities in North America," presented at the 9th American Medical Association Research Conference, Denver, Colo., 1968.
52. O. Auerbach, A. P. Stout, E. C. Hammond, *et al.,* "Smoking Habits and Age in Relation to Pulmonary Changes," *New Engl. J. Med.,* **269,** 1045 (1963).
53. G. Freeman, N. J. Furiosi, and G. B. Haydon, "Effects of Continuous Exposure of 0.8 ppm NO_2 on Respiration of Rats," *Arch. Environ. Health,* **13,** 454 (1966).
54. G. Freeman, S. C. Crane, R. J. Stephens, *et al.,* "Environmental Factors in Emphysema and a Model System with NO_2," *Yale J. Biol. Med.,* **4,** 567 (1968).
55. G. Freeman, unpublished data. Contract No. PHS 86-65-27 under the auspices of the National Air Pollution Control Administration, 1969.
56. W. H. Blair, M. C. Henry and R. Ehrlich, "Chronic Toxicity of Nitrogen Dioxide; II. Effect on Histopathology of Lung Tissue," *Arch. Environ. Health,* **18,** 186 (1969).
57. J. H. Riddick, K. I. Campbell, and D. L. Coffin, "Effects of Chronic Nitrogen Dioxide Exposure on Dogs; I. Histopathology of the Lung," presented at the Joint Meeting of the American Society of Clinical Pathologists and the College of American Pathologists, Chicago, Ill., September, 1967.

58. G. Freeman, S. C. Crane, R. J. Stephens, *et al.*, "Pathogenesis of the Nitrogen Dioxide-Induced Lesion in the Rat Lung: a Review and Presentation of New Observations," *Amer. Rev. Respirat. Diseases,* **98,** 429 (1968).
59. G. Freeman, personal communication, 1970.
60. G. C. Buell, Y. Takiwa, and P. K. Mueller, "Lung Collagen and Elastin Denaturation *in vivo* Following Inhalation of Nitrogen Dioxide," presented at the 59th Annual Air Pollution Control Association Meeting, San Francisco, Calif., June 1966.
61. O. Auerbach, E. C. Hammond, D. Kirman, *et al.,* "Emphysema Produced in Dogs by Smoking," in *Inhalation Carcinogenesis* (AEC Symposium Series 18), U.S. AEC, Oak Ridge, Tenn., 1970, pp. 375–387.
62. American Cancer Society, *1970 Cancer Facts and Figures,* American Cancer Society, New York, 1970, pp. 9, 17.
63. Shimkin, M. B., *Science and Cancer,* PHS Publication No. 1162, U.S. DHEW, National Cancer Institute, Washington, D.C., 1964.
64. E. L. Winder and D. Hoffman, "Air Pollution and Lung Cancer," prepared discussion, *Proc. Nat. Conf. Air Pollution,* PHS Publication No. 1022, DHEW, Washington, D.C., 1963, pp. 144–148.
65. "Air Quality Data from the National Air Surveillance Networks and Contributing State and Local Networks," U.S. DHEW, PHS, EHS, National Air Pollution Control Administration, 1966.
66. W. Haenszel, D. B. Loveland, and M. G. Sirken, "Lung Cancer Mortality as Related to Residence and Smoking Histories; I. White Males," *National Cancer Inst. J.,* **28,** 947 (1962).
67. P. Stocks and J. M. Campbell, "Lung Cancer Death Rates among Nonsmokers, Pipe and Cigarette Smokers; an Evaluation in Relation to Air Pollution by Benzopyrene and other Substances," *Brit. Med. J.,* **2,** 923 (1955).
68. P. Stocks, "Relations between Atmospheric Pollution in Urban and Rural Localities and Mortality from Cancer, Bronchitis, and Pneumonia with Particular References to 3,4 Benzo(a)pyrene, Beryllium, Molybdenum, Vanadium, and Arsenic," *Brit. J. Cancer,* **14,** 379 (1960).
69. D. J. B. Ashley, "Environmental Factors in the Aetiology of Lung Cancer and Bronchitis," *Brit. J. Prevent. Social Med.,* **23,** 258 (1969).
70. W. C. Hueper, P. Kotin, E. C. Tabor, *et al.,* "Carcinogenic Bioassays on Air Pollutants," *Arch. Pathol.,* **74,** 89 (1962).
71. U. Saffiotti, F. Cefis, and L. H. Kolb, "Bronchiogenic Carcinoma Induction by Particulate Carcinogens," *Proc. Amer. Cancer Res.,* **5,** 55 (1964).
72. I. J. Selikoff, *et al.,* "Asbestos Exposure, Smoking, and Neoplasia," *J. Amer. Med. Assoc.* **204,** 106 (1968).
73. U. Saffiotti, F. Cefis, L. H. Kolb, *et al.,* "Experimental Studies of the Conditions of Exposure to Carcinogens for Lung Cancer Induction," *J. Air Pollution Control Assoc.,* **15,** 23 (1965).
74. L. M. Shabad, L. M. Pylev, and T. S. Kolesnichenko, "Importance of the Deposition of Carcinogens for Cancer Induction in Lung Tissue," *J. Nat. Cancer Inst.,* **33,** 135 (1964).
75. L. N. Pylev, "Effect of Dispersion of Soot in Deposition of 3,4-benzpyrene in Lung Tissue of Rats," *Hyg. and Sanit.,* **32,** 174 (1967); translated from *Gigiena i Sanit.*
76. S. Laskin, "Carcinogenic Effects of SO_2 and Topical Approaches by Hydrocar-

bons," in *Inhalation Carcinogenesis* (AEC Symposium Series 18), U.S. AEC, Oak Ridge, Tenn., 1970, pp. 321–351.
77. J. R. Dixon, *et al.*, "Role of Trace Metals in Chemical Carcinogenesis-Asbestos Cancers," presented at the 2nd Annual Conference on Trace Substances in Environmental Health, 1968.
78. A. M. Ray and T. H. Rockwell, "Exploratory Study of Automobile Driving Performance under the Influence of Low Levels of Carboxyhemoglobin," Systems Research and Industrial Engineering, Ohio State University Report IE-i, August 1967.
79. W. E. Ehrich, S. Bellet, and F. H. Lewey, "Cardiac Changes from CO Poisoning," *Amer. J. Med. Sci.*, **208,** 511 (1944).
80. F. H. Lewey and D. L. Drabkin, "Experimental Chronic Carbon Monoxide Poisoning of Dogs," *Amer. J. Med. Sci.*, **208,** 502 (1944).
81. M. Abramowicz and E. H. Kass, "Pathogenesis and Prognosis of Prematurity," *New Engl. J. Med.*, **275,** 938 (1966).
82. P. Astrup and H. G. Pauli, eds., "A Comparison of Prolonged Exposure to Carbon Monoxide and Hypoxia in Man," *Scand. J. Clin. Lab. Invest.*, **24,** 1 (1968).
83. R. R. Beard and G. A. Wertheim, "Behavioral Impairment Associated with Small Doses of Carbon Monoxide," *Amer. J. Public Health,* **57,** 2012 (1967).
84. R. D. Steward, J. E. Peterson, E. D. Baretta, *et al.*, "Experimental Human Exposure to Carbon Monoxide," reported under Contract CRC-APRAC, Project CAPM 3-68, Coordinating Research Council, Inc., 1969.
85. E. J. Catcott, "Effects of Air Pollution on Animals," in *Air Pollution,* World Health Organization, 1961.
86. C. A. Mills, *Air Pollution and Community Health,* Christopher, Boston, Mass., 1954.
87. C. R. Cole, R. L. Farrell, and R. A. Griesemer, "The Relationship of Animal Disease to Air Pollution," report of Contract SAPH 69436, U.S. DHEW, National Air Pollution Control Administration, 1964, Ohio State Research Foundation, Columbus, Ohio.
88. C. R. Dorn, "Survey of Animal Neoplasms in Alameda and Contra Costa Counties; I. Methodology and Description of Cases; II. Cancer Morbidity in Dogs and Cats from Alameda County," *J. Nati. Cancer Inst.*, **40,** 295 (1968).
89. H. E. Stokinger and D. L. Coffin, "Biological Effects of Air Pollution," in *Air Pollution,* Vol. I, 2nd ed., A. C. Stern, Ed., Academic Press, New York, 1968, pp. 445–546.
90. P. H. Philips and J. W. Suttie, "The Significance of Time in the Intoxication of Domestic Animals by Fluoride," *Arch. Environ. Health,* **21,** 343 (1960).
91. J. Bronch and N. Grieser, "Fluorine and Fluorine Tolerance in Fodder of Domestic Animals," *Berl. Muench. Tieraerztl. Wochschr.*, **77,** 401 (1964).
92. D. J. Birmingham, "An Outbreak of Arsenical Dermatosis in a Mining Community," *Arch. Dermatol.*, **91,** 457 (1965).
93. W. D. Harkins and R. E. Swain, "The Chronic Arsenical Poisoning of Herbivorous Animals," *J. Amer. Chem. Soc.*, **30,** 928 (1908).
94. H. E. Swann, D. Brunol, and O. J. Balchum, "Pulmonary Resistance Measurement of Guinea Pigs," *Arch. Environ. Health,* **10** (1965).
95. H. E. Swann, D. Brunol, L. G. Wayne, *et al.*, "Biological Effects of Urban Air Pollution; II. Chronic Exposure of Guinea Pigs," *Arch. Environ. Health,* **11,** 765 (1965).

96. R. F. Bils, "Ultrastructure Alterations of Alveolar Tissue of Mice; I. Due to Heavy Los Angeles Smog," *Arch. Environ. Health,* **11,** 765 (1965).
97. R. G. Smith, A. J. Vorwald, A. L. Reeves, *et al.,* "A Summary of Health Effects Resulting from Long-term Exposure of Animals to Urban Air," presented at the Annual Meeting of the American Industrial Hygiene Association, Chicago, Ill., May 1967.
98. "House Martins Return to London," *Sci. News,* **95,** 569 (1968).
99. *London Times,* December 12, 1873.
100. R. A. McCormick, "Air Pollution Climatology," in *Air Pollution,* Vol. I, 2nd ed., A. C. Stern, Ed., Academic Press, New York, 1968, p. 306.
101. G. C. Holzworth and J. A. Maga, "A Method for Analyzing the Trend in Visibility," *J. Air Pollution Control Assoc.,* **10,** 430 (1960).
102. G. S. Callendar, "On the Amount of Carbon Dioxide in the Atmosphere," *Tellus,* **10,** 243 (1958).
103. R. Revelle and H. E. Suess, "Carbon Dioxide Exchange between Atmosphere and Ocean, and the Question of an Increase of Atmospheric CO_2 during the Past Decades," *Tellus,* **9,** 18 (1957).
104. W. Bischof and B. Bolin, "Space and Time Variations of the CO_2 Content of the Troposphere and Lower Stratosphere," *Tellus,* **18,** 155 (1966).
105. B. Bolin and C. D. Keeling, "Large-scale Atmospheric Mixing as Deduced from the Seasonal and Meridional Variations of Carbon Dioxide," *J. Geophys. Res.,* **68,** 3899 (1963).
106. C. D. Keeling, "The Concentration and Isotopic Abundance of Carbon Dioxide in the Atmosphere," *Tellus,* **12,** 200 (1960).
107. J. Pelletier, unpublished data, quoted by B. D. Tebbens, in *Air Pollution,* Vol. I, 2nd ed., A. C. Stern, Ed., Academic Press, New York, 1968.
108. F. Moller, "On the Influence of Changes in the CO_2 Concentration in Air on the Radiation Balance of the Earth's Surface and on the Climate," *J. Geophys. Res.,* **68,** 3877 (1963).
109. *Public Law 88–206,* December 17, 1963, as amended by *Public Law 89–272,* October 20, 1965.
110. Public Law 88-206 89-272 as amended by Public Law 91-604, December 31, 1970.

Environmental Pollution by Mercury

J. M. Wood
Biochemistry Department,
School of Chemistry,
University of Illinois,
Urbana, Illinois

I. INTRODUCTION

A decade ago two independent large-scale incidences of poisoning were observed in Japan and Sweden. In Japan many people belonging to the fishing population around Minamata Bay were affected seriously by what is now called Minamata disease and which is characterized by sometimes irreversible or fatal neurological disorders. In Sweden certain bird populations decreased drastically, and, at the same time, many birds suffering from cramps could be found. Eventually, these two unrelated incidents were traced back to the same origin—poisoning by ingestion of methylmercury compounds.

The poisoning of Swedish birds was finally attributed to (*1*) unusually high mercury levels in dead birds (1), (*2*) relatively high mercury content in some Swedish agricultural products (especially eggs) compared with those of other European countries (2), and (*3*) the extensive use of methylmercury-dicyanodiamide (Panogen®) as a fungicide in Swedish agriculture (3).

Alkylmercury generally causes characteristic neurotoxic symptoms and, in contrast to other organic mercury compounds, alkylmercury is metabolized slowly in the mammalian body (4). Therefore, cumulative alkylmercury poisoning results as small quantities are ingested sequentially. The use of methylmercury as a fungicide is now banned in Sweden (February 1966), and since this legislation, mercury content in birds and eggs has decreased drastically.

Minamata disease was traced back to the mercury-containing effluent of a large chemical plant (5). The neurologic syndrome of the disease was identical to that described previously for alkylmercury poisoning, and CH_3-Hg-S-CH_3 was isolated from shellfish in the Minamata Bay area (6). The Minamata factory was using inorganic mercury salts or mercury metal as catalysts, and when spoiled, these catalysts were partially deposited in a factory dump and partially lost into the sewage system. It has been suggested by Kurland that inorganic mercury issued with sewage effluent into the sea may be alkylated by "plankton or other marine life" (5). However, when it was discovered that the spent catalyst of an acetaldehyde reactor in the plant contained approximately 1% methylmercury, it was concluded that biological methylation of mercury did not make any significant contribution to the total methylmercury ingested by shellfish in the bay.

In Sweden great interest has been focused on the occurrence of mercury in various organisms. Studies were prompted because of the disastrous effects of the use of methylmercury as a fungicide, and they were facilitated by the supersensitive analytical methods for mercury developed in another context by Westermark and Sjöstrand (7,8). Extensive investigations have been made on the mercury content of fish (9). Fish from inland and coastal waters contain much higher levels of mercury than those caught in the ocean (i.e., several ppm to 0.01–0.1 ppm, respectively). In some cases, this rise in mercury content may be correlated with industrial water pollution. It is evident that airborne mercury from industrial areas may pollute distant waters (10). The most disturbing aspect of mercury content in fish is that mercury exists predominantly as recalcitrant methylmercury (11,12), even in areas where inorganic mercury or phenylmercury are predominant in sewage effluent. Clearly, inorganic mercury and phenylmercury (which is readily degraded to Hg^{2+}) must have been methylated either in the aquatic environment or in the fish.

With methylmercury compounds, groups occupying the cationic coordination positions are of minor importance in biological systems allowing exchanges of the following type:

$$RHgX + Y^{\ominus} \rightleftharpoons RHgY + X^{\ominus}$$

A good example of this is the use of organomercurials to titrate free cysteine sulfhydryl groups in purified enzymes. Similarly, organomercurials are used frequently to establish the involvement of sulfur as the catalytic site of some enzymes by use of enzyme kinetics.

After Westöö demonstrated that liver homogenates can methylate inorganic mercury (13), Jensen and Jernlöv demonstrated that mercury is methylated in sludge taken from aquaria (14). Whether this alkylation reaction was chemical or enzymatic in nature was difficult to establish. In methanogenic bacteria, alkyl-B_{12} compounds serve as alkylating agents for inorganic mercury (15), arsenic (16), and selenium and tellurium (17). Clearly, methylcobalamin (methyl-B_{12}) may serve as a chemical alkylating agent in the aquarium sludge ecosystem, in N^5 methyltetrahydrofolate-homocysteine transmethylase (B_{12}-dependent) in liver (18), and in all microorganisms capable of synthesizing methylcobalamin (19). It should be pointed out that the synthesis of methylcobalamin in the microbial ecosystems is enzymatic in origin, and in order for this synthesis to progress, mercury levels must be low enough not to inhibit seriously these enzymes.

II. THE MAJOR SOURCES OF MERCURY POLLUTION OF THE ENVIRONMENT

In medieval England, mercury metal was used as an agent to induce abortion and to commit suicide. In some Far Eastern countries small quantities of mercury metal are used as a potent diuretic. In *Alice in Wonderland,* the Mad Hatter is a perfect example of a person suffering from chronic mercurial poisoning. The "Hatters" in London used mercuric oxide to treat velvet fabrics, and after constant skin contact, this mercury salt would pass through the skin into the bloodstream. After a number of years the mercury would accumulate to a level where the "Hatter" would become a mental defective. The tragedy of the Mad Hatter also affected some dentists who succumbed to mercury poisoning as a result of constant exposure to mercury vapor when mixing the mercury-silver amalgam for fillings. These two classic examples of mercury poisoning are restricted to the symptoms developed after excessive contact with *inorganic* mercury (mercury metal or mercury salts); this does *not* threaten the health and welfare of the general public.

With the discovery that *inorganic* mercury is converted to deadly poisonous *methylmercury* by microorganisms which use *methyl-Vitamin* B_{12}, it was realized that this methylmercury in the environment along with the methylmercury used as a fungicide in agriculture seriously threatens the health and welfare of the general public. It has now been established that methylmercury is beginning to accumulate in a number of foods common to the diet of the people of the United States.

Mercury pollution can be divided conveniently into two categories: (*1*) intentional and (*2*) unintentional pollution. Intentional pollution involves the widespread distribution of methylmercury in agriculture, representing about 3% of the total mercury used in the United States. Unintentional pollution involves mercury losses in a variety of forms by industry, hospitals, power plants, and institutional research laboratories.

Industrial chemists did not realize that mercury losses in industrial effluents would lead to the formation of methylmercury by microorganisms to establish a food chain for the eventual accumulation of this deadly poisonous compound in fish. Twenty-six thousand tons of chlorine are manufactured daily in the United States and 25% of this chlorine is manufactured through the electrolysis of brine by using mercury metal as an electrode. Consequently, significant quantities of mercury metal and mercury salts have been widely distributed in inland and coastal waterways in the United States and Canada. To give an example of the extent of mercury losses, 200,000 lb of inorganic mercury have been deposited in the St. Clair River System in the last 20 years. The chlorine industry is not the only contributor to this pollution problem since significant quantities of mercury compounds are lost in the following processes:

1. *Plastics industry* (in the manufacture of vinyl chloride). Losses of both mercury salts and a small amount of methylmercury by-product.
2. *Paper industry.* Losses of mercury metal and mercury salts as impurities in the caustic soda used in bleaching. Phenylmercuric acetate (PMA) was used as a slimicide by the paper industry. This practice has now been stopped, but significant levels of PMA still pollute waterways because of the large quantities used.
3. *Electronics industry.* Losses due to the increased use of mercury metal. For example, the "long life" alkaline batteries have a total mercury content of 8% (we usually throw them away).
4. *Hospitals.* Losses due to the use of mercury salts in histology laboratories.
5. *Power companies.* Losses of mercury metal from coal.

Organomercurials are still used as potent diuretics and as bactericidal agents in ointments, e.g., Mercurochrome. However, with increased emphasis on the use of pesticides and insecticides to increase crop productivity, vast quantities of organomercurials are in use widely in advanced agricultural systems, particularly in the United States. This usage enables mercury to enter a variety of crops, meats, and eggs through food chains in the environment. Seeds are particularly vulnerable to fungus deterioration, and treatment with liquid alkylmercury compounds provides a very effective method of preventing seeds from rotting when planted. Methylmercury dicyanodiamide (Panogen) is one of the most popular seed treatments. It is very toxic to humans and was banned in Sweden in 1966 for this reason. Large amounts of Panogen were exported to the United States from Sweden each year where its sale competed effectively with that of the same compound in the United States. In 1970 the export of Panogen was banned in Sweden. In addition to treatment of seeds, alkylmercury compounds are used extensively on potato, tomato, and apple crops. In a survey of mercury levels in plants in Britain, Canada, New Zealand, and Scandinavia, evidence has accumulated to show that translocation of mercury does occur, causing accumulation in harvested crops from pretreated seeds and plants (20). It is of interest to note that at the time Sweden banned the use of methylmercury compounds, the United States was using *200 times the amount sold in Sweden* (21).

III. THE METABOLISM OF INORGANIC MERCURY COMPOUNDS

Bioconversions of alkylmercury compounds are limited by their toxicity although exchanges of a variety of anions at the cationic coordination site probably occur. Since these compounds are usually uncharged, membrane transport presents no problems; thus alkylmercury is readily assimilated and concentrated in a variety of animals including man.

One example of alkylmercury assimilation was provided by Dr. Carl Rosén (3). Seed dressings treated with Panogen yielded crops in which alkylmercury accumulated via translocation. Hens fed grain from pretreated crops concentrated mercury in their liver and eggs. People, especially those partial to eggs, in turn concentrated mercury in their liver where it has a biological half-life of about 70 days (22). Hence over a period of time mercury levels were concentrating exponentially in tissues including the brain (10–15%). Excretion of mercury is mainly in the feces; thus low levels of mercury would be returned to ecosystems containing anaerobic bacteria (in sewage) where they could be methylated to methylmercury and to volatile dimethyl mercury to allow recycling.

Bioconversions of inorganic mercury to organomercurials definitely occur and provide a very interesting and challenging problem for biologists and chemists. Jensen and Jernlöv reported that anaerobes in sludge from aquaria can methylate inorganic mercury (14). They postulated that this reaction could happen in one of two ways:

$$(a)\quad Hg^{2+} \xrightarrow[2R-CH_3]{} (CH_3)_2Hg \xrightarrow[X]{} CH_3Hg^{+} + XCH_3$$

or

$$(b)\quad Hg^{2+} \xrightarrow[R-CH_3]{} CH_3Hg^{+} \xrightarrow[R-CH_3]{} (CH_3)_2Hg$$

Using cell extracts of a methanogenic bacterium (15), Wood *et al.*, were able to give some support for a combination of both (*a*) and (*b*). The formation of monomethylmercury from dimethylmercury is dependent on the pH of the environment, because under mild acidic conditions the following general reaction occurred:

$$R\text{-}Hg\text{-}R' + HX \rightarrow R\text{-}Hg\text{-}X + R'H$$

The methanogenic organism used in this study was isolated by Bryant *et al.* (23) from the symbiotic mixed culture known as *Methanobacillus omelianskii.* This original symbiotic mixed culture was isolated from canal mud at Delft, Holland (24), and was grown with ethanol and carbon dioxide. The role of the first organism, called S organism (25), was to provide H_2, which in turn provides the 8 electrons necessary for the reduction of CO_2 to CH_4 by the second organism, called MOH, for example.

S. *organism.*

$$CH_3CH_2OH + H_2O \rightarrow CH_3COOH + 2H_2$$

MOH *organism.*

$$4H_2 + CO_2 \rightarrow CH_4 + 2H_2O$$

Symbiotic relationships between H_2 evolvers and H_2 utilizers are abundant in anaerobic ecosystems such as lake sediments and in the rumen.

Cell extracts of MOH grown with 80% H_2 and 20% CO_2 catalyze the formation of CH_4 from a variety of substrates which include, in addition to CO_2, C_3 of serine (26), C_1 of pyruvate (27) N^5 methyltetrahydrofolate monoglutamate (28) methylcobalamin (29), methyl-factor B and methyl-factor III (30) and (31). The last three substrates are excellent methyl donors in CH_4 formation. The overall reaction requires ATP and H_2 as the source of electrons.

For example:

$$\mathrm{CH_3\text{-}Co(R)} \xrightarrow[H_2]{ATP} \mathrm{\dot{C}o(R)} + CH_4$$

R = 5, 6 Dimethylbenzimidazole
or 5 hydroxybenzimidazole
or water

This reaction is inhibited by mercuric acetate, sodium arsenate, tellurate, or selenate. During this inhibition, methyl groups are still transferred from methylcobalamin as can be observed by the conversion of the (Co III) methylcobalamin (red) to paramagnetic (Co II) $B_{12\text{-}r}$ (brown). This methyl transfer leads to the synthesis of methyl or dimethylmercury (depending on the concentration of Hg^{2+} used as inhibitor), dimethylarsine, and, judging by odor, methyl selenide, and telluride, respectively (17).

More recent studies in our laboratory have confirmed the nonenzymatic and enzymatic methylation of mercury; furthermore, we have elucidated the reaction mechanism for these two reactions (32). Hill *et al.* (33) studied the nonenzymatic methylation of mercuric ions by methylcobalamin, and they concluded that this reaction proceeds by electrophilic attack by Hg^{2+} on methylcobalamin. We have confirmed this mechanism and have been able to demonstrate that $CH_3^{\ominus}$ is transferred by using a spin-label to determine the valency of the cobalt atom during catalysis (34). This nonenzymatic mechanism will occur only in the presence of Hg^{2+}; reduction of Hg^{2+} to Hg_2^{2+} or to Hg^0 results in inhibition of catalysis. The following general reaction occurs:

Nonenzymatic.

$$\mathrm{CH_3\text{-}\overset{+3}{Co}} + Hg^{2+} \xrightarrow[H_2O]{X^{\ominus}} \mathrm{H_2O\text{-}\overset{+3}{Co}} + CH_3HgX$$

$X^{\ominus}$ = anion

Since methylcobalamin is a common coenzyme in both anaerobic and aerobic bacteria, significant quantities of this coenzyme are present in sediments (35). The amount of methylcobalamin present in sediments will depend on the microbial population in sediments, and this relationship regulates the rate of synthesis of methylmercury by this mechanism.

Under anaerobic conditions, Hg^{2+} is reduced to Hg^0, and, depending on the concentration of Hg^0, dimethyl- and monomethylmercury are formed as products. In this reaction studies with isotopes (36) and with spin-labeled methylcobinamide (37) indicates that methyl groups are transferred as radicals ($CH_3^{\bullet}$).

Enzymatic reaction.

$$2\ CH_3\text{-}Co^{+3} + Hg^0 \xrightarrow{-400 \text{ to } -700 \text{ mV}} (CH_3)_2\ Hg + 2\ \dot{Co}^{+2}$$

The results summarized in this section provide us with a general scheme for the synthesis of methylmercury compounds in the environment.

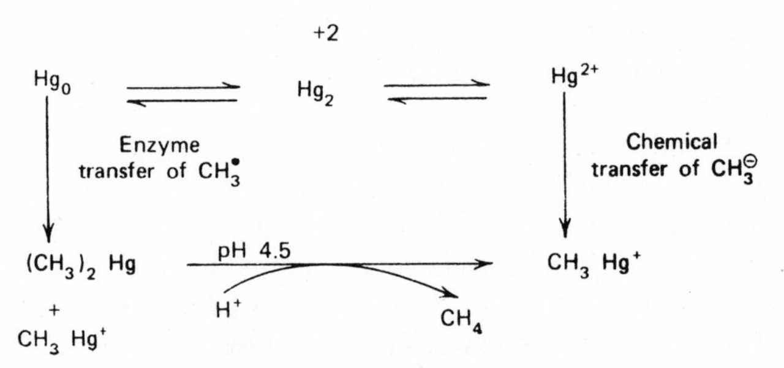

Any microorganism which is capable of synthesizing methylcobalamin will have the potentiality to synthesize methylmercury.

Furthermore, if sediments are polluted with arsenic salts, then methylcobalamin utilizing bacteria provide conditions for the synthesis of the deadly poisonous nerve gas dimethylarsine.

Pathway for the formation of dimethylarsine.

$$HO\text{-}As^{+5}(OH)(=O)\text{-}OH \xrightarrow[2e]{H_2 \rightarrow 2H^+} As^{+3}(=O)\text{-}OH \xrightarrow{CH_3B_{12} \rightarrow B_{12}\text{-}r} HO\text{-}As^{+3}(CH_3)(=O)\text{-}OH$$

$$HO\text{-}As^{+3}(CH_3)(=O)\text{-}OH \xrightarrow[H_2 \rightarrow 2e + 2H^+]{CH_3\text{-}B_{12} \rightarrow B_{12}\text{-}r} HO\text{-}As^{+1}(CH_3)(=O)\text{-}CH_3$$

$$HO\text{-}As^{+1}(CH_3)(=O)\text{-}CH_3 \xrightarrow{2H_2 \rightarrow 4e + 4H^+} (CH_3)_2As^{-3}\text{-}H$$

The conclusions that can be drawn from this research support the view of Johnels *et al.* (9) that volatile dimethylmercury (and dimethylarsine) may pollute the atmosphere as the result of inorganic contamination of anaerobic ecosystems. Depending on the pH of the environment, dimethylmercury, which is quite soluble in water, may be converted to the nonvolatile methylmercury which can be concentrated through food chains to fish living in both freshwater and marine environments.

Methylation of inorganic mercury in the liver has been demonstrated by Westöö (13) and may be accomplished by methyl transfer from bound methylcobalamin in the enzyme N_5-methyltetrahydrofolate-homocysteine transmethylase (38) (Brodie *et al.*). A mechanism for this reaction is proposed in the following scheme:

Proposed mechanism for the methylation of mercury in the liver.

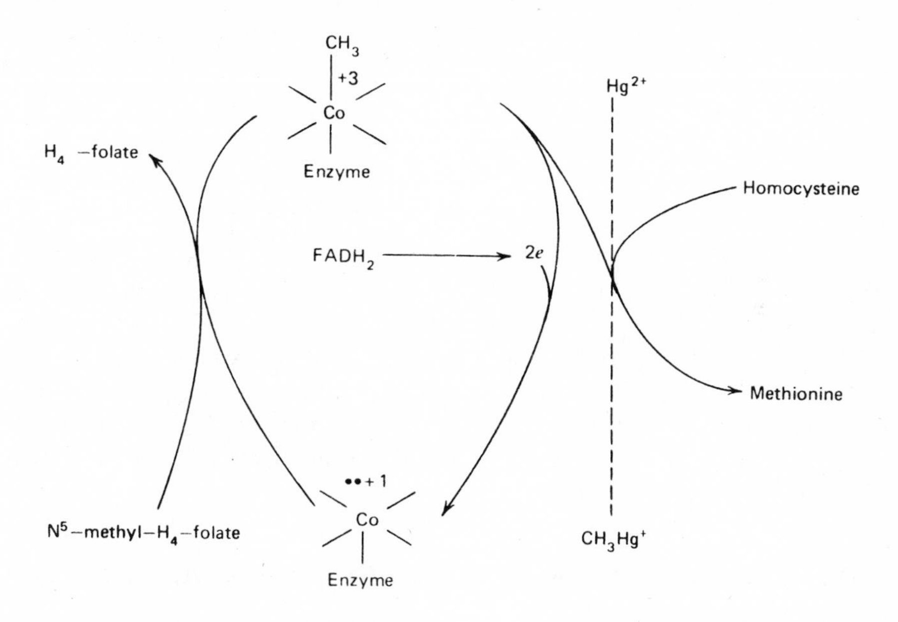

In this particular case the methyl donor would be N^5-methyltetrahydrofolate, but the methyl transfer is still mediated via cobalt. The B_{12}-dependent synthesis of methionine is found in most anaerobic bacteria, in some fecal organisms (i.e., *Escherichia coli* B) and in the mammalian liver.

The synthesis of methylmercury thiomethyl (CH_3-Hg-S-CH_3) by shellfish is easily explained by the following sequence of reactions:

Proposed mechanism for the formation of methylmercury thiomethyl.

$$\underset{\text{Cysteine}}{HS{-}CH_2{-}CH(NH_3^+)COO^\ominus} + Hg^{2+} \longrightarrow Hg^+{-}S{-}CH_2{-}CH(NH_3^+)COO^\ominus + H^+$$

$$CH_3{-}\overset{+3}{Co} + Hg^+{-}S{-}CH_2{-}CH(NH_3^+)COO^\ominus \xrightarrow[O_2]{H_2O} H_2O \rightarrow \overset{+3}{Co} + CH_3{-}Hg{-}S{-}CH_2{-}CH(NH_3^+)COO^\ominus$$

$$CH_3{-}\overset{+3}{Co} + CH_3{-}Hg{-}S{-}CH_2{-}CH(NH_3^+)COO^\ominus \xrightarrow[O_2]{H_2O} CH_3{-}S{-}Hg{-}CH_3 + \underset{\text{Alanine}}{NH_3^+{-}CH(CH_3){-}COO^\ominus} + H_2O \rightarrow \overset{+3}{Co}$$

In this reaction sequence, Hg^{2+} reacts with cysteine to give the cysteine-Hg^+ complex; this complex is an analog for homocysteine and would be methylated by the methionine enzyme (39). A second methylation step for the methylmercury cysteine complex would result in the formation of methylmercury thiomethyl. Numerous experiments designed to look for the methylation of lead by the enzyme systems have failed to yield positive results. Both Pb^{2+} and Pb^{4+} salts were tested and all results were negative.

IV. THE MAGNITUDE OF MERCURY POISONING IN JAPAN

The first major incident of methylmercury poisoning, as a result of industrial effluent, occurred in Minamata, Japan. In 1953 it was noticed that cats became seriously ill, screamed incessantly, and some killed themselves by diving from the cliffs into the ocean. Later a number of fishermen and their families became seriously ill. Initially it was thought that this epidemic was caused by an infectious agent (bacteria or virus) which was transmitted from cats to humans, and this syndrome was called Minamata disease. An investigation made by the Japanese

Health Authorities established a link between the quantities of fish eaten by afflicted families and that consumed by their cats. Subsequently, it was discovered that the effluent from a vinyl chloride plant contained high levels of mercury salts and small contaminating levels of methylmercury. When cats were fed on the factory effluent, they developed exactly the same symptoms as the cats fed Minamata Bay fish. Thus it was concluded that methylmercury was concentrating through food chains and leading ultimately to high concentrations in Minamata Bay fish. However, the vinyl chloride plant at Minamata used its own expert to investigate Minamata disease; the results of the industry survey indicated that methylmercury could not be the causative agent since fish caught in other coastal areas of Japan contained equally high levels of mercury as those taken from the Minamata Bay area. As a result of the factory study, the Minamata case was closed. Between 1953 and 1960, 116 people were irreversibly poisoned (43 fatalities) by Minamata disease. Irreversible mercury poisoning is defined as the point at which brain cell damage is extensive, and the victim remains a mental defective for the rest of his life. The pollution in Minamata Bay was considered so severe that fishing in this area was banned during the winter of 1956 because 42 cases of mercury poisoning were reported that year (40). Two years later shellfish were taken that contained 27 to 102 ppm mercury on a dry-weight basis. During 1958 the effluent was diverted into the Minamata River, and one year later shellfish in the river showed 9 to 24 ppm mercury, on a wet-weight basis, but most of the fish were found dead (41).

In 1965 a second incident occurred at the coastal town of Niigata (40). Again a vinyl chloride plant was implicated. Cats committed suicide and 120 people were affected (26 irreversibly and 5 fatally). It was established that mercury levels were high in the factory effluent, and a connection between fish location, consumption, and affliction was definitely established. Further scrutiny of the Minamata industry report to the Japanese Health Authorities revealed that in their survey of methylmercury levels of fish, they only analyzed fish taken near the vinyl chloride and chlorine plants. Therefore, this industrial report was not valid.

Experimental data from the Niigata disaster clearly demonstrated that the consumption of fish containing 5–6 ppm mercury daily would probably be lethal (42).

Mercury pollution is still a serious problem in Japan; however, these serious incidents have provided significant pressure to reduce the use of organomercurials and to monitor carefully industrial effluents. Yet pollution has been discovered recently in the Omata and Oyabe rivers. Organomercurials were replaced by other fungicides in rice crop treatment in Japan in 1968.

V. THE REASONS FOR LEGISLATION AGAINST THE USE OF ORGANOMERCURIALS IN SWEDEN

The mercury pollution problem in Sweden began about 1950 when severe poisonings were noticed in seed-eating birds, carrion, and predators. The discovery that the cause was the Panogen treatment of seeds initiated action against the chemical industry and agricultural users. Three young Swedish scientists, Ackerfors, Nilsson, and Rosén, from the Department of Radiobiology at the University of Stockholm began the long battle, risking libel and other personal hardships, to finally win their case and legislation was enacted in February 1966. In 1964 the initial step was taken by the Plant Protection Institute who demanded that the amount of mercury seed dressing used be reduced to half. In the fall of 1965 bird populations continued to be affected significantly by mercury poisoning. An interesting feature was that the drastic decrease in the use of methylmercury in 1965 had no effect on crop yields in Sweden that year.

During the public controversy over the concentration of mercury in eggs, by the food chain previously discussed (3), intensive research discovered that high levels of methylmercury (11,12,43–47) accumulated in fish both in inland and in coastal waters. Mercury content in fish was dependent on the age or weight, species, and the amount of biological activity in the environment. For example, fish in the ocean had accumulated much less mercury than those from coastal waters and some lakes and rivers.

No control of mercury pollution from industrial wastes in Sweden is yet in effect, and the release of inorganic mercury into the environment followed by its alkylation and accumulation in fish still poses a serious threat to this fish-eating nation.

VI. THE MAGNITUDE OF MERCURY POLLUTION IN THE UNITED STATES AND CANADA

In 1970, waterways in 33 states in the United States contained elevated levels of methylmercury, and many areas were closed indefinitely to fishing. The U.S. Department of Justice is initiating legal action against a number of industries in the United States for their indiscriminate elimination of mercury compounds. An illustration of the impact of industrial mercury pollution was the fact that mercury concentration in fish taken in Lake St. Clair in 1935 gave 0.07–0.11 ppm methylmercury. Fish taken from the same area in 1970 had 7.0 ppm mercury concentration. During this 35-year period of industrial growth a 100-fold increase in methylmercury levels in fish has occurred, rendering them now totally unfit for human consumption. Our inland and coastal waterways are so polluted that elevated levels of mercury are found in tuna and swordfish. However, the industrial practice of deep-sea dumping of wastes may have contributed greatly to this problem.

Currently, the pollution in the Saskatchewan River System poses the most serious health threat on the North American Continent. Fish have been taken from Lake Wabigoon that contain 24 ppm methylmercury, and the consumption of only a few fish polluted at this level would be fatal. Incredibly, fatalities caused by this dangerous situation have not yet been recorded.

The increased usage of methylmercury compounds in agriculture in the United States and Canada caused the hunting season for pheasant and Hungarian partridge to be canceled in Alberta in 1969, and elevated methylmercury levels were reported in California, Montana, and Ohio. In 1970 judges in Chicago and Washington reversed the restrictions placed by the United States Department of Agriculture on the further distribution of these deadly compounds by Noram. The basis for the decision by the two judges in support of continued use of methylmercury was that the only recorded human fatalities were the result of improper usage of these fungicides. Levels of methylmercury compounds applied to date have not yet had time to enter sufficient food chains to eliminate the human race. How long will it be before this happens?

VII. PATHOLOGICAL AND CLINICAL FEATURES OF MERCURY POISONING IN MAN

Owing to their nonpolar nature, alkylmercury compounds are uniformly distributed throughout the body, as membranes cause no special barrier. The symptoms of alkylmercury poisoning are fatigue, headache, and irritability followed by tremors, loss of feeling in the fingers and toes, blurred vision, and poor muscular coordination. After having difficulties with speech and hearing the victim tends to become an introvert, finally reaching a point at which he may need to be admitted to a mental hospital. Muscular wasting and further neurological disorders ensue. Significant amounts of alkylmercury passes through the brain membranes and accumulates causing disintegration of brain cells.

Passage of organomercurials across the placenta and concentration of them in the fetus causes a special problem. A widow of one of the Minamata victims gave birth to a seemingly healthy child which was breastfed and never given polluted fish. At 3 months of age the child suffered a convulsion, all four limbs becoming spastic; and until age 2½ when the child died, never having developed mentally (48).

Obviously the most noticeable cases of mercury poisoning originate in areas where people have been exposed to quite large doses. In the United States statistics indicate that from 3 to 11 cases of death from mercury poisoning have occurred each year since 1957. No data are available on the effect of prolonged exposure to lower levels of mercury or the level of mercury that humans can tolerate before it initiates permanent damage. It seems incredible that no standards were established for the mercury content of food in the United States until 1970. The only evidence that is available comes from experiments with animals and indicates that greater than 8 ppm mercury concentrations in brain tissue cannot safely be tolerated.

VIII. ANALYTICAL METHODS EMPLOYED FOR THE DETECTION AND QUANTITATION OF MERCURY IN TISSUES AND SEDIMENTS

A variety of analytical methods have been used to determine total mercury concentrations in foods, water, and tissues. Many of the methods used have serious defects, since analytical chemists have not recognized that methylmercury is a dangerous poison, and that this compound represents 95 to 100% of the accumulated total mercury in foods in most instances. Flameless Atomic Absorption Analysis is the most popular method used by the FDA, FWQA, and USDA. This method is accurate for *inorganic* mercury, but for methylmercury analysis it can be inaccurate as much as 100%. The neutron activation analysis technique developed by Jervis (50) has proved to be the most reliable method available at this time; however, it is seldom used because it is time consuming and expensive. It is difficult to understand why *time* should be given preference to *accuracy* when the health and welfare of the general public is concerned.

IX. POSSIBLE SOLUTIONS FOR THE MERCURY POLLUTION PROBLEM

Large quantities of mercury in various forms have been deposited in our environment. Only a small percentage of the total mercury deposited has been converted to methylmercury by microorganisms. For example, in the St. Clair River System it is estimated that the total concentration of mercury in the total fish population is approximately 100 lb. Since 200,000 lb of mercury have been deposited in that area, it is clear that the mercury pollution problem has only just started. If 95% of this mercury could be removed from the sediments in this system, it would still take many years before fish would be suitable for human consumption in this area. We face two alternative situations:

1. Find an acceptable method for removing mercury compounds from sediments.
2. Find an acceptable method for removing methylmercury from fish protein.

Considering the first situation, we find the following alternatives:

a. Dredging.

b. Cover the polluted sediments and create a new layer of biological activity.

c. Slow down the formation of methyl-Vitamin B_{12} by microorganisms. This would slow down the rate of formation of methylmercury to a point at which it is equal to or less than the rate of methylmercury metabolism in fish.

Any or all of these three approaches may be used depending on the specific conditions of the lake or river being considered. A concentration of 0.3 ppm for mercury in sediments does not appear to cause significant increases for methylmercury in fish. Sediments which contain greater than 0.3 ppm should be dealt with in the near future.

The removal of methylmercury from fish protein (the second situation) poses an extremely difficult problem. Methylmercury does not form good coordination complexes with nitrogen-containing bases. The methylmercury pyridine complex is very unstable (49); therefore it is unlikely that coordination complexes will be formed with lysine, histidine, or arginine. Attempts to remove methylmercury from protein by lyophilization, frying, boiling, or drying at high temperatures have proved unsuccessful. The only way to quantitatively remove methylmercury from protein is by treatment with concentrated acids. Undoubtedly, the major complex formed between methylmercury and protein is the covalent complex with cysteine residues.

For example:

$$-\mathrm{N(H)}-\mathrm{C(H)}(\mathrm{CH_2}-\mathrm{S}{:}\mathrm{Hg}{:}\mathrm{CH_3})-\mathrm{C}(=\mathrm{O})\text{------Protein}$$

$$\downarrow \mathrm{HCl\ (conc)}$$

$$-\mathrm{N(H)}-\mathrm{C(H)}(\mathrm{CH_2}-\mathrm{SH})-\mathrm{C}(=\mathrm{O})\text{------Protein}$$

$$+$$

$$\mathrm{CH_3\ Hg\ Cl}$$

Clearly, the removal of methylmercury from protein is not very practical.

X. DISCUSSION

Fortunately methylmercury has a biological half-life in humans from 70 to 74 days, but as little as 6 ppm in brain cells can cause irreversible brain damage. It is not known how the subclinical effects of lower levels may affect the nervous system.

Unintentional pollution by mercury must be controlled by checking the effluent wastes from industries daily and by carefully recording the quantities of mercury used and lost each year. However, recycling of solvents and careful regeneration of catalysts by the chemical industry must come eventually.

In conclusion, the increased addition of carbon, nitrogen, and phosphorus compounds to our waterways has increased microbial activity. When these microorganisms are confronted with mercury, they have the ability to detoxify their own environment at the expense of ours. The future for microorganisms looks good. It is time for us to take inventory of this problem before the situation passes beyond repair.

Acknowledgments

I wish to express my gratitude to Dr. Carl G. Rosén, Department of Radiobiology, University of Stockholm, Sweden, for making me aware of this problem during his two years at the University of Illinois. He provided the enthusiasm and many ideas required to initiate this project. Dr. R. S. Wolfe collaborated in some of the research reviewed in this chapter. He made his laboratory and facilities available to me to complete the research with methanogenic bacteria. Finally, I wish to thank graduate students F. Scott Kennedy, M. Penley, and B. C. McBride for their efforts in the laboratory. Some of the research reported in this chapter was supported by grants from the National Science Foundation, GB 8335, GB 8304, and 4481; also by the U.S. Public Health Service, AM 12599 and W00045.

References

1. K. Borg, H. Wanntorp, K. Erne, and E. Hanko, *J. Appl. Ecol.,* (Suppl) **3,** 171 (1966).
2. G. Westöö, B. Sjöstrand, and T. Westermark, *Var Föda,* **5** (1965).
3. C. G. Rosén, H. Ackerfors, and R. Nilsson, *Svensk Kem. Tidskr.,* **78,** 8 (1966).
4. J. C. Gage, *J. Ind. Med. Brit.,* **21,** 197 (1964).
5. L. T. Kurland, S. N. Faro, and H. Siedler, *World Neurol.,* **1,** 370 (1960).
6. M. Fujiki, *Jumamoto Igk Z.* **37,** 10 (1963).
7. T. Westermark and B. Sjöstrand, *Intern. J. Appl. Radiation Isotopes,* **9,** 1 (1960).
8. R. Christell, L. G. Erwall, K. Ljunggren, B. Sjöstrand, and T. Westermark, *Intern. Conf. Modern Trends in Activation Analysis Proc.,* College Station, p. 380 (1965).
9. A. J. Johnels, M. Olsson, and T. Westermark, *Var Föda,* **7,** (1967).
10. A. J. Johnels, T. Westermark, W. Berg, P. I. Persson, and B. Sjöstrand, *Oikos,* **18,** 323 (1967).
11. K. Norén, and G. Westöö, *Var Föda,* **2,** (1967).
12. G. Westöö, *Acta. Chem. Scand.,* **21,** 1790 (1967).
13. G. Westöö, *Var Föda,* **19,** 121 (1967).
14. S. Jensen and A. Jernlöv, *Nordforsk,* **14,** 3 (1968).
15. J. M. Wood, F. Scott Kennedy, and C. G. Rosén, *Nature,* **220,** 173 (1968).
16. B. C. McBride and R. S. Wolfe, *Bacteriol. Proc.,* **85,** 130 (1969).
17. B. C. McBride and R. S. Wolfe, unpublished data.
18. R. T. Taylor and H. Weissbach, *Arch. Biochem. Biophys.,* **129,** 728 (1969).
19. B. A. Blaylock, *Arch. Biochem. Biophys.* **124,** 314 (1968).
20. N. A. Smart and M. K. Lloyd, *J. Sci. Food Agr.,* **14,** 734 (1963).
21. S. Novick, *Environ.,* 3 (May 1969).
22. L. Eckman, *World Med.,* **79,** 450 (1968).
23. M. P. Bryant, E. A. Wolin, M. J. Wolin, and R. S. Wolfe, *Arch. Mikrobiol.,* **59,** 20 (1967).
24. H. A. Barker, *Antonie van Leenwenhoek J. Microbiol. Serol.,* **6,** 20 (1940).
25. C. A. Reddy, M. P. Bryant, and M. J. Wolin, *Bacteriol. Proc.,* **142,** 164 (1969).

26. J. M. Wood, A. M. Allam, W. J. Brill, and R. S. Wolfe, *J. Biol. Chem.,* **240,** 4564 (1965).
27. E. A. Wolin, M. J. Wolin, and R. S. Wolfe, *J. Biol. Chem.,* **238,** 2882 (1963).
28. J. M. Wood and R. S. Wolfe, *Biochem. Biophys. Res. Commun.,* **19,** 306 (1965).
29. M. J. Wolin, E. A. Wolin, and R. S. Wolfe, *Biochem. Biophys. Res. Commun.,* **12,** 464 (1964).
30. J. M. Wood, M. J. Wolin, and R. S. Wolfe, *Biochem.,* **5,** 2381 (1966).
31. J. M. Wood and R. S. Wolfe, *Biochem.,* **5,** 3598 (1966).
32. M. Penley, R. DeSimone, and J. M. Wood, *Biochem.,* to be published, 1971.
33. H. A. O. Hill, J. M. Pratt, S. Ridsdale, F. R. Williams, and R. J. P. Williams, *Chem. Commun.,* **341** (1970).
34. P. Y. Law, D. G. Brown, E. L. Lien, B. M. Babior, and J. M. Wood, *Biochem.,* (1971), in press.
35. A. Lezuis and H. A. Barker, *Biochemistry.*
36. M. Penley and J. M., Wood, *Biochemistry* to be published (1971).
37. P. Y. Law and J. M. Wood, *Federation Proc.* (1971).
38. J. Brodie, *Biochemistry,* **9,** 4295 (1970).
39. J. M. Wood, *J. Clin. Exp. Med.* (1971), in press.
40. G. Lofröth, *Environment,* 10, (May 9, 1969).
41. Sh. Kitamura, *Jumamoto Igk,* **37,** 494 (1963).
42. G. Westöö, *Acta. Chem. Scand.,* **20,** 2131 (1966).
43. G. Westöö, *Var Föda,* **19,** 1 (1967).
44. A. G. Johnels, *Var Föda,* **19,** 67 (1967).
45. A. G. Johnels, *Acta Oceol. Scand.,* **18,** 323 (1967).
46. G. Westöö and K. Noren, *Var Föda,* **19,** 135 (1967).
47. N. Grant, *Environment,* 18, (May 9, 1969).
48. F. Berglund and M. Berlin, *Chemical Fallout,* M. W. Miller and G. C. Berg, Eds., C. C. Thomas Publisher, Springfield, Ill., 1969.
49. S. Tejning, paper presented to the National Institute of Public Health, Stockholm, Feb. 2, 1968.
50. R. J. Jervis, D. Debrum, W. Le Page, and B. Tiefenbach, Progress Report, University of Toronto, NHG 605-7-510 (1970).

Motor Vehicle Emissions in Air Pollution and Their Control

JOHN A. MAGA
Executive Officer,
Air Resources Board,
Sacramento, California

I. INTRODUCTION

Smoke, odorous compounds, and carbon monoxide have been recognized as troublesome pollutants discharged from motor vehicles for many years. Smoke and odors were considered nuisances; carbon monoxide was of concern because of high concentrations that might occur in the vehicle passenger compartment, in garages, and at busy street intersections. In some instances, laws against excessive smoke and odors emitted from vehicles were enacted by state and local agencies. Little consideration was given, either by the industry or by air pollution control agencies, to the possibility that automobiles might cause community-wide air pollution affecting the entire population of large metropolitan areas.

The case against motor vehicles as a major factor in community-wide pollution was first made shortly after 1950, when it was reported that emissions from this source were a cause of photochemical air pollution (1). These reports stimulated and expanded research on the mechanism of photochemical reactions, the concentration of contaminants in automobile exhaust and in the atmosphere, and the contribution of motor vehicle emissions to the air pollution problem. Information obtained from these investigations firmly established that automobiles were the major source of carbon monoxide and photochemical air pollution in Los Angeles; provided the basis for the first motor vehicle emission standards in 1959 (2); and led to the motor vehicle control program undertaken by California in 1960. Continuing investigations demonstrated that other communities in California and in the United States suffered from air pollution caused by motor vehicles (3). This led to federal legislation in 1965 for the control of automotive emissions.

Air pollution created by motor vehicles was caused by factors which contribute to many environmental pollution problems: growth of metropolitan areas, increase in population, and an affluent society. In the case of air pollution, an affluent society enabled almost every family to own and operate cars. Motor vehicle registration in the United States from 1910 to 1968 is shown in Figure 1. Note the large increase in the number of motor vehicles; this increase has been much greater than

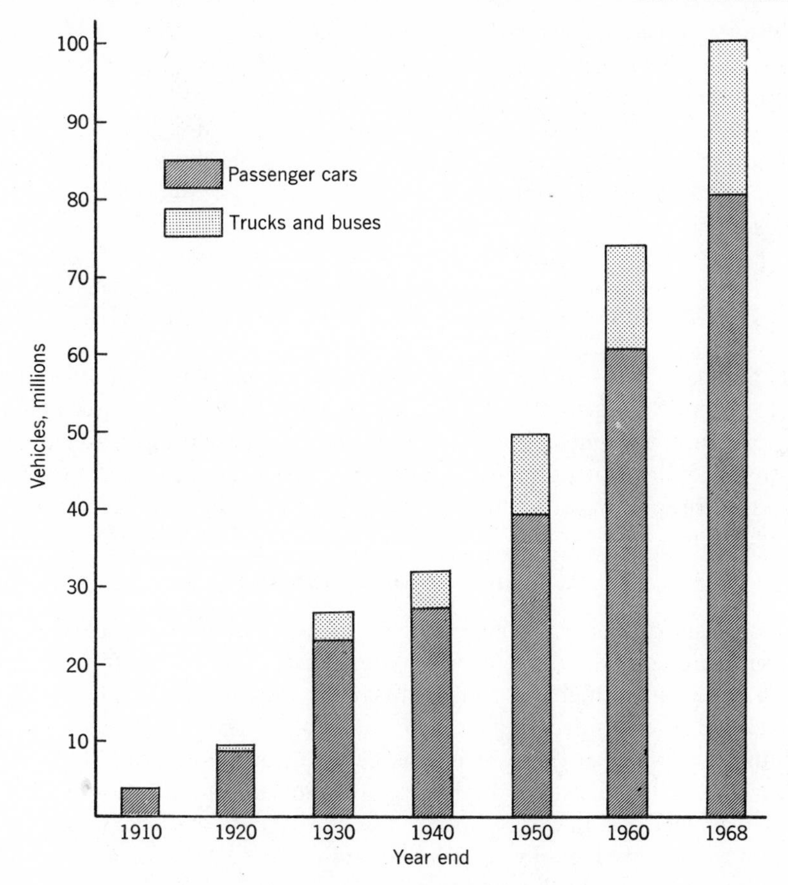

Fig. 1. Registered motor vehicles in the United States. From *Automotive Industries*, March 15, 1969.

that in population. There are now more than 100 million motor vehicles registered in the United States; this is about twice the number registered in 1950. A large percentage of the vehicles are passenger cars; in 1968, for example, there were only about 16 million trucks and buses. Most of these were small trucks, such as pickups, which are operated much like passenger cars.

The shift in population from rural to urban areas and the increase in population has created a number of large metropolitan areas. The growth of these regions has concentrated a large number of people and vehicles in a small percentage of the land area. Not only has this resulted in more severe air pollution episodes, but also it exposed a large population to the effects of the pollutants. In Los Angeles County, for example, the number of vehicles increased from 1.2 million in 1940, to the present 4 million, and the population from 2.8 million to 8 million. The number of vehicles in the county now exceeds that of all but seven states. It was in the mid-1940s that photochemical air pollution became the subject of widespread complaint from residents of the Los Angeles area.

II. MOTOR VEHICLE CREATED AIR POLLUTION

Air pollution caused by motor vehicles can be classified into three types: (*1*) nuisances in the vicinity of single vehicles; (*2*) direct effects from exposure to contaminants from a large number of vehicles; and (*3*) photochemical air pollution. The severity of these problems depends on the nature of the contaminant, quantity of emission, and meteorological conditions.

A. Nuisances near Vehicles

Nuisances experienced near vehicles usually involves smoke, odors, and irritating compounds from vehicle exhaust. Concentrations of compounds sufficiently high to cause nuisances are usually experienced in the vicinity of the operating vehicle and do not extend throughout the community. Because these problems occur near individual vehicles, the nuisances can be noted on highways and small communities with few vehicles as well as in large cities. Smoke and odors are much more common with diesel engines, although gasoline engines in a poor state of repair may cause similar problems.

B. Direct Effects from Many Vehicles

Direct effects of emissions from large numbers of vehicles which represent area-wide problems occur along heavily traveled roads and under meteorological conditions where contaminant concentrations can reach levels sufficiently high to have an effect on humans and plants. Carbon monoxide, an exhaust contaminant occurring in the order of several percent from gasoline-powered engines, is a good example of a compound which affects humans. When inhaled, it combines with the hemoglobin of the blood to interfere with the transport of oxygen.

A number of studies have been made regarding carbon monoxide concentration in the vicinity of heavy traffic (4,5). Concentrations of carbon monoxide ranged from low values to over 100 ppm, depending on traffic and wind. Peak concentration over short periods of time (less than 5 min) were measured as high as 120 ppm. The 20–30-min average concentration during maximum traffic was about 10–40 ppm. During these periods, carbon monoxide concentrations in the air throughout the community was about 10 ppm (6).

The highest concentrations of carbon monoxide within communities occur when light winds and surface-based temperature inversions occur together, conditions usually associated with high pressure systems in the winter months. At these times the carbon monoxide concentrations may be much higher than the 10 ppm reported previously. The seasonal pattern of high carbon monoxide concentrations occurring during the cold months of the year is shown in Figure 2. The contrasting seasonal variation in oxidant, a measure of photochemical air pollution which is discussed later, is shown. It can be seen that oxidant is highest in the warm months. The seasonal influence on carbon monoxide, as well as the high concentrations which were measured in several American cities, is shown also in Table I. It can be seen that the maximum concentrations were usually measured in the months of October to January.

Ethylene is another exhaust contaminant that has direct effects. This compound can damage numerous ornamental plants, such as orchids, carnations, and snapdragons (7). Elevated concentrations of ethylene follow a seasonal pattern similar to that for carbon monoxide. Damage

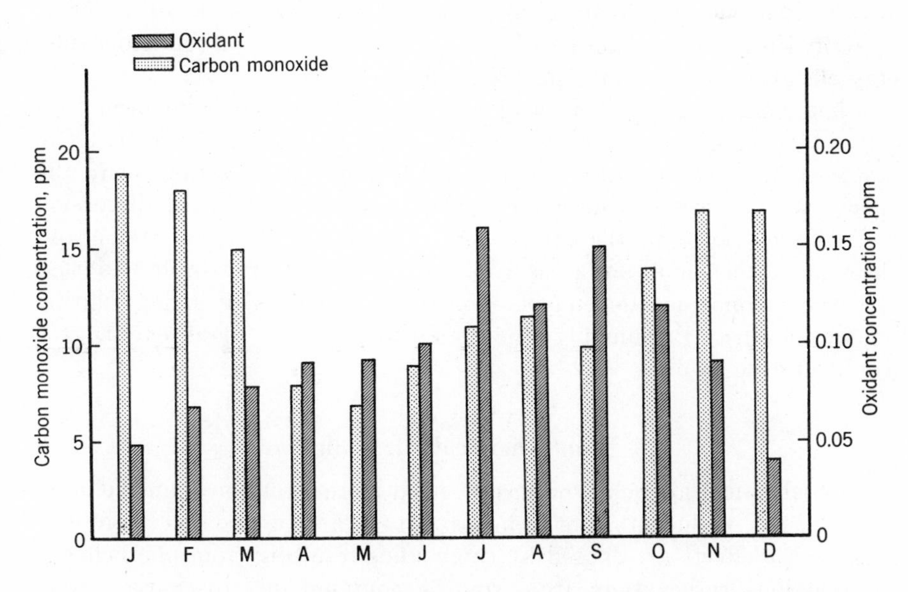

Fig. 2. Seasonal variation of atmospheric carbon monoxide and oxidant concentration (monthly averages of the maximum hourly readings in downtown Los Angeles, 1968). From *State of California Air Resources Board Air Monitoring Records.*

TABLE I

Concentration of Carbon Monoxide in Several American Cities as Measured in 1966[a]

City	Concentration of carbon monoxide, ppm	
	5-min concentration	Maximum hourly concentration
Chicago, Ill.	66 (April)	45 (April)
Cincinnati, Ohio	32 (Nov.)	20 (Oct.)
Denver, Colo.	63 (Jan.)	49 (Nov.)
Los Angeles, Calif.	39 (Jan.)	36 (Jan.)
Inglewood, Calif.	65 (Dec.)	50 (Dec.)
Philadelphia, Pa.	47 (Oct.)	40 (Oct.)
St. Louis, Mo.	68 (Nov.)	29 (Nov.)
Washington, D.C.	47 (Oct.)	38 (Oct.)

[a] From the National Air Pollution Control Administration, and Los Angeles County Air Pollution Control District.

of ornamental plants from ethylene is observed frequently during the winter months in the San Francisco Bay area, where these plants are grown extensively for commercial purposes. Ethylene is one of the principal hydrocarbons in auto exhaust. In contrast to carbon monoxide, ethylene also enters into the photochemical reaction.

Other compounds such as lead and carcinogens also have been suggested as being important because of their direct toxic effects. The exact significance of these compounds has not been established; the effects are, at present, the subject of scientific and political discussion and debate. Lead in the atmosphere over routes of heavy traffic and throughout metropolitan areas also follows a pattern similar to that of carbon monoxide—dependent upon traffic and usually highest during winter months. Presumably, the concentration pattern of carcinogens would also be similar.

C. Photochemical Air Pollution

Photochemical air pollution has received more attention than the other types of air pollution from vehicles. It is a more complex problem because the effects are caused by compounds resulting from photochemical reactions rather than from specific contaminants discharged from vehicles. The photochemical reactions involve organic compounds and oxides of nitrogen and produce a number of new compounds which in turn may be toxic, irritate eyes and the respiratory system, damage vegetation, and impair visibility. In photochemical reactions, ozone accumulates in the atmosphere; the amount of ozone produced has been regarded as a measure of the severity of the air pollution episode.

Automobiles are the sources of large quantities of both organic compounds and oxides of nitrogen. The principal organic compounds discharged from automobiles are the unburned or partially burned hydrocarbons. However, not all hydrocarbons are of equal importance in the compounds produced or in the rate of product formation. Some, mainly paraffins, react very slowly; olefins react quickly and are of greatest importance. Aromatic compounds, as a group, fall in between the paraffins and olefins in their ability to participate in the photochemical reactions. This represents a simplified picture of the participation of hydrocarbons in the reaction and that there is overlapping of the three groups. For example, some aromatics are more reactive than some olefins, and some paraffins can produce large amounts of ozone during irradiation of several hours.

Oxygenated hydrocarbons, such as aldehydes, participate in the photochemical reaction. These oxygenated hydrocarbons have not been considered to be as important as the unburned hydrocarbons because the oxygenated compounds are present in much lower quantities in uncontrolled exhausts, usually about one-tenth of the proportion of hydrocarbons. However, as hydrocarbons are controlled, the importance of oxygenated compounds will increase unless they also are controlled.

D. Meteorology

Meteorology has a strong influence on air pollution. The seasonal influence on carbon monoxide concentrations is discussed in Section II.B. Because photochemical reactions require sunlight, this type of air pollution is a daytime phenomenon and occurs in the warm seasons of the year. This seasonal pattern was shown in Figure 2. Atmospheric conditions particularly favorable to these reactions are associated with subsidence inversions. Such inversions occur frequently along the California coast and are a major factor in the differences in occurrence of photochemical air pollution found in California and that of most other cities in the United States. In contrast to carbon monoxide, photochemical air pollution episodes in many California cities, particularly Los Angeles, are much more frequent and severe than in other sections of the country. During severe attacks, ozone concentrations in the Los Angeles atmosphere exceed 0.5 ppm.

The daily variations of carbon monoxide (CO), nitric oxide (NO), and oxidant in downtown Los Angeles are illustrated in Figure 3. Con-

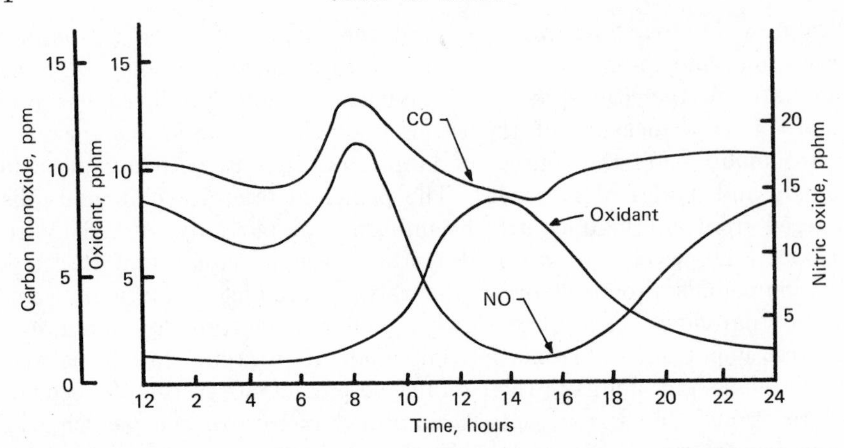

Fig. 3. Daily variations of the average hourly concentrations of carbon monoxide, nitric oxide, and oxidant in downtown Los Angeles, 1967. From *Air Resources Board Air Monitoring Records.*

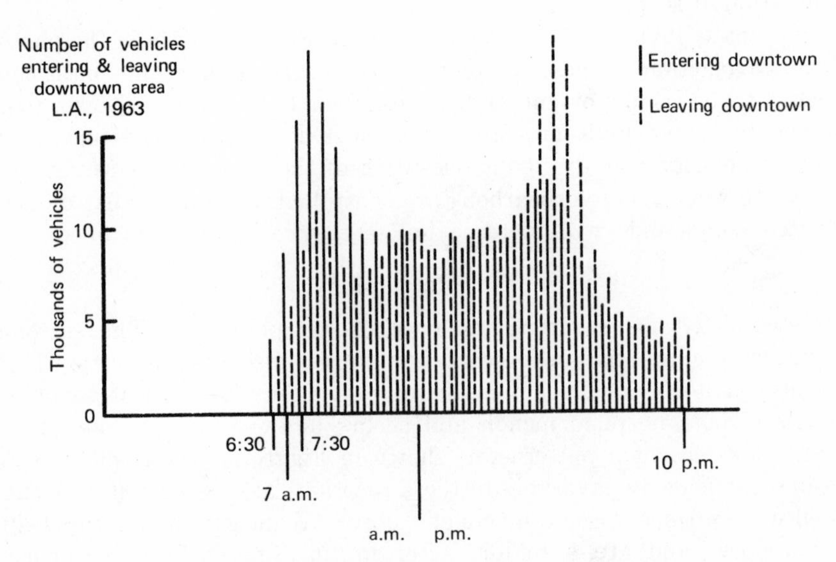

Fig. 4. The traffic pattern in downtown Los Angeles. From Traffic Counts, City of Los Angeles Department of Traffic.

centrations of carbon monoxide and nitric oxide peak in the morning and increase again at night. As might be expected, this pattern is similar to that for traffic entering and leaving the downtown area because these two compounds are discharged in large quantities in auto exhaust. The traffic pattern is shown in Figure 4. The lower concentration of carbon monoxide in midday is largely due to decreased traffic and increased atmospheric dispersion. These factors result in lower nitric oxide in midday; in addition, much of the nitric oxide is consumed in the photochemical reactions. However, oxidant, which is a measure of photochemical air pollution, accumulates with increased sunlight following the morning peak concentration of auto emissions and decreases to low values with the decrease in sunlight. The difference between the minimum and maximum values during the early morning hours indicates that contaminants have been added recently to the atmosphere. Because of the peak concentration of vehicle emissions at that time and because the photochemical reactions follow the early morning traffic, control of motor vehicle emissions must be given special attention for this traffic period.

III. POLLUTANTS AND THEIR SOURCE

A number of contaminants arising from several sources—exhaust, crankcase, fuel tank, and carburetor—are attributable to motor vehicle operation. The kinds and quantity of pollutants will vary according to the source and other factors such as fuel and the design and operational characteristics of the engine.

A. Gasoline Engines

1. *Exhaust*

The exhaust contributes the greatest quantity as well as the largest number of contaminants to the air. Compounds from this source are characteristic of products resulting from incomplete combustion and include organic compounds, oxides of nitrogen, carbon dioxide, carbon monoxide, water vapor, hydrogen, sulfur compounds, lead, scavenger compounds, and particulate matter. Some of these—carbon dioxide, water vapor, and hydrogen—are not considered to be contaminants in the air pollution problems of cities. Carbon monoxide, hydrocarbons, and oxides of nitrogen are usually defined as the most important pollutants. But concern has also been expressed from time to time about partially oxygenated hydrocarbons, particulate matter, and lead.

The amount of contaminants in the exhaust depends upon many factors, including the design of the engine and how it is operated and maintained. For example, air-fuel ratio (A/F), compression ratio, spark timing, valve overlap, surface to volume ratio of the combustion chamber, and choke affect the emission of hydrocarbons, carbon monoxide, and oxides of nitrogen. The effects of one of the more important of

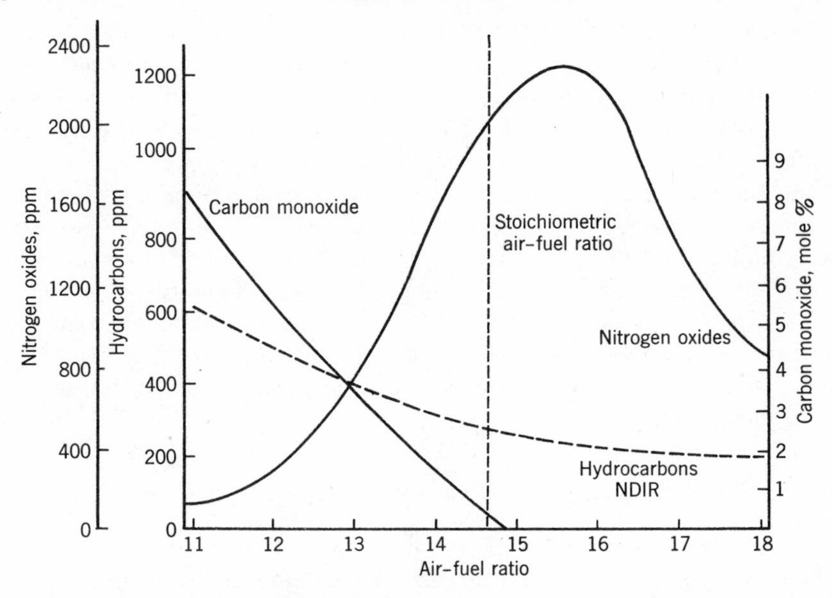

Fig. 5. Air-fuel ratio versus hydrocarbons, carbon monoxide, and oxides of nitrogen

these factors, A/F, are shown in Figure 5. As can be seen, the A/F ratio has a marked influence on emissions. About 14.7 lb of air are required to completely burn 1 lb of fuel (A/F = 14.7/1) under ideal conditions. Because peak power is developed at A/F ratios of less than this amount, gasoline-powered vehicles usually operate at average A/F ratios of about 13.5/1. This results in incomplete combustion which is characterized by high concentrations of carbon monoxide and unburned hydrocarbon in exhaust. Figure 5 shows that increased A/F ratio, within the range at which gasoline engines operate, reduces the concentration of hydrocarbons and carbon monoxide, but increases the oxides of nitrogen. This fact complicates control efforts directed at reducing simultaneously all three of these contaminants by adjusting the A/F ratio in present engines.

Modes of operation also have direct impact on contaminant concentration in exhaust. For example, idle and deceleration are characterized by high concentrations of carbon monoxide and hydrocarbons, but very low concentrations of oxides of nitrogen. On the other hand, steady-state cruising produces relatively low concentrations of hydrocarbons and carbon monoxide, but high concentrations of oxides of nitrogen. Some of the explanation for this can be found in Figure 5. Cruise conditions are with air-fuel ratios approaching that needed for complete

TABLE II

Exhaust Contaminants Concentrations by Mode of Operation[a]

Operating mode, mph	Carbon monoxide, Vol. %	Hydrocarbons NDIR-hexane, ppm	Oxides of nitrogen, ppm as NO_2
Accelerate 0–25	3.6	665	1250
Cruise 50	2.8	445	890
Decelerate 30–15	5.0	1650	—[b]
Cruise 15	4.1	560	205
Accelerate 15–30	2.8	500	1700
Cruise 50	2.4	320	1580
Decelerate 50–20	4.2	3495	—[b]
Idle	5.8	895	—[b]

[a] Los Angeles Test Station Project, 1961–1963.
[b] Very low, not measured.

combustion of gasoline; deceleration or idle are fuel-rich conditions that result in less efficient combustion. Oxides of nitrogen are produced at the high temperatures associated with efficient combustion. Table II illustrates the variation in the concentration of contaminants under various operation modes.

The quantity of contaminants emitted from exhaust is also related to these modes because of the volume of the exhaust. The volume for small passenger cars at idle is under 5 ft^3/min. Large trucks operating a full load may have over 200 ft^3/min of exhaust. Although the idle and deceleration modes are characterized by high concentrations of carbon monoxide and hydrocarbons, the volumes of these modes are small as compared to cruise conditions. If one wishes to determine the quantity of emissions, he must specify the operating modes and measure both the concentration and volume of exhaust.

2. *Crankcase*

Crankcase emissions are caused by the escape of gases from the cylinder during the compression and power strokes. The gases pass the piston rings and are called "blowby." Blowby is mainly (about 75–80%) the unburned fuel-air mixture but contains some exhaust products. Because these gases closely resemble the carbureted fuel-air mixture, hydrocarbons are the main pollutants. Concentration of hydrocarbons in blowby are high, varying from 8000 to 15,000 ppm.

The volume of blowby gases is greatest at full-load conditions (high airflow) and lowest at idle and deceleration (low airflow). It also increases with engine wear as the seal between the piston and the cylinder wall becomes less effective. The measurement of blowby gases from several hundred automobiles under conditions representative of driving in Los Angeles showed that the average volume was about 1 ft^3/min and varied from about 0.5 to 5 ft^3/min (9).

3. *Evaporation*

a. Fuel Tank

Fuel tank losses result from the evaporation of fuel and the displacement of vapors when fuel is added to the tank. Evaporation of fuel is much the larger source of losses from the tank.

The amount of fuel vapors evaporated depends on the composition of the fuel and its temperature. The fuel temperature increases with an increase in ambient temperature and as the tank is warmed by the air passing under it during the operation of the vehicle and by the hot pavement when the vehicle is parked. Obviously, evaporative losses are high when the fuel tank is exposed for prolonged periods to ambient temperatures near or above the boiling point of the fuel.

The amount of vapors lost when fuel is added to the tank (filling losses) is equal to the volume of fuel added. For example, if 10 gal of gasoline are added, then 10 gal of vapor are displaced into the atmosphere. Evaporative losses are mainly hydrocarbons, although some of the additives that were originally present in the fuel are also evaporated.

b. Carburetor

The carburetor is also a source of evaporative losses. Emissions occur mostly during the period after the engine is turned off and when the carburetor bowl is subjected to heating from the hot engine. This is known as a "hot soak." The amount and composition of the vapors depend on the vapor pressure of the fuel, volume of the bowl, and temperature of the engine prior to shut-down. It has been estimated that 10 g of hydrocarbons are emitted during an average soak period (10).

Table III shows the relative contribution of hydrocarbons, carbon monoxide, and oxides of nitrogen from uncontrolled automobiles under operating conditions in Los Angeles. As discussed previously, exhaust is the largest source of these three contaminants, as well as the only important one for oxides of nitrogen and carbon monoxide.

B. Diesel Engines

The emissions from diesel engines differ from those of gasoline engines in several important respects. Basic differences between the two types

TABLE III

Relative Magnitude of Emissions of Hydrocarbons, Carbon Monoxide, and Oxides of Nitrogen from Uncontrolled Gasoline-Powered Automobiles

	Pollutant as percent of total of each contaminant emitted		
Source	Hydrocarbons	Carbon monoxide	Oxides of nitrogen
Exhaust	60–65	98–99	98–99
Crankcase	20–25	1–2	1–2
Carburetor	5–10	0	0
Fuel tank			
Evaporative	5–10	0	0
Filling	1–2	0	0

of engines, the fuels used, fuel handling systems, methods of igniting the charge in the cylinders, and air-fuel ratios are examples of differences between the two engines.

The air in diesel engines is compressed before the fuel is introduced as a spray into the cylinders by means of injectors. The closed-injection system eliminates the carburetor losses experienced in the conventional gasoline engine. Compressing air rather than an air-fuel mixture, as in the case of gasoline engines, reduces the hydrocarbons in blowby to very small amounts as air escapes past the rings instead of an air-fuel mixture.

Diesel engines normally operate with air-fuel ratios in excess of the stoichiometric ratio. This promotes more efficient combustion of the fuel, thereby producing less unburned hydrocarbons and carbon monoxide in the exhaust. Because peak temperatures in the combustion chamber are high, large quantities of oxides of nitrogen are formed.

Diesel fuel is not as volatile as gasoline. Therefore, fuel tank evaporative losses are much less in diesel vehicles. Diesel fuels do not contain lead additives so that this compound is not discharged from these engines. Table IV compares the emissions from gasoline and diesel engines under comparable operating conditions, adjusting the values for excessive air used for diesels. It can be seen that carbon monoxide concentrations in diesels are much lower; hydrocarbons about one-half; and oxides of nitrogen somewhat higher than similar contaminants from gasoline engines. In addition, the different fuel produces different types of unburned hydrocarbons which are not similar in reactivity to those from gasoline engines.

TABLE IV

Comparison of Typical Emissions of Hydrocarbons, Carbon Monoxide, and Oxides of Nitrogen from Uncontrolled Gasoline- and Diesel-Powered Vehicles[a]

Source	Gasoline-powered automobiles	Diesel-powered vehicles
Exhaust		
Hydrocarbons[b], ppm	500–1200	200–500
Carbon monoxide, %	2.5–4.5	0.1–0.3
Oxides of nitrogen[c], ppm	300–2000	2000–3000
Crankcase		
Hydrocarbons, g/day	Approx. 100	Very low[d]
Fuel tank and carburetor		
Hydrocarbons, g/day	Approx. 80	Very low[d]

[a] From the California State Department of Public Health, and the Air Resources Board.

[b] As measured by nondispersive infrared analyzer (NDIR) and expressed as hexane.

[c] Expressed as nitrogen dioxide.

[d] Probably less than 10% of that from gasoline engines.

Emissions from diesel engines cause two well recognized air pollution problems—smoke and odors—which normally are not associated with automobiles. Smoke results from overfueling; the cause and specific compounds that are responsible for the odor are not well established.

In addition to the difference in fuel and design of diesel engines, there are other factors which make the contribution of these engines to community-wide air pollution minor compared to the gasoline engine. The most important factor is the small number of diesel-powered vehicles in use. Diesel engines are found, almost exclusively, in large trucks and buses and comprise only about 0.5% of all vehicles. Much of the driving of diesel trucks is on highways between communities. Thus not all emissions would be involved in the air pollution problems of cities. For example, diesels comprise only about 0.5% of the registered vehicles in California and burn about 6% of the total fuel. It has been estimated, however, that they only burn 1.5% of the vehicle fuel used in Los Angeles County (11).

Although diesels now are sources mainly of local nuisance rather than major contributors of community-wide air pollution, it is not to be assumed that there would not be community-wide air pollution if all vehicles were powered by diesel engines. Oxides of nitrogen concentrations would be increased, some photochemical reaction would probably still occur, and millions of operating diesels would very likely produce widespread smoke and odor problems.

IV. CONTRIBUTION OF MOTOR VEHICLES TO AIR POLLUTION

Air pollution in large metropolitan regions is a complex problem whose nature and severity varies according to the number, kind, and size of the sources and to the meteorological conditions. Contaminants from many sources are intermingled within the air mass over the community and may be transported considerable distances so that it is often difficult to relate the effects felt to the source of the contaminant. Photochemical reactions further complicate the problem because effects now experienced are not caused by the pollutant as discharged, but by the reaction products.

The solution to a problem in any area must be based on the knowledge of the kind and quantity of the pollutants, as well as the source. Automobiles are present in large numbers, have similar engines, and are operated in much the same manner. They, therefore, are a common and major source of air pollution in all metropolitan areas. Contaminants from nonvehicular sources are much more variable from community to community and to a large extent depend upon the type of fuel burned—coal, oil, natural gas—and the industries present. Many common pollutants, such as fly ash, sulfur dioxide, and dust, are almost always nonvehicular in origin.

The major contaminants from automobiles—carbon monoxide, hydrocarbons, and oxides of nitrogen—are also discharged from other sources. For example, hydrocarbons are emitted during the production, refining, distribution, and marketing of petroleum products, and from organic solvents used in painting, degreasing, and cleaning operations. Oxides of nitrogen are formed in all high-temperature combustion of fossil fuels. Carbon monoxide is produced in the regeneration of catalysts during petroleum refining. Table V shows the relative emission of the above contaminants from motor vehicles and other sources for several American cities during the period 1963–1966 before exhaust control systems were required on vehicles in these cities.

In each of these localities automobiles discharged a very high percentage of the carbon monoxide and were the source of much of the hydrocarbons emitted. In all the cities, large quantities of oxides of nitrogen were discharged from other sources, and in St. Louis and Washington, D.C., these other sources were the major contributors of these compounds.

TABLE V

Relative Emission of Hydrocarbons, Carbon Monoxide, and Oxides of Nitrogen from Motor Vehicles and Nonvehicular Sources

	Hydrocarbons % from		Carbon monoxide % from		Oxides of nitrogen % from	
Place	Motor vehicles	Other	Motor vehicles	Other	Motor vehicles	Other
Los Angeles county (1965)[a]	72	28	97	3	62	38
St. Louis interstate area (1963)[b]	62	38	97	3	32	68
San Francisco Bay area (1965)[c]	51	49	78	22	57	43
Washington, D.C., metropolitan area (1965–1966)[d]	75	25	98	2	39	61

[a] Los Angeles County Air Pollution Control District.

[b] R. Venzia and G. Ozolins, Interstate Air Pollution Study-Air Pollutant Emission Inventory, U.S. Department of Health, Education, and Welfare, PHS, December 1966.

[c] Bay Area Air Pollution Control District.

[d] Technical Report, Washington, D.C., Metropolitan Area Air Pollution Abatement Activity, National Center for Air Pollution Control, November 1967.

The relative contribution of motor vehicles and all other sources also depends on the extent of control efforts. For example, if nonvehicular sources had not been controlled in Los Angeles, the percentage of hydrocarbons and carbon monoxide from motor vehicles would be decreased from 72 and 97%, respectively, to about 60 and 80%.

V. LEGAL BASIS FOR CONTROL OF MOTOR VEHICLE EMISSIONS

Traditionally, programs for the control of air pollutants have been viewed as a responsibility of local government. However, motor vehicle emissions control has required a different legislative approach. Automobiles are mobile, common sources of air pollution in all communities, mass produced in assembly plants serving large areas or several states, and usually registered and licensed by the states.

Present programs to control the emissions from motor vehicles are largely the result of laws enacted by the California Legislature and the U.S. Congress. The first laws in this area were enacted by the state of California in 1959 and 1960. The laws required standards for air quality and motor vehicle emissions and established a state agency to test control systems and to approve those vehicles that met the emission standards. The California activities led to the first standards prescribing limits to the pollutants that could be discharged from motor vehicles and to the first program requiring control systems to be installed on cars.

In 1965 the U.S. Congress required the Secretary of Health, Education, and Welfare to set emission standards and to certify control systems for all new motor vehicles. With the exception of California, states and other political subdivisions are prohibited from adopting motor vehicle emissions standards for new vehicles and from certification of control systems for such vehicles.

California can enforce more strict standards if the Secretary finds that these are needed to meet compelling and extraordinary conditions in that state and if they are technologically feasible. Thus federal standards will provide a uniform approach nationwide for new cars. California has been permitted to establish a number of more stringent emission standards through waivers granted by the Secretary of the Department of Health, Education, and Welfare.

The federal standards are directed at the manufacturer of the motor vehicles and do not apply to used cars. Once the equipped vehicles are sold, they are under the jurisdiction of the states. Therefore, states must become active in the enactment and enforcement of laws to prevent the removal of control systems and to require their continuing effective operation. Many questions still remain on the type of inspection, maintenance, and enforcement program which will be needed to ensure success.

VI. EMISSION STANDARDS

Motor vehicle emission standards have evolved over the past decade and are still undergoing changes in the degree of control required, pollutants included, and method of expression (12,13).

From the previous discussion, it is clear that such standards must consider a number of factors: (*1*) several kinds of problems exist—nuisances from smoke and odors, direct effects of toxic compounds, and photochemical reactions; (*2*) pollutants are discharged from four points—exhaust, crankcase, fuel tank, and carburetor; and (*3*) there are a number of pollutants.

A. Test Cycles

Levels specified in emission standards must be related to the test procedures and vehicle operating conditions used to determine the emissions. One of the more important parts of the test is selection of the vehicle operating modes to be sampled.

A comprehensive study of driving modes and emissions was made in Los Angeles in 1956 by the Coordinating Research Council (14). In this study, vehicle operating conditions were expressed in terms of 11 modes which were related to vehicle speed. Vehicle operating parameters for these were determined by analyzing composite samples from automobiles representing the vehicle population in Los Angeles. The findings of this survey became the basis for the establishment of motor vehicle emission standards in California in 1959.

The 11-vehicle operating modes were not suitable for testing on a dynamometer. Therefore, it was necessary to modify these modes to obtain a cycle pattern which could be used conveniently to operate a vehicle on a chassis dynamometer. As shown in Table VI, this continuous dynamometer cycle included 10 modes, 7 of which were used with appropriate weighting factors as the basic test procedures. The seven-mode method has been utilized by California for certifying exhaust control systems. This same procedure was incorporated in 1968 in the Federal Vehicle Control Regulations.

The basic exhaust emission test is conducted on a chassis dynamometer and was developed to represent an urban trip of about 21 min and at an average speed of 22 mph. It includes the emissions for a "cold start" and requires that the above cycle be repeated nine times.

The evaporative emission test is designed to represent emissions from typical driving and parking sequences. It includes the daily fuel tank

TABLE VI

Dynamometer Test Cycle–Light-Duty Vehicles

Mode, mph	Time in mode, sec	Cumulative time, sec	Weighting factor
Idle	20	20	0.042
0–25	11.5	31.5	0.244
25–30	2.5	34.0	Not used
30 Cruise	15	49	0.118
30–15	11	60	0.062
15 Cruise	15	75	0.050
15–30	12.5	87.5	0.455
30–50	16.5	104	Not used
50–20	25	129	0.029
20–0	8	137	Not used

TABLE VII

Heavy-Duty Vehicle Driving Cycle[a]

Mode	Manifold vacuum, in. Hg	Time in mode	Weighting factor
Idle	—	70	0.036
Cruise	16	23	0.089
Part throttle acceleration	10	44	0.257
Cruise	16	23	0.089
Part throttle deceleration	19	17	0.047
Cruise	16	23	0.089
Full load	3	34	0.283
Cruise	16	23	0.089
Closed throttle	—	43	0.021

[a] Procedures require that all modes except idle be operated at constant rpm. Idle is operated at the manufacturer's recommended speed.

loss, operating loss during the exhaust emission test, and "hot soak" loss from the carburetor after the engine has been stopped.

Heavy-duty vehicles (over 6000 lb gross vehicle weight) are not operated in the same manner as are passenger cars. One of the major differences is that the heavier vehicles normally are operated at a higher load factor. This can have a significant bearing on the emission standards, test procedures, and control options. A greater variety in the operating schedule for heavy-duty vehicles exists, ranging from their intermittent use as recreational vehicles to more or less continuous operation on delivery routes.

The test cycle for these vehicles was also developed to duplicate operating conditions under urban area driving conditions. In contrast to light-duty vehicles, the tests for a heavy-duty vehicle are run on an engine dynamometer and includes idle and 8 operating modes at a constant speed of 200 rpm as shown in Table VII.

B. Emission Standards

1. *Gasoline-Powered Vehicles*

a. Crankcase and Evaporative Emissions

The emission standards for these sources are concerned only with hydrocarbons. The standard for crankcase emissions permits no discharge of hydrocarbons, and, in effect, it requires 100% control. Those for

TABLE VIII

Emission Standards for Crankcase and Evaporative Losses[a] for Light-Duty Vehicles

Uncontrolled vehicle	Emission standards	Remarks
Crankcase emissions HC,[b] 120 g	No emissions	Federal
Evaporative losses/test HC, 50 g	HC, 6 g/test	1970 Models—California 1971 Models—Federal and California

[a] Fuel tank and carburetor standards apply only to light-duty vehicles.
[b] HC—hydrocarbons.

evaporative emissions require about a 90% control. Standards prescribing such high degrees of control were possible because of the availability of very effective control systems. Table VIII shows the average emissions from an uncontrolled vehicle and from those permitted under current emission standards.

b. Exhaust Emissions

Light-Duty Vehicles. Exhaust emission standards have had to consider more than one contaminant. Moreover, it was not always possible at the beginning to require a very high degree of control because the control systems were not available. The exhaust emission standards, particularly in California, have established control goals for future years based on the judgment of the technological feasibility of systems which could be developed by the specified year.

The first exhaust emission standards were established in California in 1959 and prescribed limits of 275 ppm of hydrocarbons and 1.5% of carbon monoxide. These were used by both the federal government and California until the 1970 models, although an adjustment had been made in the values applied to small cars.

Table IX shows the current exhaust emissions standards for light-duty vehicles (under 6000 lb gross vehicle weight). The table also shows the emissions from an uncontrolled vehicle. It indicates that the standards are now expressed in grams per mile to take into account the exhaust volume. An emission of 2.2 g/mile of hydrocarbons represents a concentration of about 180 ppm in a vehicle weighing 4000 lb; 23 g of carbon monoxide would be about 1.0% in such a vehicle.

It can be seen that control of oxides of nitrogen will be required for 1971 model automobiles in California, and that the emissions standards in that state are a series of requirements which become more stringent year by year, until the 1975 model vehicles.

The emission standards for 1975 models represent very stringent control requirements and, for this reason, are often termed "low emission standards." The feasibility of such standards has been the subject of recent studies by two groups (15,16). The conclusions of both studies are that such emission limits can be achieved.

Heavy-Duty Vehicles. The exhaust emission standards for heavy-duty vehicles (over 6000 lb gross vehicle weight) are given in Table X. These emissions continue to be expressed in concentrations rather than in mass of pollutants. As yet, there is no standard for oxides of nitrogen from these vehicles. Standards for the heavier vehicles have remained in concentrations because of the same engine being used in a 7000 lb vehicle or a 20,000 lb vehicle. Also, the mass emission standards that are applied to passenger cars would impose very strict control requirements on a large truck or bus, thus raising questions of technical feasibility as well as fairness from the standpoint of work performed.

TABLE IX

Exhaust Emission Standards for Light-Duty Vehicles

Uncontrolled emissions, g/mile	Exhaust emission standards			Remarks
	Model year	g/mile		
	1970	HC	2.2	Federal and California
		CO	23	Federal and California
	1971	HC	2.2	Federal and California
		CO	23	Federal and California
		NO_x	4.0	California
HC[a] 11				
CO[b] 80				
NO_x[c] 5.7				
	1972	HC	1.5	California
		CO	23	Federal and California
		NO_x	3.0	California
	1974	HC	1.5	California
		CO	23	Federal and California
		NO_x	1.3	California
	1975	HC	0.5	California
		CO	12	California
		NO_x	1.0	California

[a] HC—Hydrocarbons, as measured by nondispersive infrared analyzer (NDIR).
[b] CO—Carbon Monoxide.
[c] NO_x—Oxides of nitrogen, expressed as nitrogen dioxide.

TABLE X

Exhaust Emission Standards for Heavy-Duty Vehicles

Uncontrolled emissions, ppm	Exhaust emission: Model year	Exhaust emission: Standard	Remarks
	1970	HC 275 ppm	Federal and California
		CO 1.5%	Federal and California
HC[a] 900 ppm			
CO 3.5%			
	1972	HC 180 ppm	California
		CO 1.0%	California

[a] HC—Hydrocarbons, as measured by nondispersive infrared (NDIR).

2. *Diesel-Powered Vehicles*

Diesel emission standards have been established by the federal government for smoke. These limit the smoke from 1970 model diesel engines to an opacity that shall not exceed 40% during acceleration and 20% during the engine lugging mode.

Most diesel vehicles comply with present hydrocarbon and carbon monoxide emissions standards for gasoline-powered heavy-duty vehicles. For example, the concentration of carbon monoxide in diesel exhaust is normally about 0.1 to 0.3%. As emissions standards for heavy-duty gasoline-powered vehicles become more stringent, however, emphasis will be placed on additional standards for diesel engines.

VII. CONTROL IMPLICATIONS OF EMISSION STANDARDS

Figure 6 shows the emissions of hydrocarbons, carbon monoxide, and oxides of nitrogen from an uncontrolled light-duty vehicle and from one complying with the 1970 and 1975 crankcase, evaporative, and exhaust emissions standards in California.

It can be seen that a new vehicle in 1975 will be discharging very much less of the hydrocarbons, carbon monoxide, and oxides of nitrogen than those discharged from a 1960 model car. The effects of the controls on hydrocarbon emissions from a large population of cars is shown in Figure 7. This figure illustrates the expected decrease in hydrocarbons in the South Coast Basin as the California emission standards go into effect and controlled vehicles replace uncontrolled ones. The plot in-

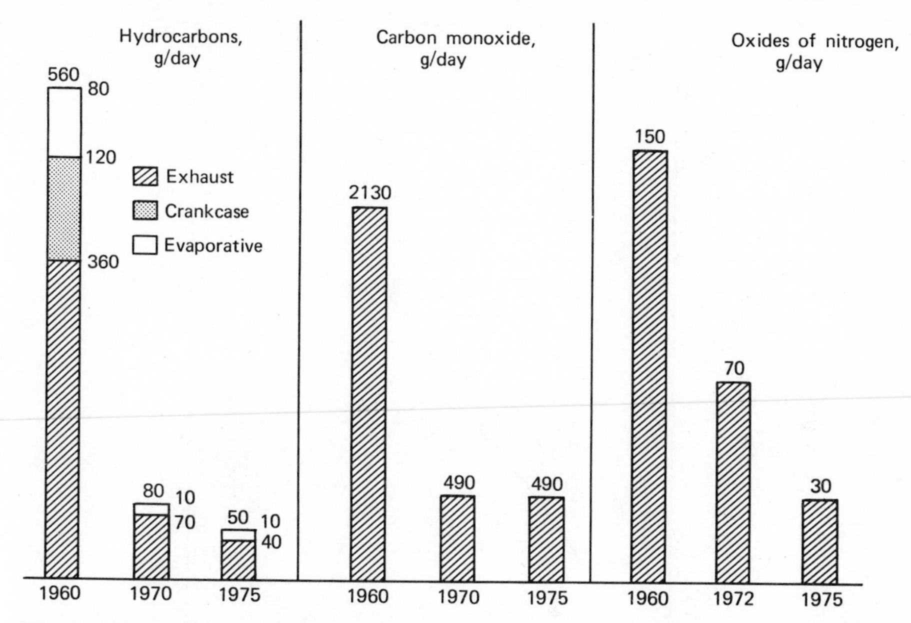

Fig. 6. Comparison of exhaust emissions from an uncontrolled light-duty vehicle and one meeting the new California emissions standards.

cludes expected increased vehicle population. Although they are not shown here, similar curves can be drawn for carbon monoxide and oxides of nitrogen.

VIII. FUTURE MOTOR VEHICLE EMISSION STANDARDS

A. Gasoline-Powered Vehicles

The wide concern over air pollution, public demand for clean air, and increased numbers of motor vehicles seems certain to require even more control of emissions from vehicles. This has already been reflected in the standards for light-duty vehicles in California. The federal government can be expected to revise its standards in the near future. Other contaminants, such as aldehydes and particulate matter, will be included in future standards.

Heavy-duty gasoline-powered vehicles will also be subjected to additional control requirements. These will include more stringent limits on exhaust hydrocarbons and carbon monoxide and standards for oxides

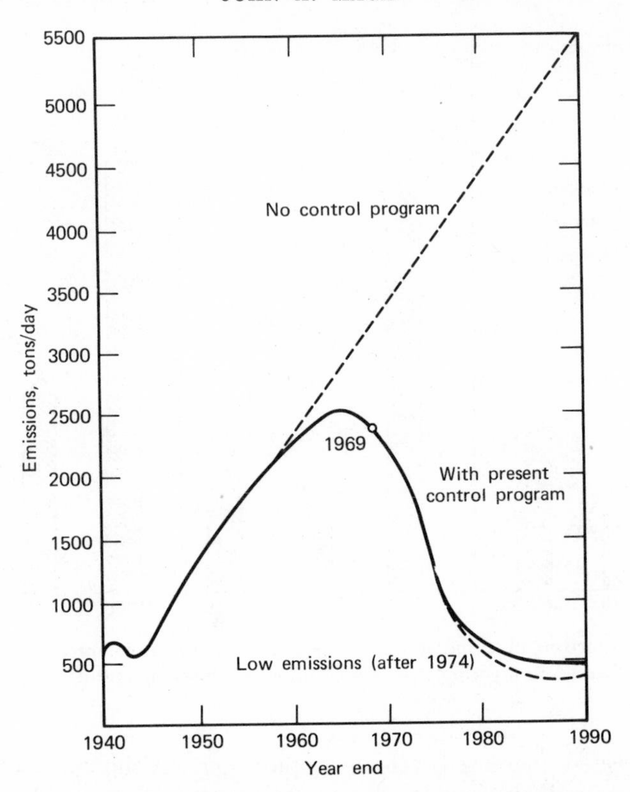

Fig. 7. Hydrocarbon emissions from motor vehicles in the South Coast Basin. From California Air Resources Board.

of nitrogen and evaporative loss. Control for the heavier vehicles will probably continue to follow those for light-duty vehicles by a few years. This permits the controls developed for the passenger cars to be extended to the large trucks and buses which comprise less than 5% of all vehicles.

B. Diesel-Powered Vehicles

Diesel exhaust emission standards for hydrocarbons, carbon monoxide, and oxides of nitrogen are required by law to be set in California by January 1, 1971. A different set of conditions must be considered in establishment of these standards. The definition of a test cycle, methods of measurement, and testing procedures cannot be translated directly from the corresponding gasoline vehicle requirements. The technological feasibility of control is also relatively unexplored, and many control techniques developed for gasoline engines are inherently applicable for diesel engines. This is particularly true for oxides of nitrogen.

IX. MOTOR VEHICLE EMISSION OR CONTROL SYSTEMS

It is possible to reduce the pollutants from motor vehicles by several methods, including: devices added to the vehicle, modification of existing engines, new propulsion systems, and change in the way people are transported. The final control of motor vehicle created air pollution will probably include all of these alternatives.

Many of the early efforts to control the exhaust emissions were directed at developing control devices which could be attached to the engine. These devices provided further combustion of the hydrocarbons and carbon monoxide in the exhaust from the engine. This early effort was followed by emphasis on minimizing the emissions discharged from the engine through carburetion, spark timing, and engine design. These courses were followed because they offered the earliest promise of success and because they did not disrupt the current patterns of engine design, manufacture, and use. The emphasis on "low emission" and "smog free" vehicles has again stimulated research on separate control devices which can be added to the vehicles and has expanded efforts to develop electric cars, steam cars, gas turbines, and other types of engines as a replacement for the present internal combustion engine.

Changes in the manner in which people are transported will not completely eliminate the use of the automobile. Mass transit systems would not, therefore, eliminate the need for controlling vehicle emissions. Changes in transportation patterns could, however, be important in reducing emissions by decreasing the use of private cars in metropolitan areas.

A. Crankcase Emissions

The control of crankcase emissions has proven to be effective and inexpensive because the blowby volume is low; only hydrocarbons are of concern, and a proven and simple control method was already available. All control systems are similar in principle. The road draft tube is removed, and the gases from the crankcase are introduced into the intake manifold or air cleaner or both. The flow of the crankcase gases must be regulated so that the proper air-fuel mixture is maintained. This is usually accomplished by a tapered valve; the flow through the valve is regulated by the intake manifold vacuum. These control systems are frequently referred to as positive crankcase ventilation.

The first crankcase control devices were installed on 1961 model cars in California. They have been placed on 1963 and subsequent model year vehicles throughout the United States. Such devices are now operating on almost 65 million vehicles in the United States.

Crankcase control devices have been the only devices installed on used cars. California requires that such systems be placed on 1955 and later model used vehicles at the time of change of ownership of cars not then equipped with these systems. The requirement is limited to the more densely populated counties where air pollution is most severe.

B. Evaporative Emissions

Systems to control hydrocarbons which are evaporated from the fuel system were first required to be on 1970 model cars in California. Similar control systems will be installed on all light-duty 1971 model vehicles sold in the United States.

The control systems collect the hydrocarbons evaporated from the fuel tank and carburetor. This requires the sealing of vents on the fuel tank and connecting the tank and the carburetor to a hydrocarbon storage unit. The unit usually is a canister of activated carbon for absorbing the hydrocarbons; the crankcase has been used for storage in some systems. The vapors are stored during the period when the car is not in operation. Upon starting the engine, some of the air flowing into the engine passes through the carbon canister, absorbing the fuel vapors and carrying them into the engine where they are burned. Vapors stored in the crankcase are introduced into the intake manifold by means of the crankcase control system. The systems must be designed to prevent changing the carburetor settings for exhaust emission control.

C. Exhaust Emissions

1. *Hydrocarbons and Carbon Monoxide*

The control of exhaust emissions has proved to be much more difficult than crankcase and evaporative emissions. More than one pollutant is involved and the exhaust flow and concentrations are highly variable with the variety of vehicle operating conditions. Control systems on 1966 model cars in California and 1968 models in the United States have been of two types: (*1*) addition of air into the exhaust manifold, and (*2*) modification of the engine.

In the air injection system, air is added into the exhaust manifold so that further combustion of the hydrocarbons and carbon monoxide may occur while the exhaust is hot. The air is added near each exhaust valve by means of a pump driven by a fan belt.

Engine modification has relied heavily on changes in the carburetion and ignition systems. As was shown in Figure 5, changes in carburetion are reflected in changes in hydrocarbons, carbon monoxide, and oxides of nitrogen. Retarding spark timing reduces oxides of nitrogen and hydrocarbons. Although engine modification systems have relied heavily on changes in carburetion and spark timing, they also have included changes in combustion chamber design to achieve low surface-volume ratios.

By the end of 1969, there were approximately 35 million vehicles in the United States with these two exhaust control systems. These systems have reduced hydrocarbons and carbon monoxide. Neither, however, has caused revolutionary changes in the design of engines now used. The emission standards which go into effect in 1975 models, however, will probably require more fundamental modifications, separate control systems, or the development of new kinds of engines.

2. *Oxides of Nitrogen*

Oxides of nitrogen control will become mandatory on 1971 model cars in California. Several methods of oxides of nitrogen control are under study and development. It appears that the 1971 model standards can be met in many vehicles by changing the current vacuum spark advance practice. The more stringent oxides of nitrogen emission standards for 1972 and 1974 model vehicles will probably require other control approaches. Exhaust recirculation has been the most widely suggested method. There are questions if this method also will enable vehicles to comply with very low emission standards for these compounds.

3. *Low Emission Vehicles*

Much research is being conducted by automobile manufacturers, the refining industry, and others on "low emission" control systems. One approach that has been the subject of considerable research is a highly efficient exhaust manifold in which higher temperatures and better mixing of additional air with the exhaust products are provided. Such a control system could include higher temperature in the reactor for oxidation of the hydrocarbons and carbon monoxide combined with rich operation and exhaust gas recirculation for the reduction of oxides of nitrogen.

Catalytic devices are also the subject of much study. Such devices could include catalytic reduction of oxides of nitrogen and oxidation of the hydrocarbons and carbon monoxide. Catalytic systems could be used as the primary control scheme or in conjunction with other control methods.

Table XI summarizes the exhaust emissions from some of the laboratory control systems aimed at achieving low emission standards.

D. Alternate Power Plants

The emphasis on very low exhaust emissions has led to suggestions that other types of motor power be used. Many types of engines have

TABLE XI

Emission from Conventional Engines Equipped with Low Emission Control Systems (g/mile)

	Sun Oil[a]	Chrysler—Esso[b]			Du Pont[c]	Ethyl[d]
		Reactor	Catalyst	Synchro[e]		
HC	0.7	<1.5	1.7	0.25	0.2	<0.7
CO	12.0	<20.0	12.0	7.0	12.0	<10.4
NO_x	0.6	<1.3	1.0	0.6	1.2	<2.5

HC—Hydrocarbons (NDIR).
CO—Carbon Monoxide.
NO_x—Oxides of Nitrogen as NO_2.

[a] P. E. Oberdorfer, discussion of a paper, "A Progress Report on the Development of Exhaust Manifold Reactors," by E. N. Cantwell, J. T. Rosenlund, W. J. Barth, F. L. Kinnear, and S. W. Ross, DuPont, presented at SAE International Automotive Engineering Congress, Detroit, January 1969.

[b] "Two Hands," Chrysler Corporation and Standard Oil Company (New Jersey), June 22, 1969 (APCA Meeting).

[c] "Ethyl Lean Reactor Car," Fact Sheet for Vehicle Demonstration, President's Environmental Quality Council, San Clemente, California, August 22, 1969.

[d] "The Search for a Low-Emission Vehicle," staff report prepared for the Committee on Commerce, U.S. Government Printing Office, Washington, D.C., U.S. Senate, 1969.

[e] Synchro-Thermal Reactor-Rich mixture, timed air injection to reactor, and exhaust gas recirculation.

been advanced, but the gas turbine, steam engine, and electric car are the most frequently mentioned. The current emphasis on alternate power plants for vehicles is almost entirely because of the concern over the air pollution problems caused by the internal combustion engine, rather than the limitations of present engines to power vehicles. Therefore, it is doubtful that there would be much interest in alternate power plants if there were no air pollution problems from present engines.

Development of alternate power plants is proceeding parallel to the development of low emission control systems for present engines. If effective and reliable low emission control systems are developed, presumably there would be less emphasis on alternate power plants as a means of controlling pollutants.

1. *Gas Turbine Engine*

The gas turbine engine has been tested in passenger cars and also in heavy vehicles. A few companies are marketing such engines for 1970 model trucks. Although this engine can be built for heavy-duty engines, it is not clear at this time if costs can be reduced so that it will be competitive with present passenger car engines or if it can be built in sizes for very light vehicles. Emissions from gas turbines have been reported to range in the "low emission" category.

2. *Battery Power*

Hydrocarbon, oxides of nitrogen, and carbon monoxide would not be discharged from battery-powered vehicles. The "Morse Report" (15) and subsequent studies have stated that this type of motor power will be restricted for at least a decade to vehicles with limited range and speed.

3. *Steam Engine*

Steam engines should be capable of being designed with low emissions. A few reports are available which indicate "low emissions." The tests, however, have not been extensive and the test conditions may not be representative of operating conditions in urban traffic. A number of projects are underway in steam power development. At this time, it is not known if the steam engine can be developed to a point where it will be a satisfactory replacement for the engine now in use.

X. AN UPDATING

New emission standards and test procedures that make significant changes in the amount of pollutants that may be discharged, in vehicle operating cycles, in sampling procedures, and in instrumentation have been adopted since this manuscript was first prepared. The extent of the changes that have been made in such a short period of time demonstrate the pace at which laws and regulations are enacted now to bring about more effective emissions control. These actions reflect the concern over air pollution by the public, legislative bodies, and governmental agencies. They also illustrate the difficulties that face both those who enforce air pollution control laws and those who must comply with the legal and administrative control requirements.

The U.S. Environmental Protection Agency prescribed new exhaust test procedures for emission values that are different than those determined by the previous test procedures. It was necessary, therefore, for that agency to adopt revised standards for hydrocarbons and carbon monoxide in 1972 and later model years. It also revised the evaporative emission standard from 6 to 2 g/test and adopted an oxides of nitrogen emission standard for 1973-model vehicles.

The most significant development, however, was the action of the U.S. Congress in legislating very stringent exhaust emission standards for motor vehicles. These and the new test procedure are discussed more fully below.

A. Light-Duty Vehicles

Late in 1970, Congress made extensive changes in the exhaust emission requirements for light-duty vehicles, when it established standards for exhaust hydrocarbons, carbon monoxide, and oxides of nitrogen. This was the first time that Congress, following the pattern set by the California Legislature in 1967, placed emission standards into law rather than relying on an administrative agency to establish the values as was the practice in the past.

The new standards were not based on a scientific study of the degree of control required to meet air quality standards. Instead, it was required that emissions of hydrocarbons and carbon monoxide in 1975-model vehicles be reduced by at least 90% from those for the 1970 models. The oxides of nitrogen in 1976 models must be reduced by this same percentage from 1971 cars. Because there were no previous federal standards for oxides of nitrogen, the degree of control of these compounds is somewhat less than those for hydrocarbons and carbon monoxide. Apparently, it was the intent of Congress to prescribe exhaust emission standards sufficiently stringent to eliminate the automobile as an important source of hydrocarbons, carbon monoxide, and oxides of nitrogen.

When the amendments to the Clean Air Act were being considered in 1970, the automobile industry vigorously opposed the emission standards on the basis that they were unnecessarily severe and could not be achieved by 1975 and 1976. Congress, as a compromise, provided that the Administrator of the Environmental Protection Agency could delay application of the standards for one year if he found that they were not attainable. He must, however, establish interim standards during that year. The law does not provide for further postponement; this could only be done by a change in the Clean Air Act.

The debate over the feasibility of the standards did not stop with their enactment into law. The automobile industry is faced with a difficult challenge if it is to meet any of the standards. The outlook for achieving the requirements for hydrocarbons and carbon monoxide is, however, more encouraging than for oxides of nitrogen. Less research has been done on the control of oxides of nitrogen, and the severe limitations placed on carbon monoxide emissions makes it more difficult to reduce the oxides of nitrogen also.

The new testing procedures established by the Environmental Protection Agency are considerably more complicated and lengthly than previous Federal and California ones and include a constant volume sampler, a continuous driving cycle, and measurement of hydrocarbons by flame ionization detectors (FID) and oxides of nitrogen by chemiluminescence (CI) analysis. In some ways, the procedures are almost as controversial as the emission standards. The new dynamometer cycle appears to be a better one for evaluating emission control systems. However, the test is very long and complicated. Also, it is not clear if the new procedures are superior to the previous ones for estimating emissions into the atmosphere.

Because of the many differences in the new and previous testing procedures, emission values determined by the two methods do not give comparable results. On the average, the quantity of hydrocarbons and carbon monoxide as measured by the new procedures are about twice as high and oxides of nitrogen about 50% higher than the 1970–1971 values measured under previous procedures.

Table XII shows the federal exhaust emission standards for the 1970-, 1975-, and 1976-model year vehicles. The emissions are all on the basis of the new test procedures.

B. Heavy-Duty Vehicles

Late in 1970, the California Air Resources Board adopted emission standards for hydrocarbons, carbon monoxide, and oxides of nitrogen from both gasoline and diesel-powered heavy-duty vehicles (18). The new standards are effective in the 1973-model year and become more stringent in the 1975-model year. This represents the first effort to control the emissions of these compounds from diesel engines and the first controls of oxides of nitrogen in gasoline heavy-duty vehicles. Early in 1971 that Board also adopted the first evaporative emission standards for heavy-duty gasoline-powered vehicles (19). These become effective in the 1973-model year.

TABLE XII
Exhaust Emission Standards for 1970-, 1975-, and 1976-Model Years

Pollutant	1970 Standards, g/mile	1975–1976 Standards, g/mile
Hydrocarbons	4.1	0.41
Carbon monoxide[a]	34	3.4
Oxides of nitrogen	4.0[a]	0.4

[a] Emissions from 1971-model year vehicles; there was no federal standard for oxides of nitrogen prior to 1973 models.

TABLE XIII
Motor Vehicle Emission Standards for Gasoline and Diesel-Powered Heavy Duty Vehicles

Pollutant	Emission standards, g/BHP-hr	
	1973-Model year	1975-Model year
Hydrocarbons plus:		
Oxides of nitrogen	16	5
Carbon monoxide	40	25

Table XIII shows the exhaust emission standards for heavy-duty vehicles. A new test procedure was required for the diesel-powered vehicles. The standards were expressed in grams per brake horsepower hour, and the hydrocarbons and oxides of nitrogen were combined into one value.

References

1. A. J. Haagen-Smit, "Chemistry and Physiology of Los Angeles Smog," *Ind. Eng. Chem.*, 1342 (June 1952).
2. "Technical Report of California Standards for Ambient Air Quality and Motor Vehicle Exhaust." State of California, Department of Public Health, Chap. XVIII, pp. 104–106.
3. "Motor Vehicles, Air Pollution, and Health." A report of the Surgeon General to the U.S. Congress, U.S. Department of Health, Education, and Welfare, pp. 113–114, June 1962.
4. A. J. Haagen-Smit, "Carbon Monoxide in City Driving," *Environ. Health* (May 1966).
5. Robert T. Brice and Joseph F. Roesler, "The Exposure to Carbon Monoxide of Occupants of Vehicles Moving in Heavy Traffic," *J. Air Pollution Control Assoc.*, **16**, 11 (November 1966).
6. Air Quality Data from the National Air Sampling Networks and Contributing State and Local Networks 1864–1965, U.S. Department of Health, Education, and Welfare, Public Health Service.
7. W. Crocker, P. W. Zimmerman, and A. E. Hitchcock, "Ethylene-Induced Epinasty of Leaves and the Relation of Gravity to It," *Contrib. Boyce Thompson Inst.*, **4**, 2 (1935).
8. U.S. Public Health Service, "Survey of Lead in the Atmosphere of Three Urban Communities," U.S. Department of Health, Education, and Welfare, PHS Publication No. 999-AP-12, January 1965.
9. "An Evaluation of Crankcase Emissions," State of California Department of Public Health, Bureau of Air Sanitation, May 31, 1963.
10. State of California Department of Public Health, report to the Advisory Committee on Air Sanitation, "Progress Report Motor Vehicle Emissions and Proposed Standards," *Evaporative Emissions*, Section V, p. 6, January 8, 1964.
11. "Profile of Air Pollution Control in Los Angeles County," Los Angeles Air Pollution Control District, 1969.

12. J. A. Maga and J. Kinosian, "Motor Vehicle Emission Standards—Present and Future," SAE Paper 660104, January 10–14, 1966.
13. J. A. Maga and G. C. Hass, "Present and Future Emission Standards for Heavy–Duty Vehicles," SAE Paper 690765, November 4–7, 1969.
14. "Los Angeles Traffic Pattern Survey," report by the Traffic Survey Panel of Automobile Manufacturers Association and Los Angeles Air Pollution Control District, 1957.
15. "The Automobile and Air Pollution," report of the Panel on Electrically Powered Vehicles, U.S. Department of Commerce, Part I, p. 1.
16. "Control of Vehicle Emissions after 1974," report to the California Air Resources Board by the Technical Advisory Committee, November 19, 1969.
17. Federal Register, Vol. 36, No. 128, July 2, 1971 Wash. D.C.
18. "California Exhaust Emission Standards, Test, and Approval Procedures for Engines in 1973 and Subsequent Model Diesel and Gasoline-Powered Motor Vehicles over 6001 Pounds Gross Vehicle Weight," California Air Resources Board, February 17, 1971.
19. "California Fuel Evaporative Emission Standards and Approval Procedures for 1973 and Subsequent Model Year Gasoline-Powered Motor Vehicles over 6001 Pounds Gross Vehicle Weight," California Air Resources Board, June 16, 1971.

Spectroscopic Methods for Air Pollution Measurement

PHILIP L. HANST
Environmental Protection Agency,
Research Triangle Park
North Carolina

I. INTRODUCTION

In the past, optics has played an important part in air pollution analysis and monitoring. However, the role of optics has not been as large as it can and should be in the future. There is now an upsurge of interest in applications of optics to air pollution studies. The interest is shown in symposia at scientific and engineering meetings, in seminars in colleges, in articles in the press and the scientific literature, and in a general demand for more information on the application of optics to pollution studies.

Portions of this study touch on the past, the present, and the future of optics in air pollution studies. Past work is treated briefly with reference to more extensive reviews. The present status of optical systems and components that are applicable to pollution studies is discussed in terms of long optical paths, radiation sources, and detection instrumentation. Future applications of optical systems in pollution measurement are implicit in many of the discussions, especially in the section devoted to radiation absorption by pollutants.

The word *for* in the title of this study was chosen deliberately. It implies a discussion of particular topics applicable to air pollution measurement rather than a complete review. The methods discussed herein generally have not yet been fully applied; but in the opinion of the author, they can and should be. In particular, long-path absorption techniques, the laser absorption technique, and the nondispersive gas filter method are set forth as promising candidates for further application.

This is not a complete work, nor an entirely accurate one, but it is hoped that it is a timely work, and that it contains as few inaccuracies as is practical in the present fast-moving situation in optics.

The author acknowledges with appreciation the assistance of William J. Henson in recording the spectra shown in Section III and the contributions of Vincent H. Early in the experimental work on the many-mirror multiple-pass cells reported in Section IV. The cooperation of Dr. Claus B. Ludwig in allowing quotations from his Convair reports and in supplying photographs and drawings for reproduction is gratefully acknowledged.

II. GENERAL DISCUSSION

A. The Role of Spectroscopy in the Study of Atmospheres

The atmospheric absorption spectrum, which extends from the ultraviolet to the far infrared, presents a vast amount of information on the composition and physical properties of the atmosphere. The use of spectra in the study of the atmosphere dates back to the work of Fraunhofer. Although most of Fraunhofer's lines were due to absorption by gases in the atmosphere of the sun, some lines (telluric) were due

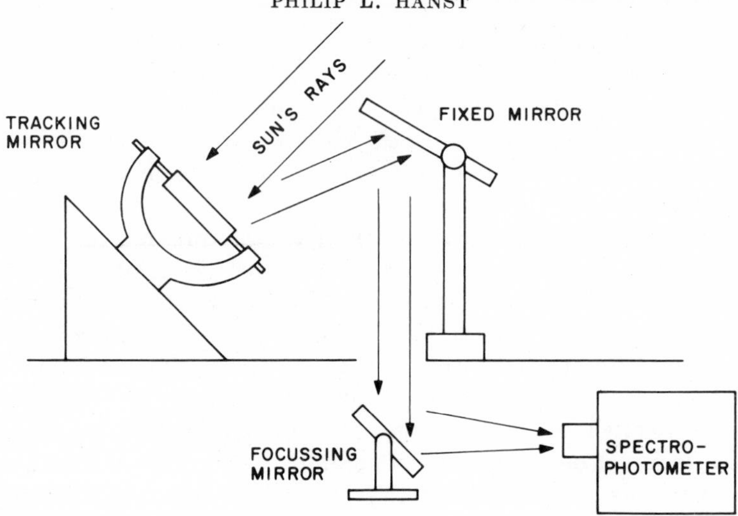

Fig. 1. Technique of atmospheric absorption spectroscopy, using the sun as a light source.

to gases in the atmosphere of the earth. As spectroscopy has evolved, absorption and emission spectra have become a major source of knowledge of the earth's atmosphere, especially of its upper reaches. With regard to the atmospheres of the sun, planets, and stars, spectroscopy is almost the only source of information. Helium was discovered in the sun by means of its spectrum before it was discovered in the earth's atmosphere. The ammonia and methane in Jupiter's atmosphere are known only by their absorption spectra. The same is true of the carbon dioxide in the atmosphere of Mars and Venus.

The earth's atmosphere has been studied by many techniques, including distillation, chemical analysis, mass spectroscopy, and chromatography. In the study of the earth's atmosphere, the spectroscopy of the electromagnetic spectrum is used less widely than these direct atmospheric sampling methods. However, the resurgence of optics during recent years promises to accentuate the potential applications of optical methods in the study of all levels of the earth's atmosphere.

Direct applications of spectroscopy in the study of the atmosphere include the Fraunhofer-type experiments, in which the sun is viewed directly through the atmosphere. The type of optical system used for such work is diagramed in Figure 1. A sun-tracking mirror, two or three auxiliary mirrors, and the spectrometer system are the essential parts of the experimental equipment.

To obtain the full amount of information available in the spectrum, the instrumental resolving power must be sufficient to permit the recording of the true shapes of the spectral lines and the true fine structure of the molecular bands. The required resolution is a function of the pressure of the parcel of air under study. In looking through the entire atmosphere, one may see very narrow lines caused by constituents at high altitudes, wider lines caused by constituents at low altitudes, and composite lines caused by constituents being uniformly distributed. Pollutants near sea level will have absorption lines approximately 0.1 cm^{-1} wide.

The pure atmosphere has been the subject of much spectroscopic analysis. Work prior to 1954 has been reviewed by Goldberg (1). A 1964 summary of atmospheric transmission studies has been given by Howard, Garing, and Walker (2). A basic reference work is *Photometric Atlas of the Infrared Solar Spectrum,* published by the Royal Belgian Observatory (3). In that work the spectrum was recorded by looking at the sun from an observing station at a height of 11,700 ft in the Alps. The wealth of detail in the spectrum is illustrated in Figure 2 by a small portion of the spectrum copied from the atlas. The majority of the absorption lines are due to water vapor, but the two strong lines marked are attributed to atmospheric methane.

The spectrum of the polluted atmosphere is much less well known than the spectrum of the clean atmosphere. Stair and Gates (4,5) both looked at the sun through the smog of Los Angeles and observed absorp-

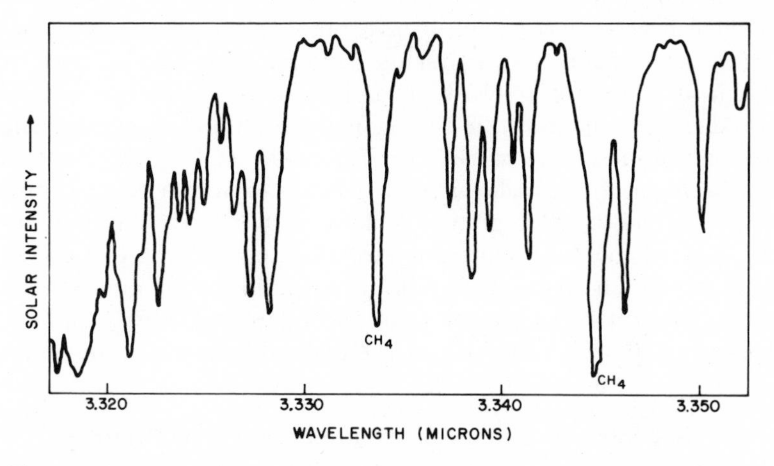

Fig. 2. A portion of the *Photometric Atlas of the Infrared Solar Spectrum.*

tion bands due to pollutants. However, their resolving powers were limited and further work is needed. Scott *et al.* (6) observed the characteristic bands of many pollutants in the Los Angeles air in 1956, using folded paths of several hundred meters. Path folding was achieved with a "White" cell (Section IV). A Nernst glower source and a commercial prism spectrometer were used. The spectra showed bands due to ozone, carbon monoxide, peroxyacyl nitrates, ethylene, acetylene, and other hydrocarbons, as illustrated in Figure 3. With higher resolution, more detail could have been observed.

An ever-present difficulty in atmospheric spectroscopy is the interference from atmospheric carbon dioxide and water vapor. These two molecules are the principal polyatomic constituents of the clean atmosphere; each has strong absorption bands in the middle infrared region. The water vapor spectrum is particularly rich in lines. Water absorption not only blanks out completely the far infrared region of the spectrum, but also appears strongly in the interesting and useful region of the middle infrared between 5.5 and 7.5 μ. Even the microwave region suffers from water interference. In early attempts to propagate radar pulses through the atmosphere, the received signals were many times weaker than anticipated. The disappointing results were found to be caused by the unfortunate choice of the operating wavelength of 1.25 cm, close to the peak absorption by a water vapor line in the microwave spectrum.

When searching for pollutants occurring at concentrations of less than 1 ppm, it is not advisable for one to attempt to remove water vapor or carbon dioxide from the air. Treatment of the air to remove a major constituent such as water will always raise the question of whether the concentrations of minor constituents have not also been altered. Generally, it is necessary to work in and around the lines and bands of CO_2 and water, and for this reason high resolution instrumentation is desired for atmospheric work. In Figure 4*a* and *c*, the atmospheric transmission spectrum is shown as it is likely to appear to an instrument of comparatively low resolution looking across a 1000 ft path at sea level. There is actually more fine structure than shown, and the ability of high resolution instrumentation to work in and around the water and CO_2 absorption is greater than the figure indicates. Parts *b* and *d* of Figure 4 show where some of the accessible bands of air pollutants fall.

B. Infrared Studies of Atmospheric Photochemistry

Special mention should be made of the role of infrared spectroscopy in the discovery of the organic peroxy nitrates in polluted air; they

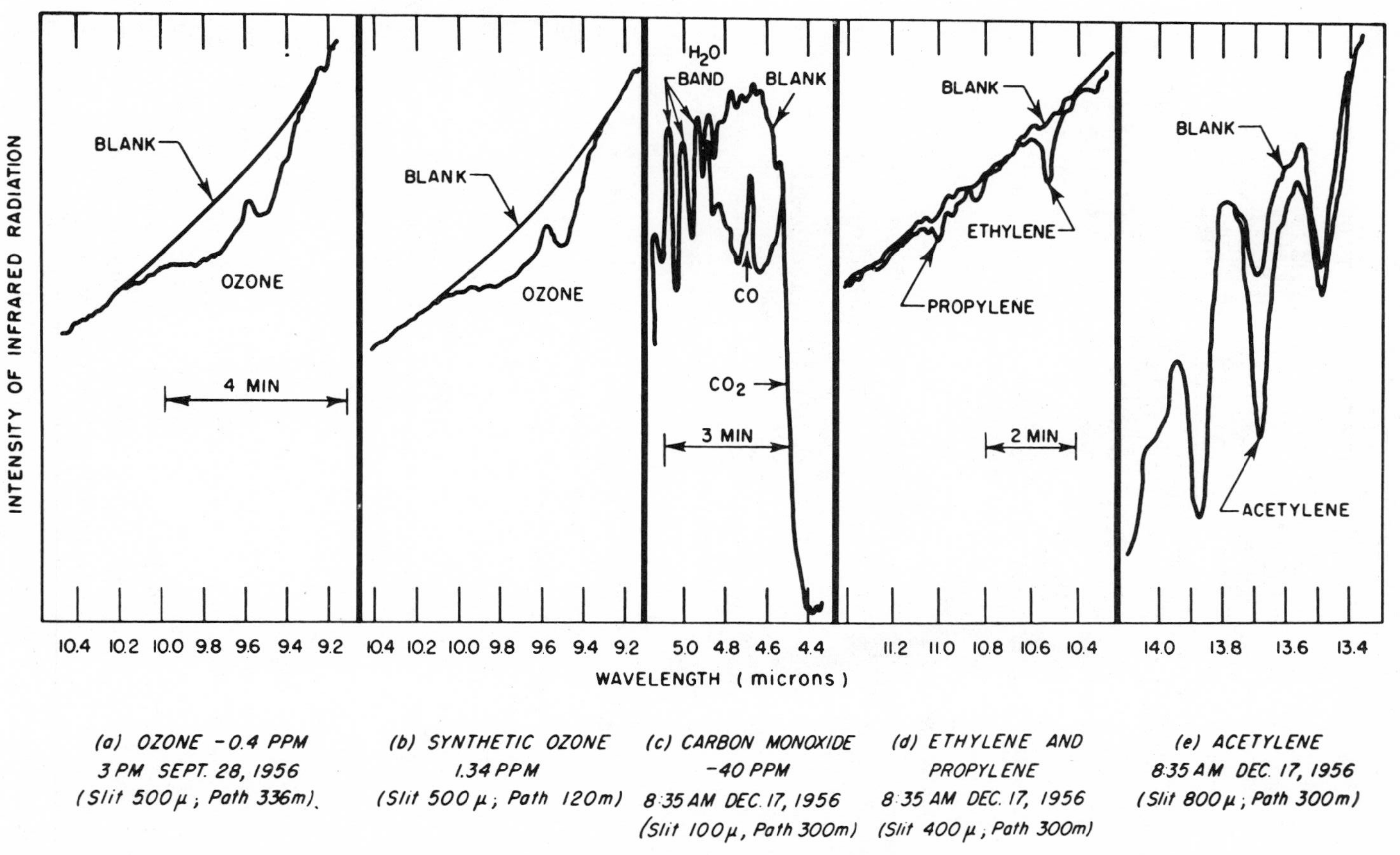

Fig. 3. Infrared bands of pollutants in South Pasadena.

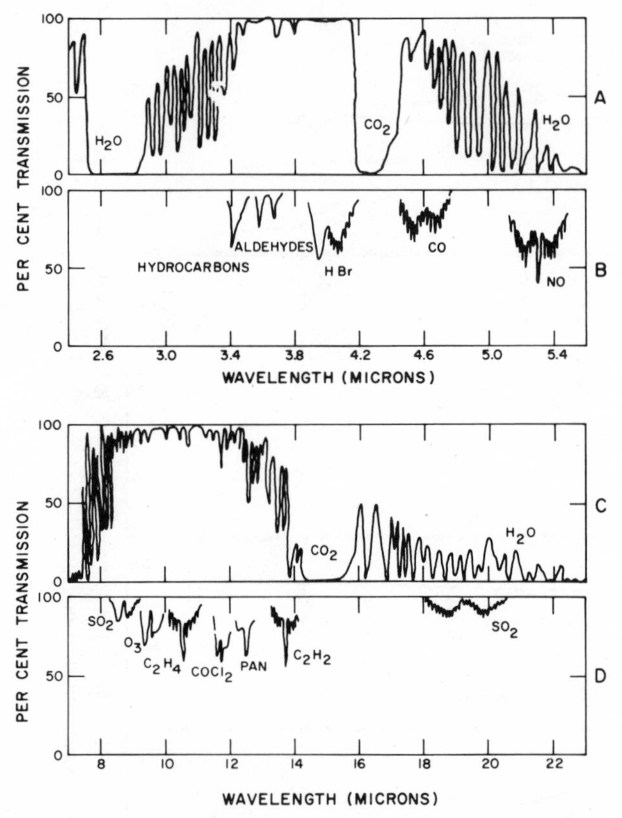

Fig. 4. Atmospheric and pollutant absorption bands.

were unknown prior to the air pollution studies. They are difficult to synthesize in the laboratory, and have no known applications in chemical technology. The part they play in photochemical smog, however, is very significant (7).

The organic peroxy nitrates were discovered in a 1955 research program at the Franklin Institute in which infrared absorption spectroscopy was being used as the analytical technique for following the photochemical reactions of hydrocarbons and nitrogen oxides in air. A long folded infrared path was used so that reactants and products at concentrations of parts per million in air could be detected. A principal objective of the work was to determine by the infrared absorption method whether or not ozone was formed in the photochemical reactions.

Ozone was detected, but in the same spectra a new set of absorption bands appeared. These bands appeared and disappeared together, so it could be assumed that they belonged to a single compound. Since the bands showed no resolvable fine structure, they most likely belonged to a fairly large polyatomic molecule. A nitrogen-containing compound was suspected, because it was necessary to account for the disappearance of nitrogen oxides during the photolysis. From known infrared spectra it could be stated that the new compound was neither an organic nitrate, a nitrite, a nitro compound, nor a nitroso compound. For some time the newly discovered molecule was merely designated compound X.

Through consideration of the chemistry of the material, with some assistance (but not much) from mass spectroscopic analysis, and through detailed consideration of the infrared spectrum, the compound X was finally identified as peroxy acetyl nitrate, abbreviated PAN. The structural formula is $CH_3\overset{\overset{\displaystyle O}{\|}}{C}—OONO_2$. The presence of the material in the Los Angeles atmosphere was verified by means of the infrared absorption spectrum by the Franklin Institute group in 1957 (6). PAN since has been shown to be both damaging to crops and irritating to the eyes. The higher homologues PBN (butyl) and PPN (propyl) have also been studied.

Recently, the related compound peroxybenzoyl nitrate (PBzN) has been discovered in photochemical reactions of air pollutants. PBzN has been shown to be much more irritating to the eyes than PAN (8). The discovery and identification of PBzN were made by means of its infrared absorption spectrum.

The infrared spectrum of PAN is shown in Figure 29, Section III. The presence of the distinctive carbonyl band at 5.75 μ is noteworthy, since it was a key to the understanding of the structural formula. The PAN story is a significant illustration of the power and the continuing importance of the infrared method in the study of gaseous mixtures.

C. Energy Limitations on Long-Path Absorption Spectroscopy

The application of absorption spectroscopy to pollutant analysis has been limited by the amounts of energy available from sources of radiation. Since a long light path is required for the detection of trace constituents, the divergence of light beams from thermal sources is likely to produce a very weak signal at the receiver. Direct solar spectroscopy is the one aspect of atmospheric analysis that has not suffered from a lack of energy but it does have the obvious disadvantage of restricting the line of sight to certain slant paths through the entire atmosphere.

Absorption spectroscopy on confined samples has depended primarily on an ingenious device known as the "White" cell (9). This is an optical arrangement of three mirrors in which the light beam is folded over a very long path without losing energy through beam divergence (Section IV). The cells have been specially fabricated in many laboratories and are available commercially. With the multiple-pass cell, however, the detection sensitivity is energy limited. To achieve maximum detection sensitivity, elaborate detection instrumentation is needed. Any development which will alleviate the problem of beam divergence or increase the source energy will significantly improve pollutant detection. Lasers overcome the energy limitation by having a large amount of power at a single wavelength and a coherence which allows extremely good collimation.

D. Applications of the Various Spectral Regions

The spectrum can be considered in terms of wavelength regions characterized by the type of physical change that takes place in the absorbing or emitting molecule and by the type of instrumentation used in studying the spectrum.

1. *Ultraviolet, 0.25–0.40 μ*

The ultraviolet is the region of absorption by electronic transitions in molecules. Combined with the electronic changes are vibrational and rotational changes which in a few cases produce a characteristic structure in the bands. While the ultraviolet band systems of different molecules are likely to overlap, correlation techniques have been developed for distinguishing one band system in the presence of others. The larger molecules do not have uv band structure at atmospheric pressure and therefore the applicability of the uv in pollution monitoring by absorption spectroscopy is limited. If Raman scattering techniques can be developed for pollution measurement, the ultraviolet region will become of much greater importance because the intensity of the Raman scattering increases with the fourth power of the radiation frequency.

2. *Visible, 0.40–0.70 μ*

The visible region of the spectrum, by definition, has very few molecular absorption bands. NO_2 is the only colored gas which is a common pollutant.

3. *Near Infrared, 0.70–2.50 μ*

This is the spectral region where the overtones of the fundamental molecular vibration-rotation bands appear. The overtones are about 100 times weaker than the fundamentals; thus this region is not generally useful for molecular detection and analysis. A notable exception is in the study of the atmospheres of the planets, where large concentrations of absorbing gas are viewed over extremely long paths. Thus the overtone region is the principal source of information on the atmospheres of Mars, Venus, and Jupiter.

4. *Middle Infrared, 2.50–25 μ*

This is the spectral region where the strong fundamental vibration-rotation bands of molecules appear. Nearly every air pollutant will have a characteristic absorption band in this region. It is sometimes called the "fingerprint" region of the infrared and is used extensively in chemical analysis. Generally the absorption bands in this region differ widely as to shape, location, and intensity distribution. Even the two members of pairs of similar molecules such as CO–NO and O_3–SO_2 have significantly different spectra. More detailed discussion and numerous examples of pollutant infrared spectra are given in the following sections.

5. *Far Infrared, 25–500 μ*

Rotational lines as well as some vibration-rotation bands appear in the far infrared. Unfortunately, very intense water vapor absorption blanks out the far infrared in atmospheric work.

6. *Millimeter and Microwave*

Molecular rotational lines appear in the microwave spectrum. Most large molecules have a microwave spectrum, and the region is very powerful in the study of molecular structure and other physical properties. However, the molecular gas under study must be at a very low pressure in order for the fine structure of the spectrum to be resolvable. At atmospheric pressure the spectra are so "smeared out" that distinguishing between molecules is impractical. Pollution measurements using long microwave absorption cells containing air samples at reduced pressure are a possibility deserving further investigation.

E. Concentrations and Distribution of Air Pollutants

The atmosphere, participating in biological processes, acquires trace concentrations of many organic and inorganic gases. It acquires trace gases through earth erosion, volcanic action, and other inanimate processes. Since excessive amounts of these trace gases are generally detrimental to human welfare, they are properly called pollutants. Transported through the atmosphere the pollutants are subject to photolysis, oxidation, and reaction with other trace gases. Reactions involving the pollutants are most severe at the highest altitudes. Above the ozonosphere the strength of the ultraviolet sunlight is sufficient to break up any molecule.

The processes of gas transport, photolysis, and chemical reaction create a condition in the atmosphere in which the concentrations of pollutants vary with altitude, weather, terrain, time of day, and other factors. This is the normal or steady-state condition of the atmosphere to which man has adapted himself in an ecologically balanced world. Unfortunately, since World War II the increasing population, increasing human consumption, and, especially, increasing use of fossil fuels are perturbing the atmospheric ecological balance. We hope that the cleansing power of the entire atmosphere still far exceeds the rate of emission of pollution, but there is no doubt that in the vicinity of large cities the cleansing power has been overtaxed.

Pollution sources, concentrations, and removal processes have been reviewed recently by Robinson and Robbins (10), as well as by Ludwig *et al.* (11), Altshuller (12), and Stephens and Burleson (13). Because of the variability in pollutant concentrations, no single concentration can be regarded as typical or normal. However, it is practical to make estimates for two opposite cases of interest: a large city and a rural area. Estimates covering most of the important pollutants, compiled from the previously mentioned sources are presented in Table I. These estimates apply to ground level; in most cases, the vertical distribution is not well known.

It is contended in Section III that each of the pollutants listed in Table I can be measured by infrared techniques. Some, such as methanic, carbon monoxide, ozone, and total hydrocarbon, can be measured at all existing atmospheric concentrations with instrumentation based on the presently existing optical technology. Others, such as sulfur dioxide, nitrogen dioxide, peroxyacetyl nitrate, olefinic hydrocarbons, and formaldehyde, will be measurable by new instrumentation to be developed through reasonable and foreseeable extensions of optical technology.

Long-term pollution buildup on a global scale is a frightening prospect necessitating the most serious and careful study. Spectroscopic methods are advantageous in global atmospheric surveillance, and remote sensing systems for pollution monitoring from satellites have been under study by the National Aeronautics and Space Administration for the past several years. A particularly relevant and informative NASA report is "Study of Air Pollutant Detection by Remote Sensors" (11). The following discussion of the prospect of global pollution buildup is quoted with permission, from a monthly report of the General Dynamics Corporation, published under their NASA contract.

TABLE I

Estimated Pollutant Concentrations

Pollutant	Principal sources	Typical city concentration, ppm	Typical rural concentration, ppm
Carbon monoxide	Auto exhaust	5.0	0.1
Sulfur dioxide	Oil burners	0.2	0.002
Nitric oxide	Combustion	0.2	0.002
Nitrogen dioxide	Combustion	0.1	0.001
Ozone	Atmospheric photochemical reactions	0.3	0.010
Methane	Natural gas, decaying organic matter	3.0	1.4
Ethylene	Auto exhaust	0.05	0.001
Acetylene	Auto exhaust	0.07	0.001
Peroxyacetyl nitrate	Atmospheric photooxidation of olefins	0.03	0.001
Olefins with three-or more carbons	Auto exhaust	0.02	0.001
Total hydrocarbons excluding methane	Auto exhaust	2.0	0.005
Ammonia	Decaying organic matter	0.010	0.010
Hydrogen sulfide	Decaying organic matter	0.004	0.002
Formaldehyde	Incomplete combustion, atmospheric reactions	0.05	0.001

The primary scavenging processes of pollutants are: (*a*) the deposition and "wash-out" of liquid and solid particles, (*b*) the adsorption and chemical reactions of species in the lower atmosphere, and (*c*) the conversion to "normal" atmospheric constituents in the upper atmosphere by thermal and photochemical reactions. If these processes are not sufficient to remove a pollutant, accumulation takes place.

In general, the interaction of air pollution with the earth's atmosphere produces both local and global effects. The local effects due to human activities are monitored by point sampling in order to determine the "air quality" in urban areas, source emissions, dispersal rates and the effects on human and animal life, and on vegetation. It is believed that the local effects will continue to be effectively observed by ground-based techniques.

Long term global effects are more difficult to classify, but involve the gradual changes in global meteorology (14,15). At present, the only methods being used to determine long-term effects are point samplings of pollutants at selected ground stations around the globe (16); inferences from geological evidence, such as the observation of lead aerosol accumulation in the top layers of oceans and ice core drillings in Antarctica; the change of the atmospheric electrical conductivity; and plant and forest damage even in mountainous regions. However, these methods are clearly insufficient because they are influenced heavily by the local weather conditions and also by the local scavenging processes taking place in the lower atmosphere. Thus, a general trend of global buildup can be established only after many years of data taking and there is no method, other than speculative calculations, to establish global dispersal patterns and scavenging processes.

The best known facts of changes in our environment are the increase of the CO_2 and aerosol concentrations (17). It is evident that the known removal processes by the photosynthesis of vegetation and the assimilation of CO_2 by rocks and oceans are not sufficient. CO_2 increases of 0.72 ppm/year in the northern and 1.3 ppm/year in the southern hemisphere are deduced from measurements at Mauna Loa, Hawaii, and Little America, South Pole (15). The increase of aerosol content is deduced from local observations of the visibility (turbidity), ocean water analysis, and electrical conductivity observations (15).

No such trends have been reported for other pollutants as yet. For a number of pollutants, known removal processes in the lower atmosphere *seem* to be sufficient so that no build-up is expected. Examples of these pollutants are the oxides of nitrogen, the fluorides, sulfides, and ammonia. However, the possibility of the formation of some stable intermediate compounds, which are not as easily measured, cannot be precluded (18).

For a number of other pollutants, the removal processes in the lower atmosphere are either very slow or not known at all so that a build-up of these pollutants in the global atmosphere is possible. At present, the methods and instrumentation used are insufficient to establish this. In the case of CO, an increase of 0.03 ppm/year is expected in the absence of any removal processes (19). In the absence of measurements sufficiently accurate to detect such concentrations, it is conceivable that such accumulation, in fact is occurring (18). Although the absolute annual increase of CO is less than the increase of CO_2, its toxic nature makes a world-wide surveillance of this pollutant very important.

The situation is similar for SO_2. Although some removal processes are known, such as $SO_2 \rightarrow SO_3 \rightarrow H_2SO_4$, it is not clear how efficient is this mechanism. In the absence of removal, the annual increase is expected to be 0.006 ppm (18).

Other pollutants, such as the stable nitrogen-containing hydrocarbons, are extremely important in air pollution surveillance. In unstable meteorological conditions (super adiabatic lapse rate), these compounds can be mixed throughout the troposphere. It has been assumed that photochemical processes are sufficient to break them up; however, no measurements have been made.

It is expected that a global surveillance will reveal non-uniform distributions of pollutants as was already indicated by the different CO_2 increases in the northern and southern hemispheres, and the different electric conductivities of the atmospheres over the Pacific and Atlantic Oceans (15).

Global surveillance should show a relationship between the pollutant content and the synoptic weather situation, and perhaps provide a forecasting system for pollution levels. Such a system would be of great value in the future if pollution abatement is not successful, and as more nations become industrialized.

III. RADIATION ABSORPTION BY POLLUTANTS

A. Infrared Absorption Spectra of Selected Pollutants

The fundamental infrared vibration–rotation bands mainly fall in the spectral region 3 to 23 μ. The band location within the spectrum and the detailed structure of the infrared bands are derived from the molecular geometry, atomic masses, and chemical bond strengths. The bands, therefore, are a most distinctive property of the molecules for use in measurement systems. This has been known for many years and there are many published infrared spectra of air pollutant molecules. However, as instrumentation has become developed more highly, spectral resolution has increased, and many of the published spectra do not show the wealth of detail which modern instrumentation can feature. For these reasons and for reference in subsequent discussions, the spectra of a group of the principal air pollutants have been recorded on a grating instrument and are presented in the following figures:

Pollutant	Figure(s)
Methane	5, 6
Ethane	7
Propane	8
Butane	9
Pentane	10
Hexane	11
Formaldehyde	12
Acetaldehyde	13
Nitrogen dioxide	14
Hydrogen chloride	15
Deuterium oxide	16
Carbon monoxide	17
Nitric oxide	18
Sulfur dioxide	19
Ozone	20
Ethylene	21, 22
Ammonia	23
Methanol	24
Propylene	25, 26
Acetylene	27, 28
Peroxyacetyl nitrate	29

Most of the principal atmospheric pollutants are included in the list above, the bands studied being those which are accessible in the presence of atmospheric water vapor and carbon dioxide. Some strong and characteristic bands are not shown because of water interference, but are accessible either in dry air or when the sample is at low pressure. These include the carbonyl bands near 5.8 μ and the NO_2 fundamental near 6.2 μ.

Spectra were recorded on a Perkin-Elmer single-beam Ebert-type grating spectrophotometer (E-14). The resolution was limited by slit widths to values between 0.5 and 1.0 cm^{-1}, as indicated by the apparent widths of lines in the spectra. Gases were obtained from Matheson lecture bottles; formaldehyde vapor was obtained by heating pure solid paraformaldehyde; methanol and acetaldehyde vapors were taken from equilibrium with the liquids; and ozone was generated by passing tank oxygen through an electric discharge and collecting the ozone in a trap cooled in liquid nitrogen.

Most of the spectra have marked on them the positions of certain gas laser lines. The application of some of the lines to pollutant measurement will be discussed in a subsequent section. The figures show wavelengths in air, which may be converted to vacuum wave numbers by application of the appropriate correction factor (20).

B. Discussion of Individual Spectra

1. *Methane*

Methane, a spherical top molecule, has a comparatively simple infrared absorption spectrum. There are two strong infrared bands, shown in Figures 5 and 6. The shorter wavelength band is the more useful for detection and measurement in the atmosphere because of its greater freedom from interference by water vapor. The distinctive fine structure of the band centered at 3.3 μ has permitted the development of nondispersive infrared detection instruments for methane. This spectral detail can also be used to advantage in the development of laser absorption methods of hydrocarbon measurement.

2. *Ethane*

Ethane, having a larger moment of inertia and a lower degree of symmetry than methane, has a much more complex spectrum. Figure 7 shows that ethane appears to have two complex bands near 3.4 μ which overlap and are difficult to resolve. The Q-branch at 3.385 μ seems to be a dividing mark between two parts of the spectrum, the

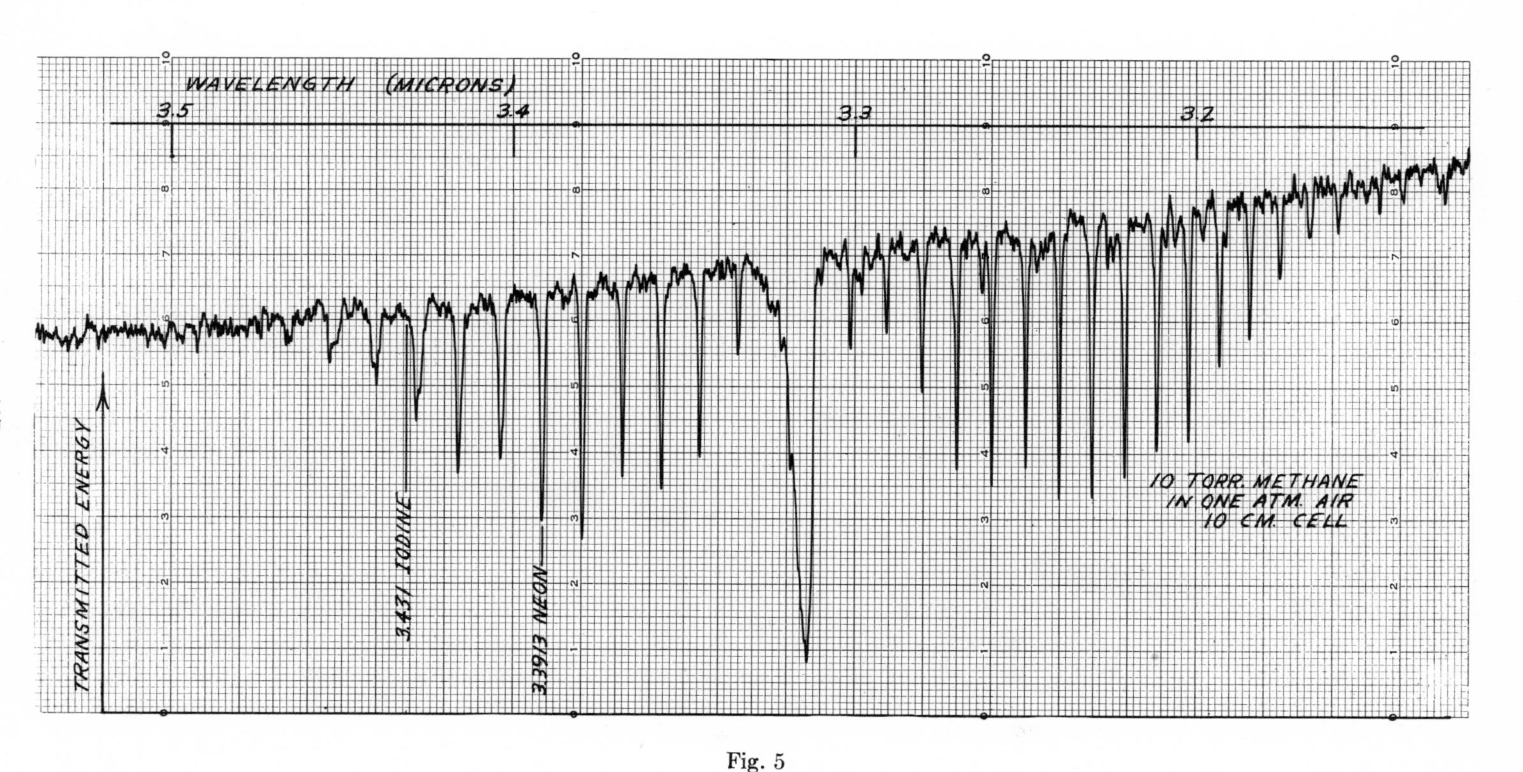

WAVELENGTH (MICRONS)
3.5
3.4
3.3
3.2
TRANSMITTED ENERGY
3.431 IODINE
3.3913 NEON
10 TORR. METHANE
IN ONE ATM. AIR
10 CM. CELL

Fig. 5

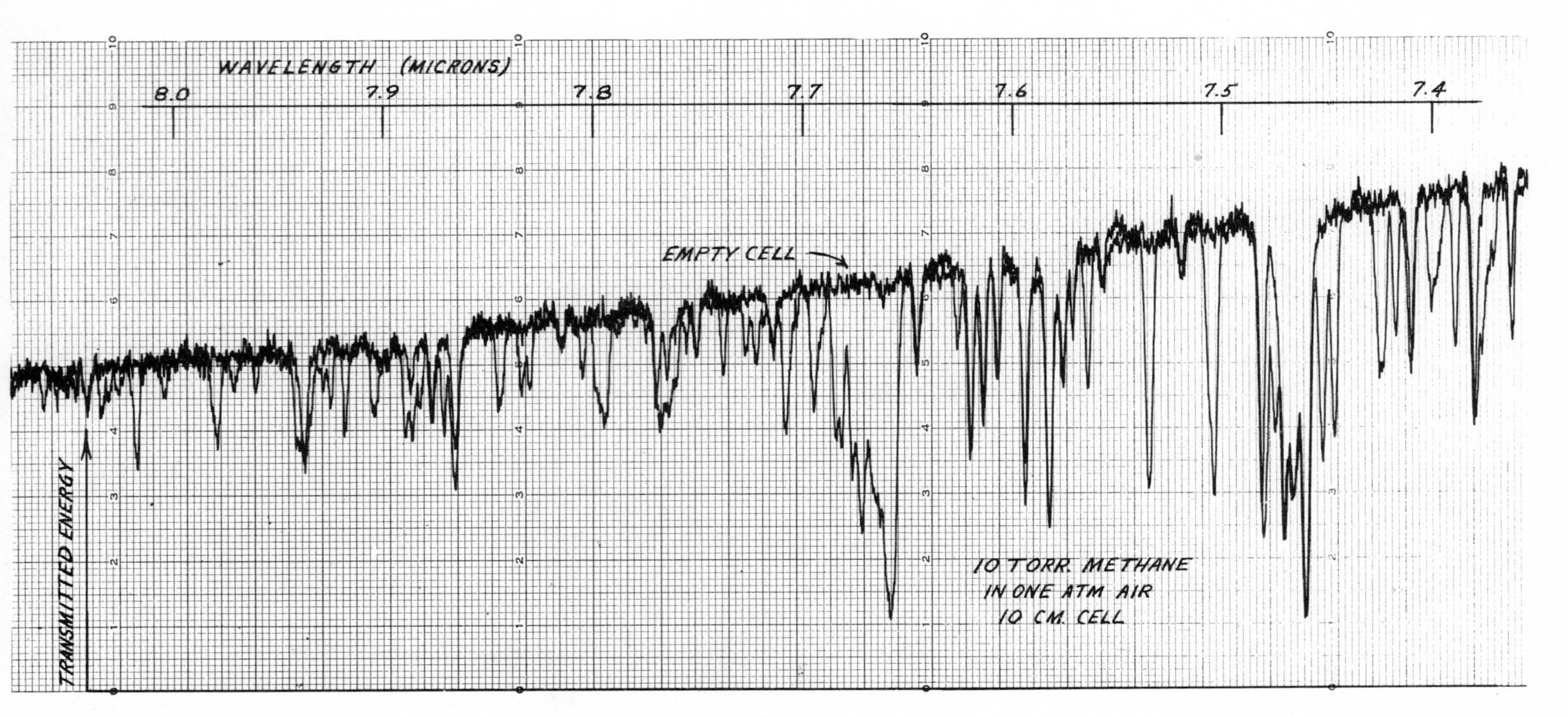
WAVELENGTH (MICRONS)
8.0
7.9
7.8
7.7
7.6
7.5
7.4
EMPTY CELL
10 TORR. METHANE
IN ONE ATM AIR
10 CM. CELL
TRANSMITTED ENERGY

Fig. 6

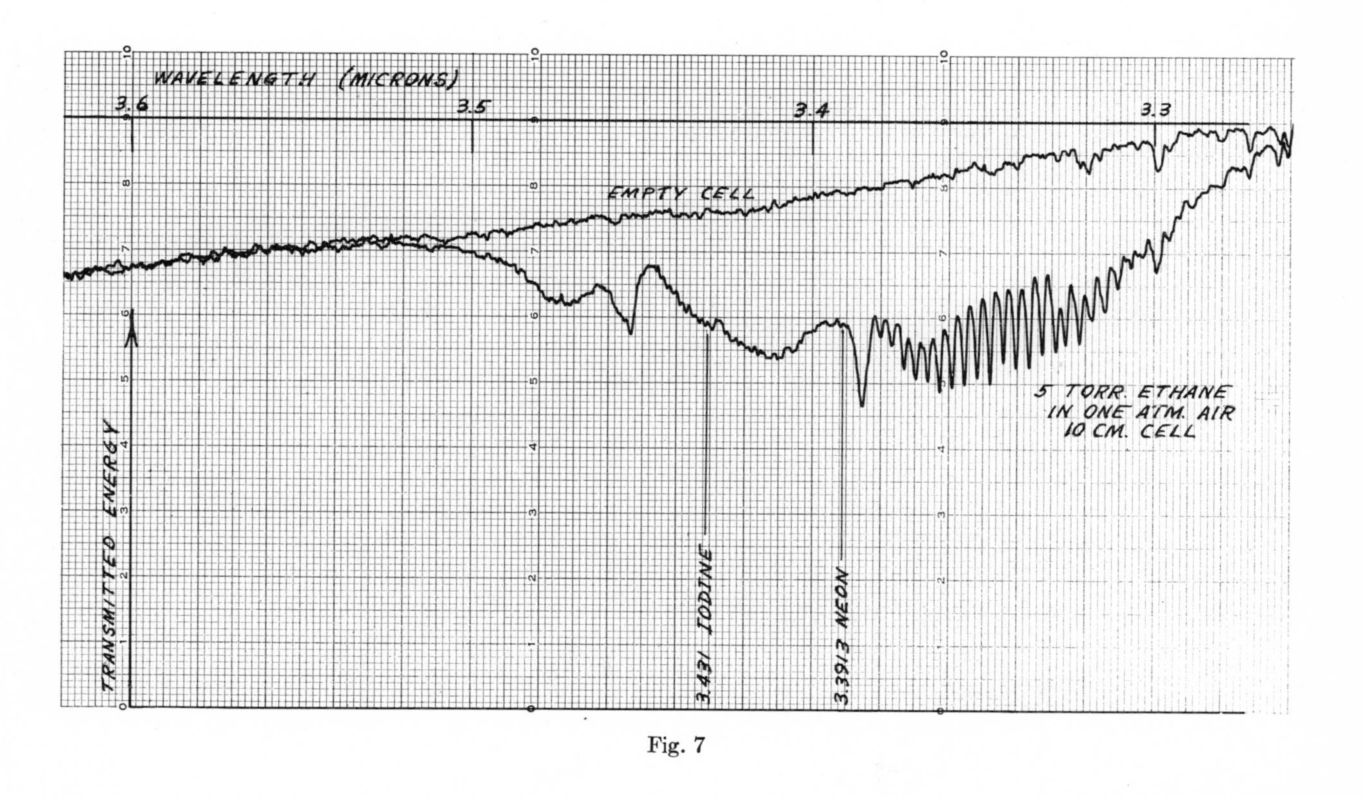
WAVELENGTH (MICRONS)
3.6
3.5
3.4
3.3
EMPTY CELL
5 TORR. ETHANE
IN ONE ATM. AIR
10 CM. CELL
TRANSMITTED ENERGY
3.431 IODINE
3.3913 NEON

Fig. 7

short wavelength part having fine structure shown in the spectrum that may be only partially resolved, and the long wavelength part having at one atmosphere pressure little or no resolvable fine structure.

All organic molecules which contain hydrogen-carbon linkages have infrared absorption near 3.4 μ. The absorbing transition can be visualized as producing a change of vibration of one hydrogen atom against the remainder of the molecule; therefore the band is designated the C–H band.

In ethane and in all heavier paraffinic hydrocarbons the C–H band is the strongest in the infrared spectrum (21). Organic molecules with two or more carbons and with a low degree of symmetry have an extremely complex spectrum in which lines are packed so close together that at atmospheric pressure there is no resolvable fine structure. This lack of fine structure can be verified with low resolution instrumentation by observing that the attenuation relationship, $\ln I_0/I = \alpha pl$, is obeyed at all points in the spectrum. In laser absorption studies the lack of fine structure is demonstrated by showing that the absorption coefficient is not strongly dependent upon total sample pressure. Nearly all pollutant molecules will show fine structure in their spectra when the sample pressure is reduced to a few torrs, as is discussed in a following section.

3. *Propane, Butane, Pentane, and Hexane*

In the heavier alkanes the 3.4-μ band (the so-called C–H band) becomes progressively more intense and more diffuse as the number of carbons is increased, as shown in Figures 8–11. In propane there is no resolvable fine structure at atmospheric pressure, but there is an envelope of a P–Q–R structure in the bands. By the time hexane is reached, the P–Q–R structure has disappeared. The integrated absorption strength of the band increases roughly in proportion to the number of C–H bonds in the molecule. Other hydrocarbon absorption bands fall near 6.8 and 7.2 μ but are considerably less attractive than the 3.4 μ-band for measurement in the atmosphere because they fall within a region of strong water vapor absorption. They are also weaker than the C–H bands.

4. *Formaldehyde*

Formaldehyde has a low degree of symmetry, but is light enough that a complex fine structure appears in its spectrum at atmospheric pressure (Fig. 12). The C–H absorption of aldehydes extends to longer wavelengths than is the case for most other organic compounds. This distinguishing characteristic can be the basis of selective infrared

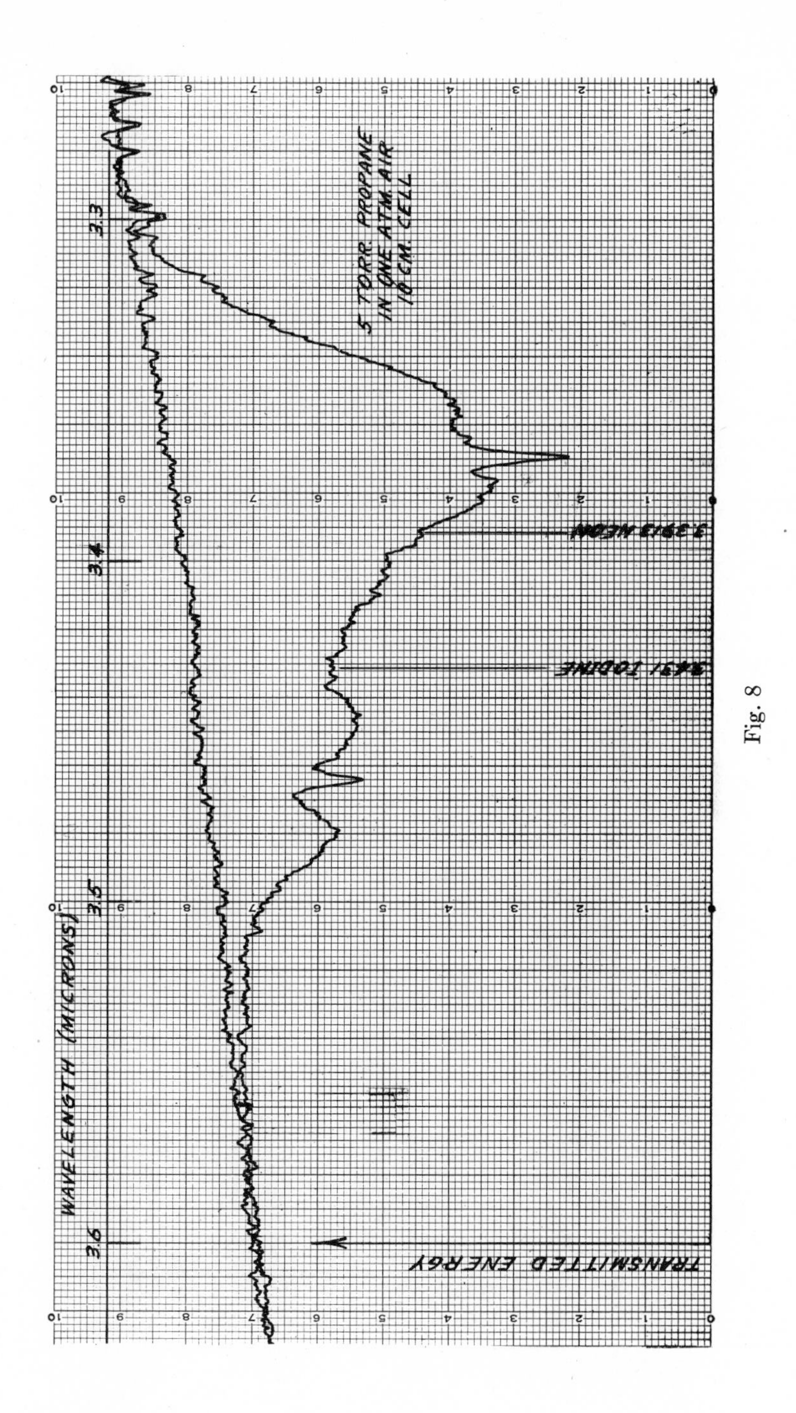

5 TORR. PROPANE
IN ONE ATM. AIR
10 CM. CELL
3.3913 NEON
3.431 IODINE
WAVELENGTH (MICRONS)
TRANSMITTED ENERGY
3.3
3.4
3.5
3.6

Fig. 8

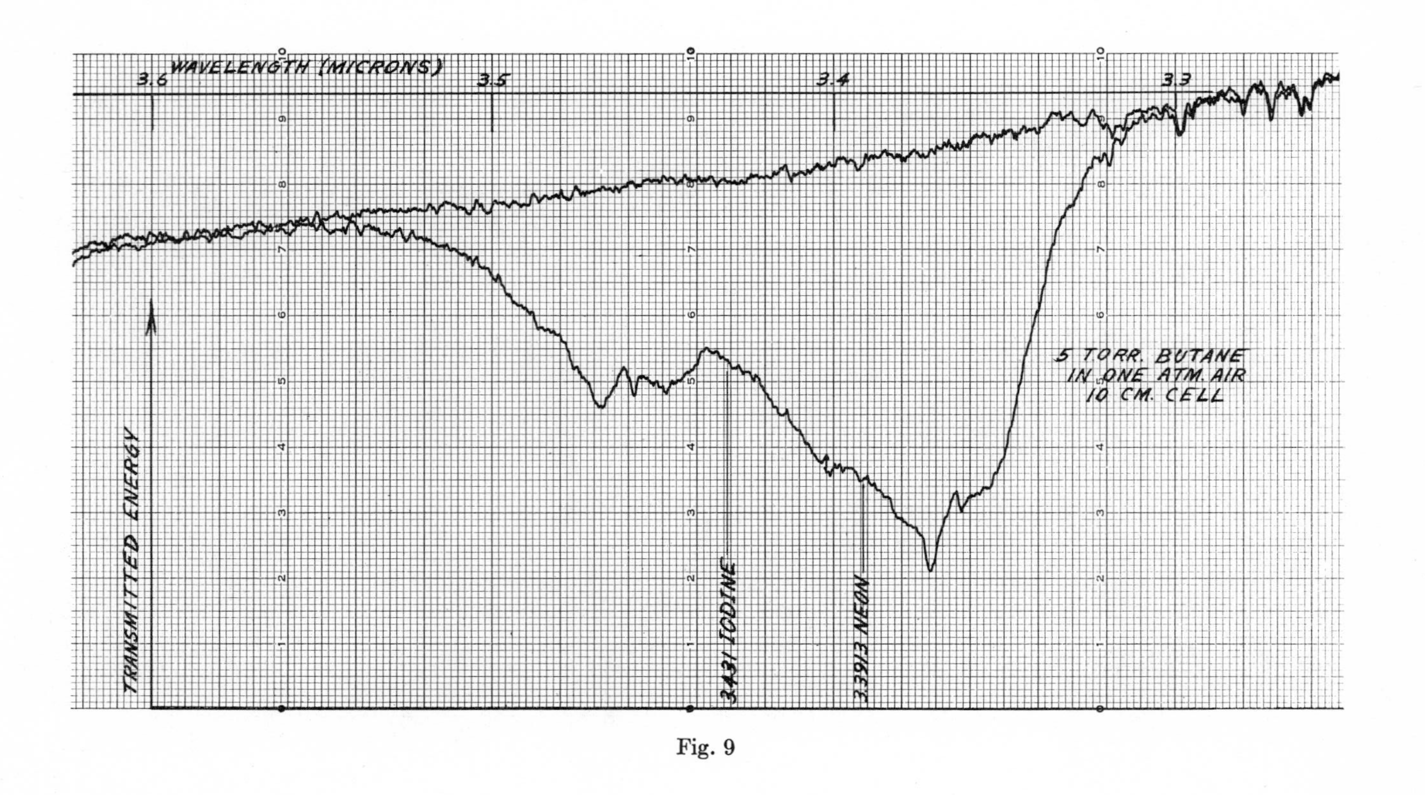
WAVELENGTH (MICRONS)
3.6
3.5
3.4
3.3
5 TORR. BUTANE
IN ONE ATM. AIR
10 CM. CELL
TRANSMITTED ENERGY
3.431 IODINE
3.3913 NEON

Fig. 9

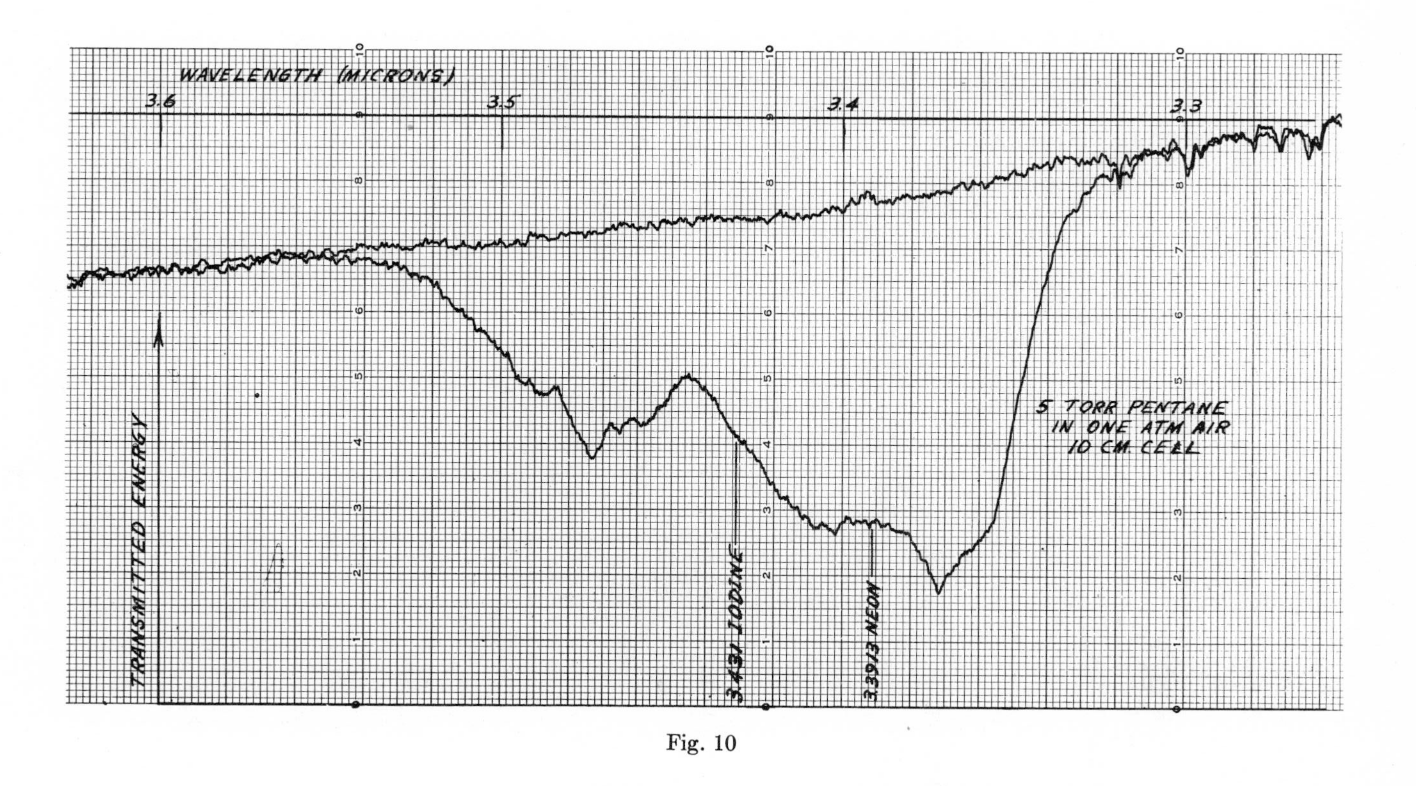
WAVELENGTH (MICRONS)
3.6
3.5
3.4
3.3
TRANSMITTED ENERGY
5 TORR PENTANE
IN ONE ATM AIR
10 CM CELL
3.431 IODINE
3.3913 NEON

Fig. 10

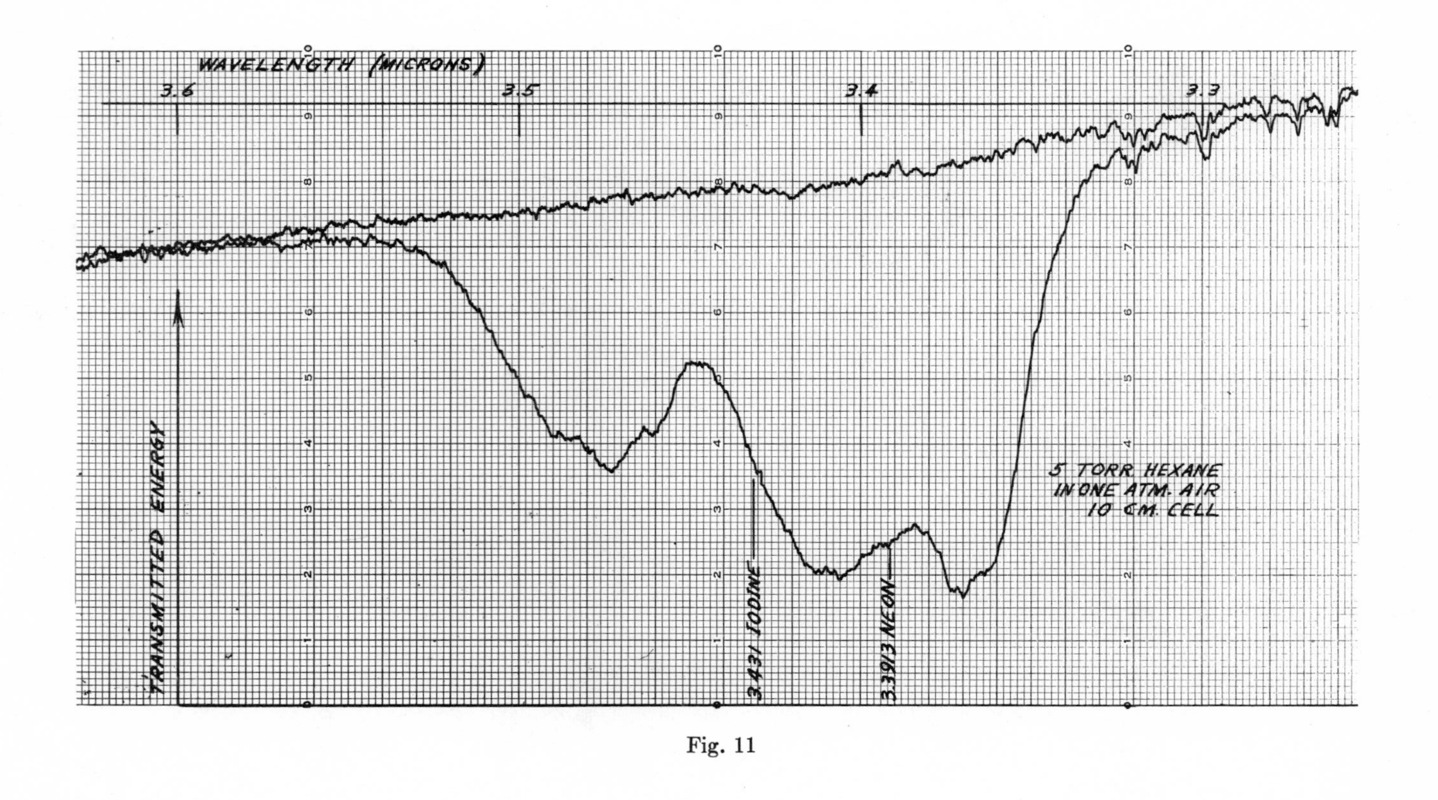

WAVELENGTH (MICRONS)
3.6
3.5
3.4
3.3
TRANSMITTED ENERGY
3.431 IODINE
3.3913 NEON
5 TORR HEXANE
IN ONE ATM. AIR
10 CM. CELL

Fig. 11

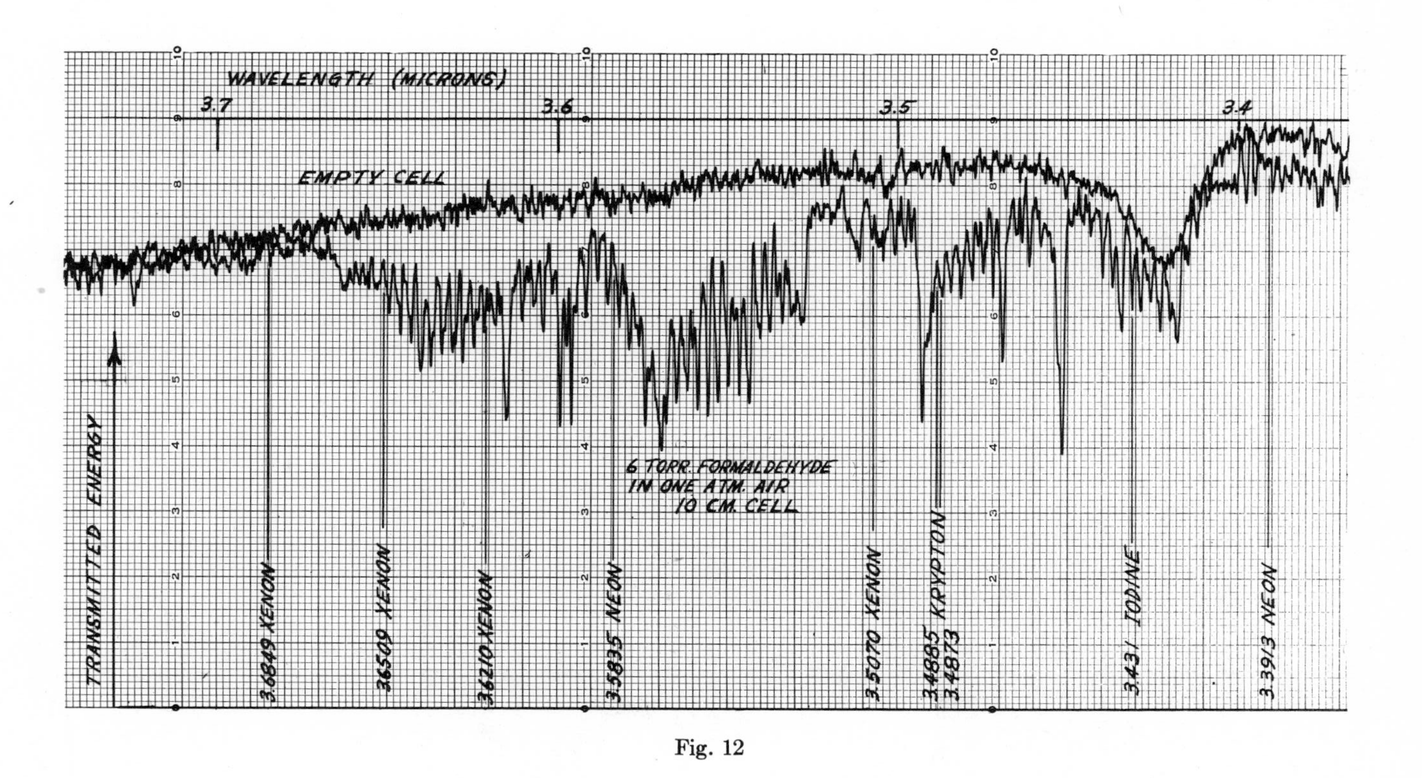
WAVELENGTH (MICRONS)
3.7
3.6
3.5
3.4
EMPTY CELL
TRANSMITTED ENERGY
6 TORR. FORMALDEHYDE
IN ONE ATM. AIR
10 CM. CELL
3.6849 XENON
3.6509 XENON
3.6210 XENON
3.5835 NEON
3.5070 XENON
3.4885 KRYPTON
3.4873
3.431 IODINE
3.3913 NEON

Fig. 12

measurement of aldehydes in the presence of other organic compounds. With formaldehyde the absorption does not extend quite as far to the long-wave side as in the case of the heavier aldehydes.

The fine structure in the formaldehyde spectrum makes the molecule an excellent candidate for the nondispersive type of cross-correlation detection system discussed in Section VI. As very few pollutants absorb in the 3.5 to 3.7-μ spectral region, the nondispersive method should be quite selective in the detection of formaldehyde. The formaldehyde vapor appeared to be thermally stable in the absorption cell, which means that it could be used in a gas filter cell.

Formaldehyde is one of the few pollutant molecules which have structured absorption bands in the near ultraviolet. However, our measurements show that the infrared bands have much greater absorptivities than the near ultraviolet bands.

5. *Acetaldehyde*

Acetaldehyde and the heavier aldehydes have C–H absorption extending to 3.7 μ and beyond, as shown in Figure 13. Fine structure does not appear in the bands at atmospheric pressure but would appear at lower pressures. Absorption between 3.5 and 3.7 μ is characteristic of aldehydes.

6. *Nitrogen Dioxide*

The strongest fundamental infrared band of nitrogen dioxide is centered at 6.2 μ and is therefore difficult to observe in air because of water vapor interference. However, a weaker band is accessible at 3.4 μ, Figure 14. The NO_2 used for recording the spectrum was in equilibrium with N_2O_4 without added gas. The total NO_2–N_2O_4 pressure was about 20 torrs, and the value of 10 torrs for the NO_2 pressure alone was obtained by reference to a spectrum of NO_2 at the lower total pressure of 3 torrs where the equilibrium is well to the NO_2 side. The 3.4 μ NO_2 band has a distinctive and useful structure at atmospheric pressure and is an excellent candidate for the gas-filter type of detection system. Atmospheric hydrocarbons will be the principal interference at 3.4 μ, but their spectra are generally continuous and will be discriminated against by a NO_2-containing gas filter or other high-resolution optical device. As in the cases of ozone and SO_2, the NO_2 spectrum has such a large number of lines packed so closely together that there will be no individual line structure except at extremely low pressures. The structure in the spectrum is due to Q-branch clustering of lines, as in the ozone case.

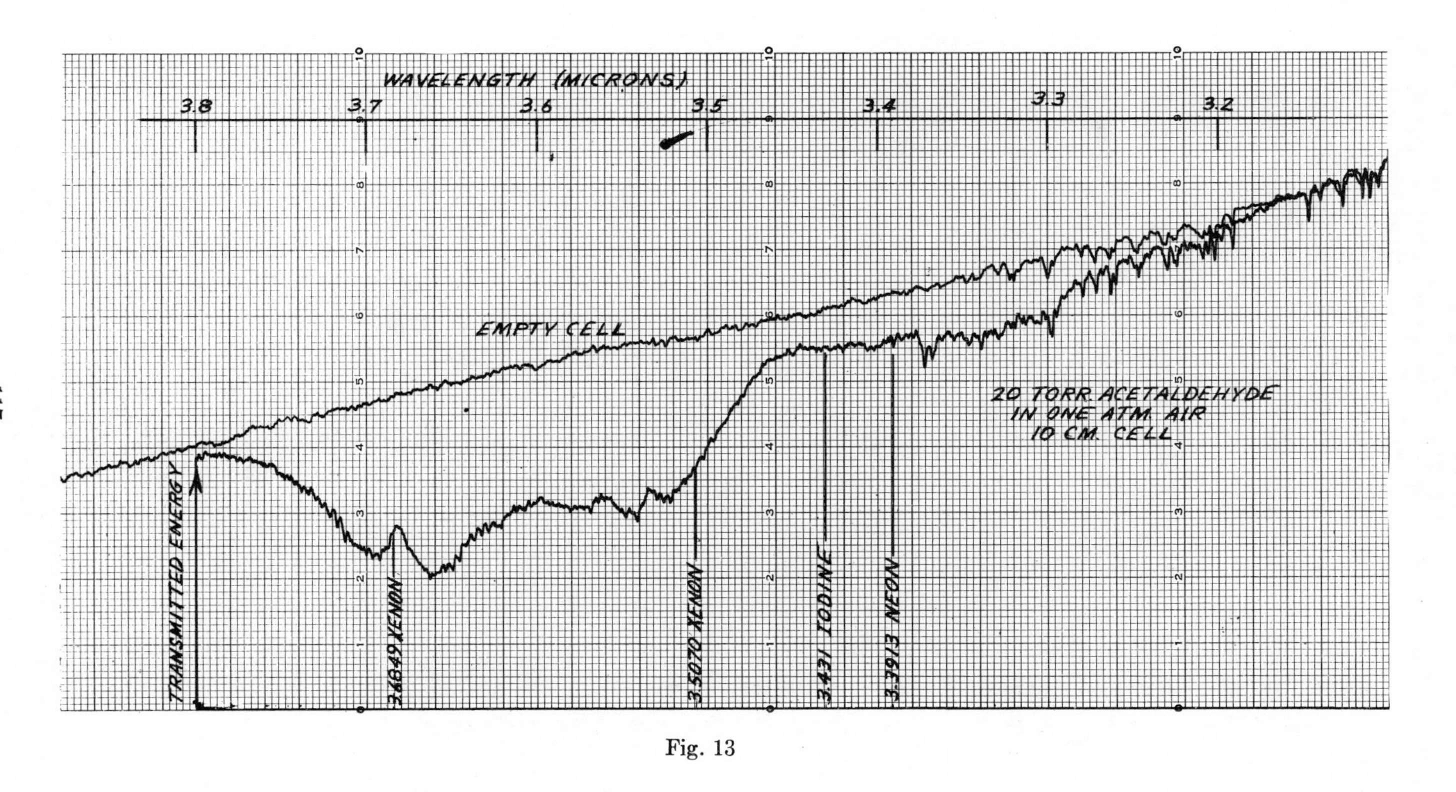
WAVELENGTH (MICRONS)
3.8
3.7
3.6
3.5
3.4
3.3
3.2
EMPTY CELL
20 TORR ACETALDEHYDE
IN ONE ATM. AIR
10 CM. CELL
TRANSMITTED ENERGY
3.6849 XENON
3.5070 XENON
3.431 IODINE
3.3913 NEON

Fig. 13

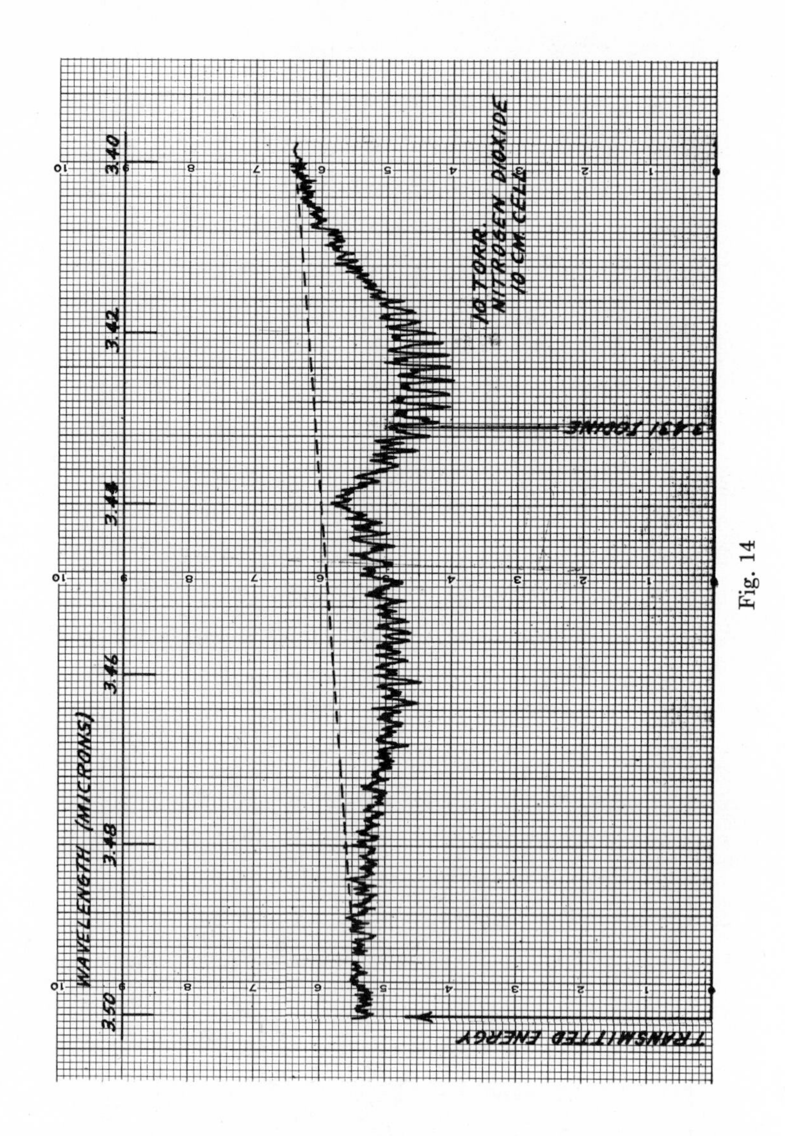

WAVELENGTH (MICRONS)
3.40
3.42
3.44
3.46
3.48
3.50
10 TORR.
NITROGEN DIOXIDE
10 CM. CELL
3.431 IODINE
TRANSMITTED ENERGY

Fig. 14

Note that the NO_2 band at 3.4 μ has approximately the same absorption strength as the well-known ultraviolet and visible absorption bands which give NO_2 its brown color. The fine structure is deeper and more detailed in the infrared bands than in the ultraviolet and visible.

7. *Hydrogen Chloride*

HCl is a stable molecule evolved in the decomposition of chlorinated plastics and other materials. Its strong distinctive absorption spectrum falls in an easily accessible region of the infrared, as shown in Figure 15. The spectrum consists of pairs of lines which are due to the presence in the gas of two isotopic species, HCl^{35} and HCl^{37}. The distinctive fine detail of the spectrum makes HCl a good candidate for nondispersive analysis. There do not appear to be any coincidences of HCl lines with known strong laser lines.

8. *Deuterium Oxide*

The detection of deuterium oxide in the atmosphere may be of interest in the vicinity of atomic energy installations. Figure 16 shows that there are characteristic D_2O lines in an open region of the spectrum.

9. *Carbon Monoxide*

The fine structure in the carbon monoxide band permits selective and quantitative measurement based on nondispersive gas filter principles (Fig. 17). These principles are used in a number of commercial CO measuring instruments. The nondispersive technique also is being applied in an instrument under development for NASA for remote sensing of atmospheric carbon monoxide (22). Possibly a laser attenuation method of CO measurement could be highly selective and sensitive.

10. *Nitric Oxide*

The nitric oxide fundamental is shown in Figure 18. Because the molecule has two states of different angular momentum (lambda doubling), there are approximately twice as many lines as in the case of CO. The band is centered at about 5.3 μ, which is a region of appreciable water vapor absorption. With an optical path of 100 m or more needed to bring out the NO lines, the atmospheric water vapor will absorb almost totally when averaged over the band. However, narrow spectral regions of fairly good transmission between the water lines may permit application of laser absorption to NO detection in humid air. The water interference becomes less serious when the sample pressure is reduced, and

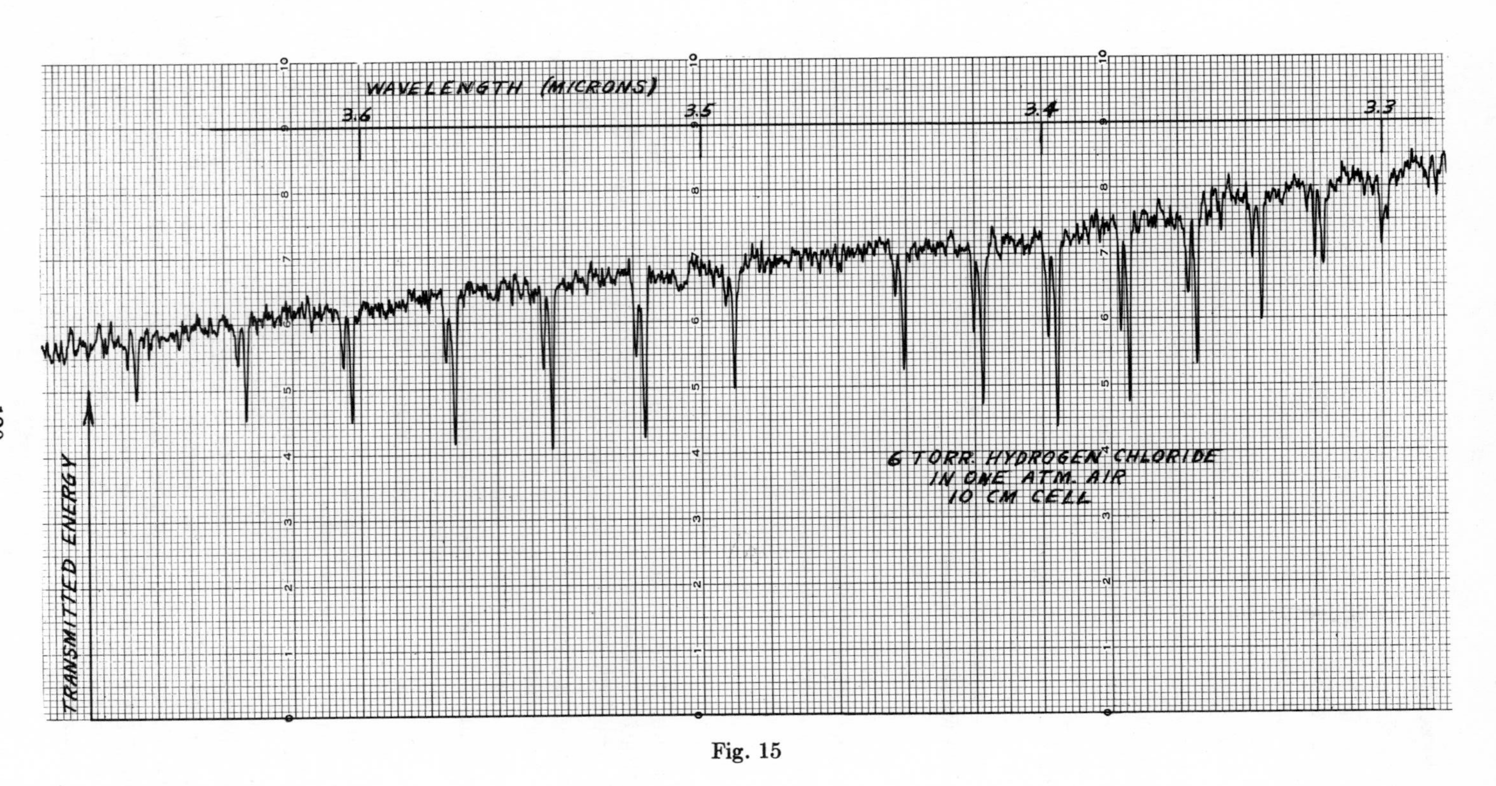
WAVELENGTH (MICRONS)
3.6
3.5
3.4
3.3
TRANSMITTED ENERGY
6 TORR. HYDROGEN CHLORIDE
IN ONE ATM. AIR
10 CM CELL

Fig. 15

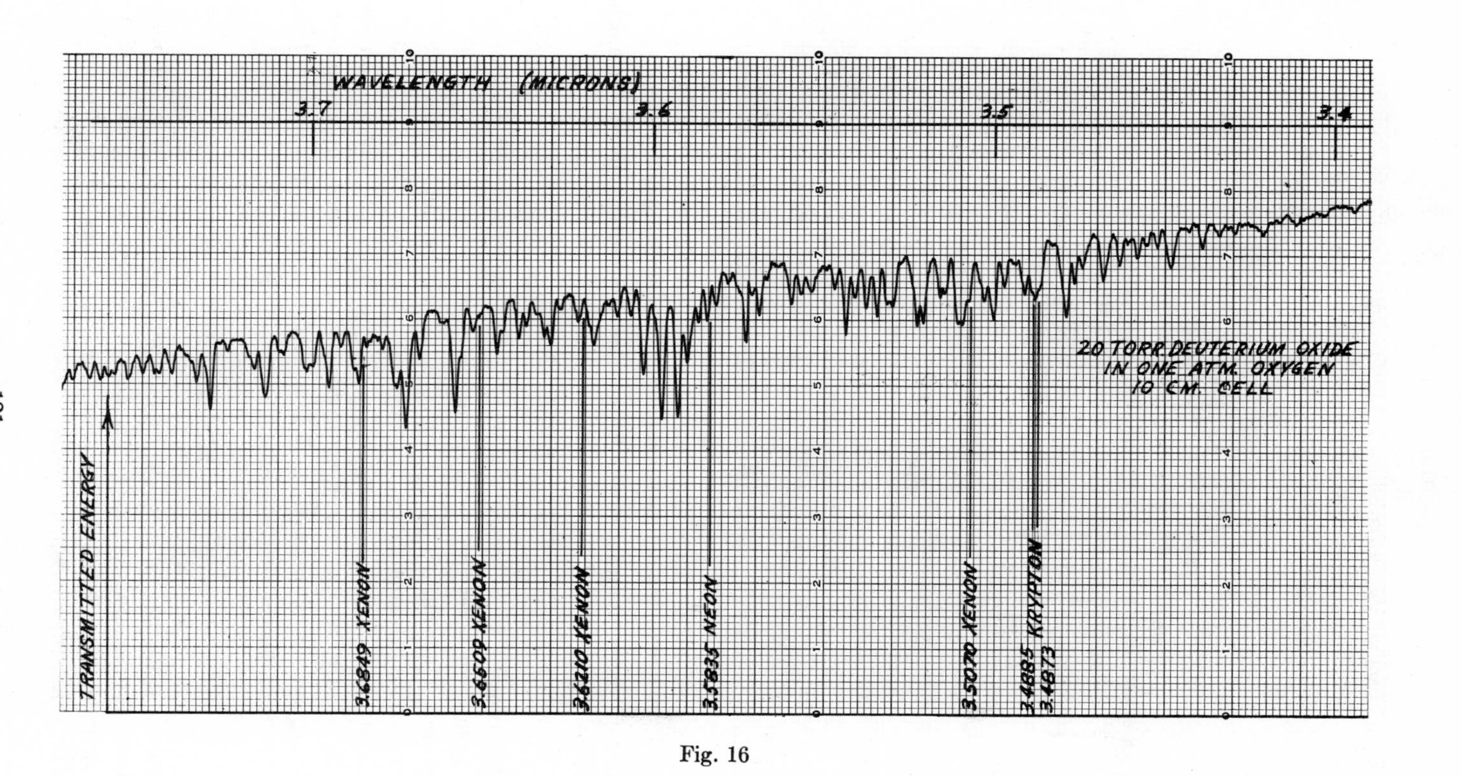

WAVELENGTH (MICRONS)
3.7
3.6
3.5
3.4
20 TORR DEUTERIUM OXIDE
IN ONE ATM. OXYGEN
10 CM. CELL
TRANSMITTED ENERGY
3.6849 XENON
3.6609 XENON
3.6210 XENON
3.5835 NEON
3.5070 XENON
3.4885 KRYPTON
3.4873

Fig. 16

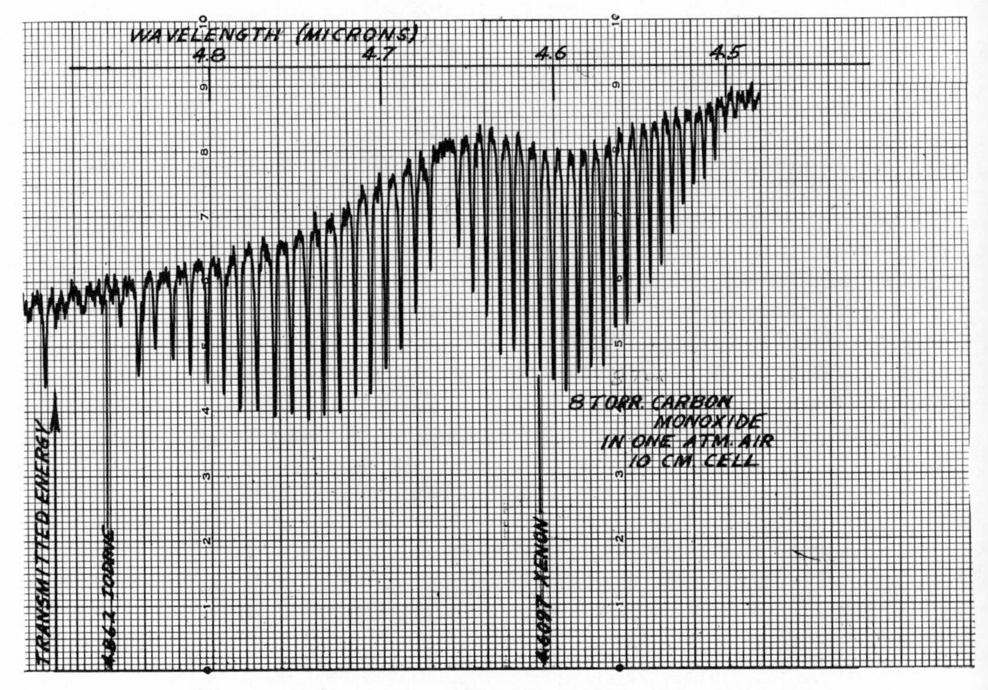

Fig. 17

with low pressure samples a gas filter cross-correlation detection method may be applied.

11. *Sulfur Dioxide*

Sulfur dioxide is a heavy molecule with low symmetry (asymmetric top). Thus its spectrum is so complicated with overlapping lines that it does not show much fine structure at atmospheric pressure. The strongest SO_2 infrared band is centered at about 7.35 μ, a region of a moderate water vapor interference. The empty cell absorption spectrum in Figure 19 illustrates atmospheric water lines superimposed on an absorption band peaked at 7.4 μ that is due to potassium nitrate contamination on the cell windows.

12. *Ozone*

Ozone, a principal constituent of photochemical smog, is an important pollutant for regular monitoring. Its distinctive principal band near 9.6 μ, Figure 20, is fortunately quite free from interference by the bands of other atmospheric constituents. Ozone is not amenable to nondis-

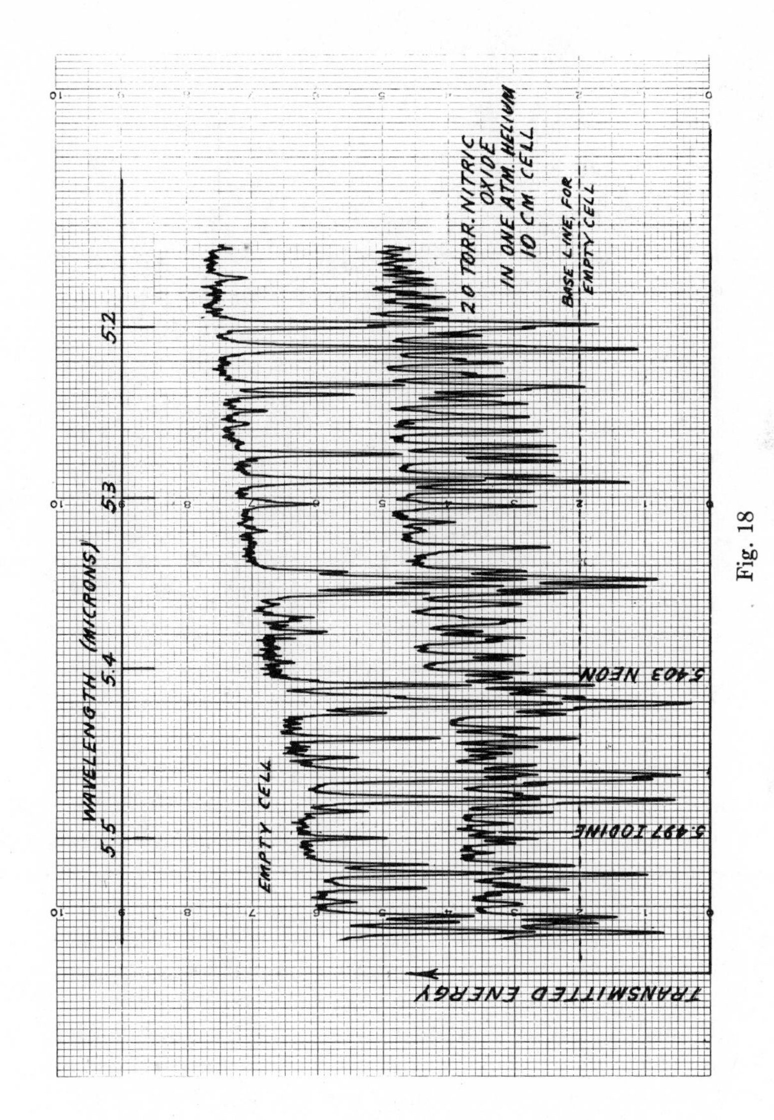
WAVELENGTH (MICRONS)
5.5
5.4
5.3
5.2
EMPTY CELL
20 TORR. NITRIC OXIDE IN ONE ATM. HELIUM 10 CM CELL
BASE LINE FOR EMPTY CELL
5.403 NEON
5.497 IODINE
TRANSMITTED ENERGY

Fig. 18

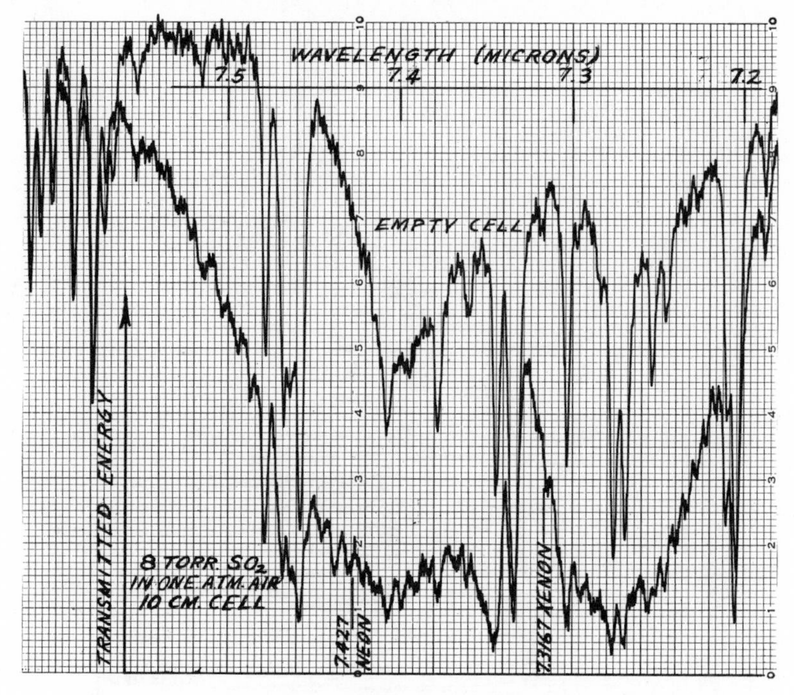

Fig. 19

persive gas-filter methods as it is unstable and would decompose in the filter cell. However, ozone is an excellent candidate for detection by the absorption of laser radiation since many strong CO_2 lines fall within the ozone band.

Laboratory measurements have shown that none of the CO_2 laser lines exhibits either an abnormally high or an abnormally low absorption coefficient when passed through ozone and that the absorption coefficients can be estimated from the recorded spectrum. These observations indicate that at atmospheric pressure the ozone lines are packed so closely together that no space remains between them, and there is not a finer structure than shown. The peaks in the spectrum are due to clusters of lines similar to the Q-branch clustering of the symmetric top molecule where each Q-branch is associated with a single value of the rotational quantum number J and running values of a second quantum number K.

The precise measurement of the amount of ozone in an absorption cell experimentally is difficult and was not attempted in the present work. The ozone absorption coefficients were taken from a previously

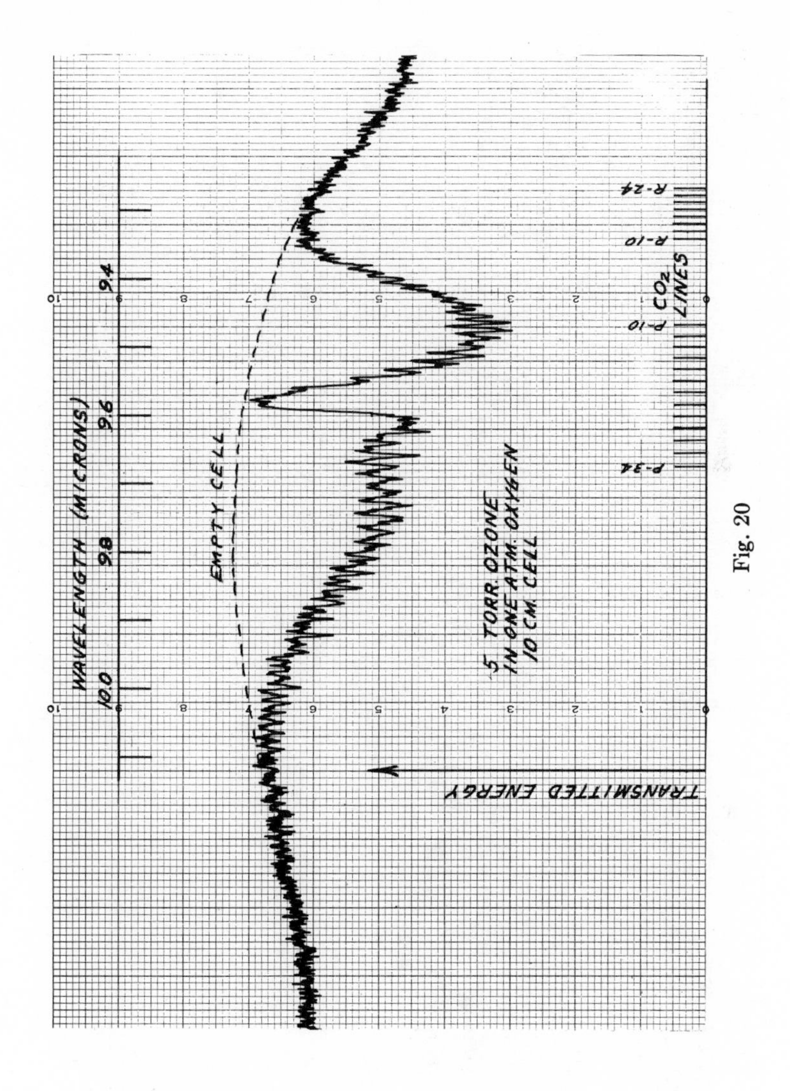
WAVELENGTH (MICRONS)
9.4
9.6
9.8
10.0
EMPTY CELL
.5 TORR. OZONE
IN ONE ATM. OXYGEN
10 CM. CELL
TRANSMITTED ENERGY
CO_2 LINES
R-24
R-10
P-10
P-34

Fig. 20

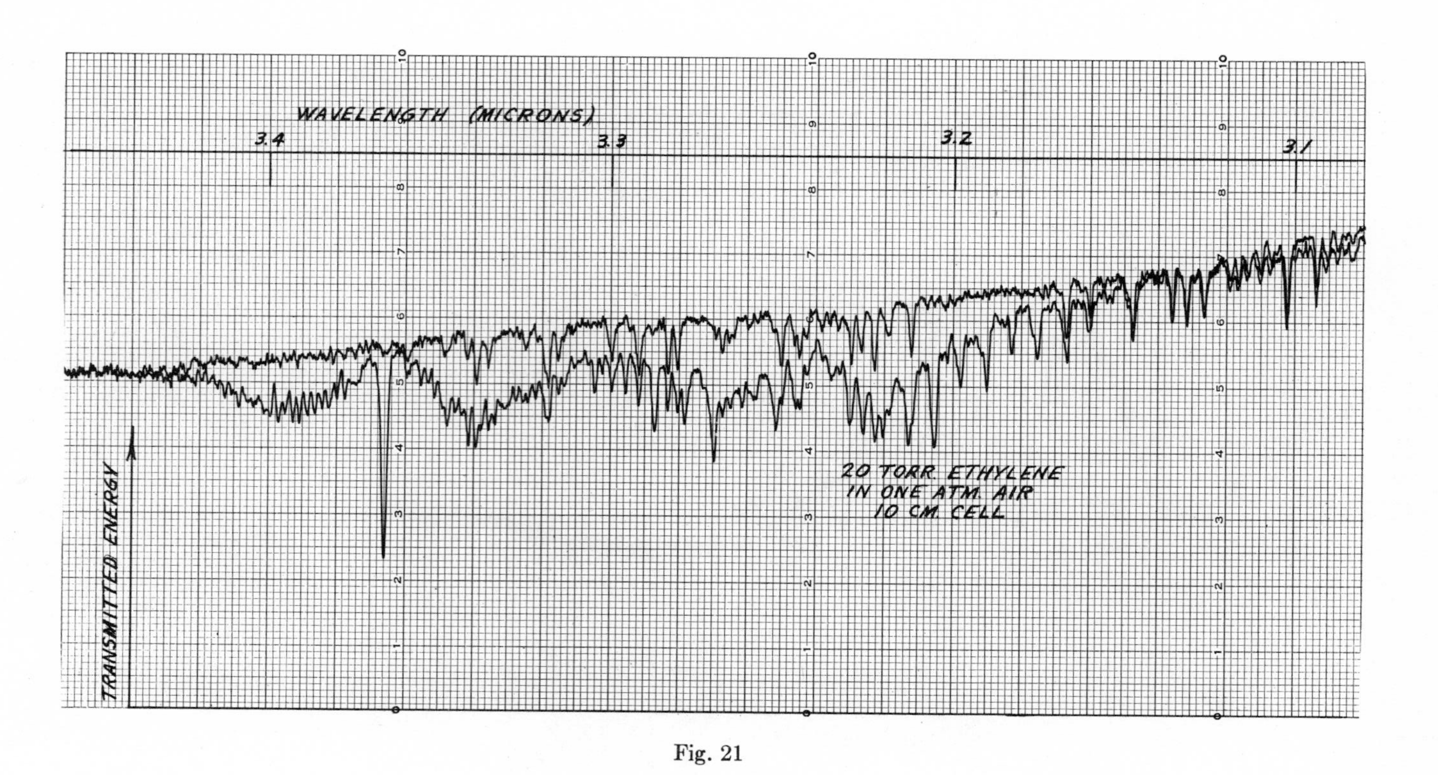
WAVELENGTH (MICRONS)
3.4
3.3
3.2
3.1
20 TORR. ETHYLENE
IN ONE ATM. AIR
10 CM. CELL
TRANSMITTED ENERGY

Fig. 21

published paper, with application of a 10% correction to allow for the limited resolution of the previously published spectrum (23).

13. *Ethylene*

Ethylene has two accessible bands, one near 3.4 μ and the other at 10.5 μ, Figures 21 and 22. The spectrum is irregular but with much fine structure at atmospheric pressure. The 10.5 μ band encompasses many CO_2 laser lines.

14. *Ammonia*

The ammonia spectrum shows two large clusters of lines plus many other doublets and single lines spread over the 8 to 13 μ region. The two clusters and some surrounding detail are shown in Figure 23.

15. *Methanol*

The spectrum of methanol (Fig. 24) is included as being representative of the spectra of the many polyatomic molecules which under certain special conditions may be of concern as air pollutants. Because the

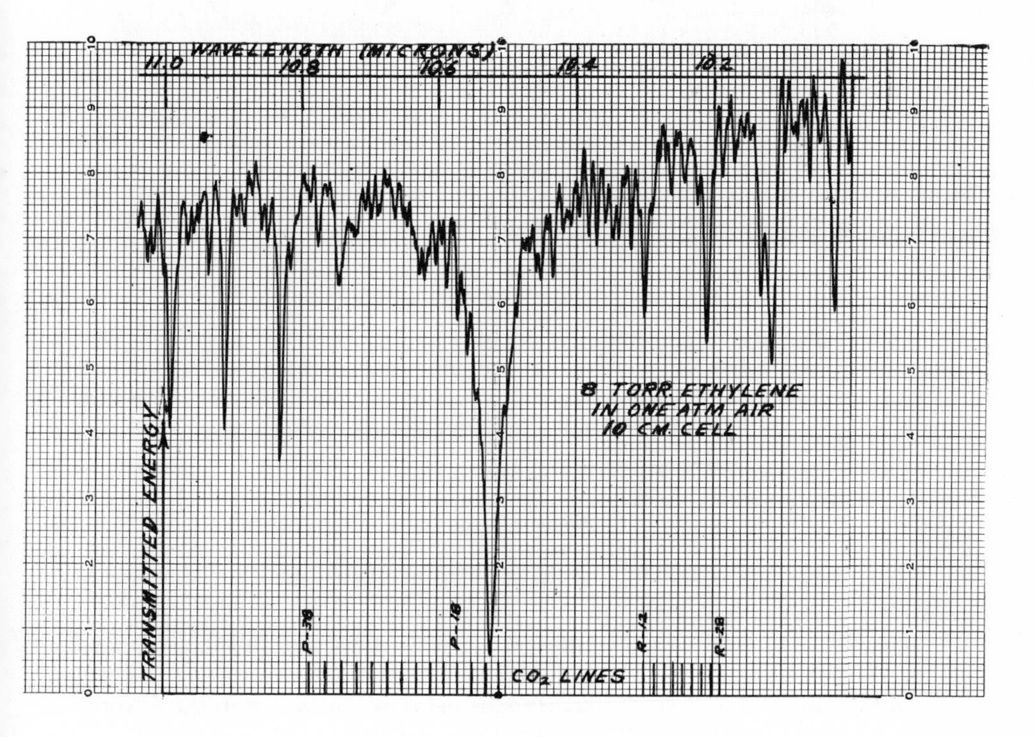

Fig. 22

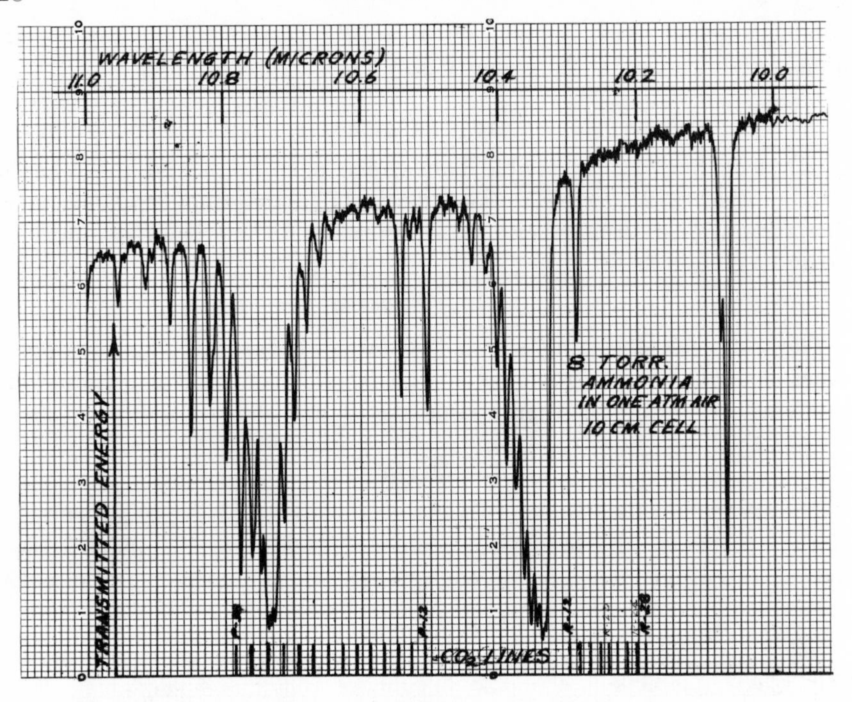

Fig. 23

molecule is nearly a symmetric top, there is a considerable amount of structure in the spectrum even though methanol is a fairly large 6-atom molecule.

16. *Propylene*

Propylene has a characteristic Q-branch and other structure at two places in the infrared, Figures 25 and 26. The spectra of the heavier alkenes, such as the butenes and the pentenes, will have less structure but can, nevertheless, be expected to have distinctive characteristics, especially in the band centered at 11 μ.

17. *Acetylene*

Acetylene, a prominent constituent of auto exhaust, is a linear molecule with a regular fine structure in its spectrum. Two bands, shown in Figures 27 and 28, are of interest in atmospheric studies. The long wavelength band has a Q-branch at 13.70 μ which is the strongest absorption feature of any of the bands studied. From the spectrum the absorption coefficient probably can reach a maximum of about 90 cm^{-1}

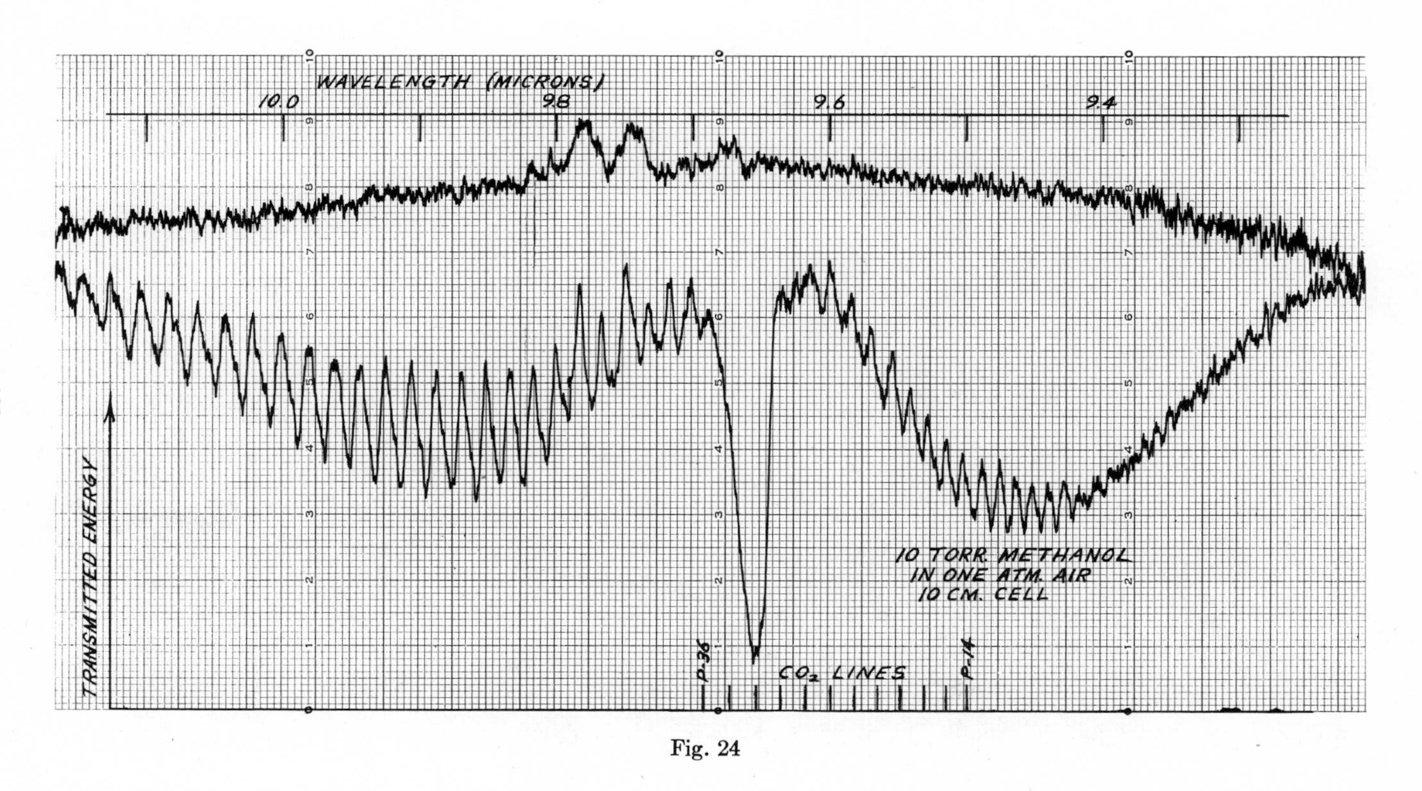
WAVELENGTH (MICRONS)
10.0
9.8
9.6
9.4
TRANSMITTED ENERGY
10 TORR METHANOL
IN ONE ATM. AIR
10 CM. CELL
CO_2 LINES
P-36
P-14

Fig. 24

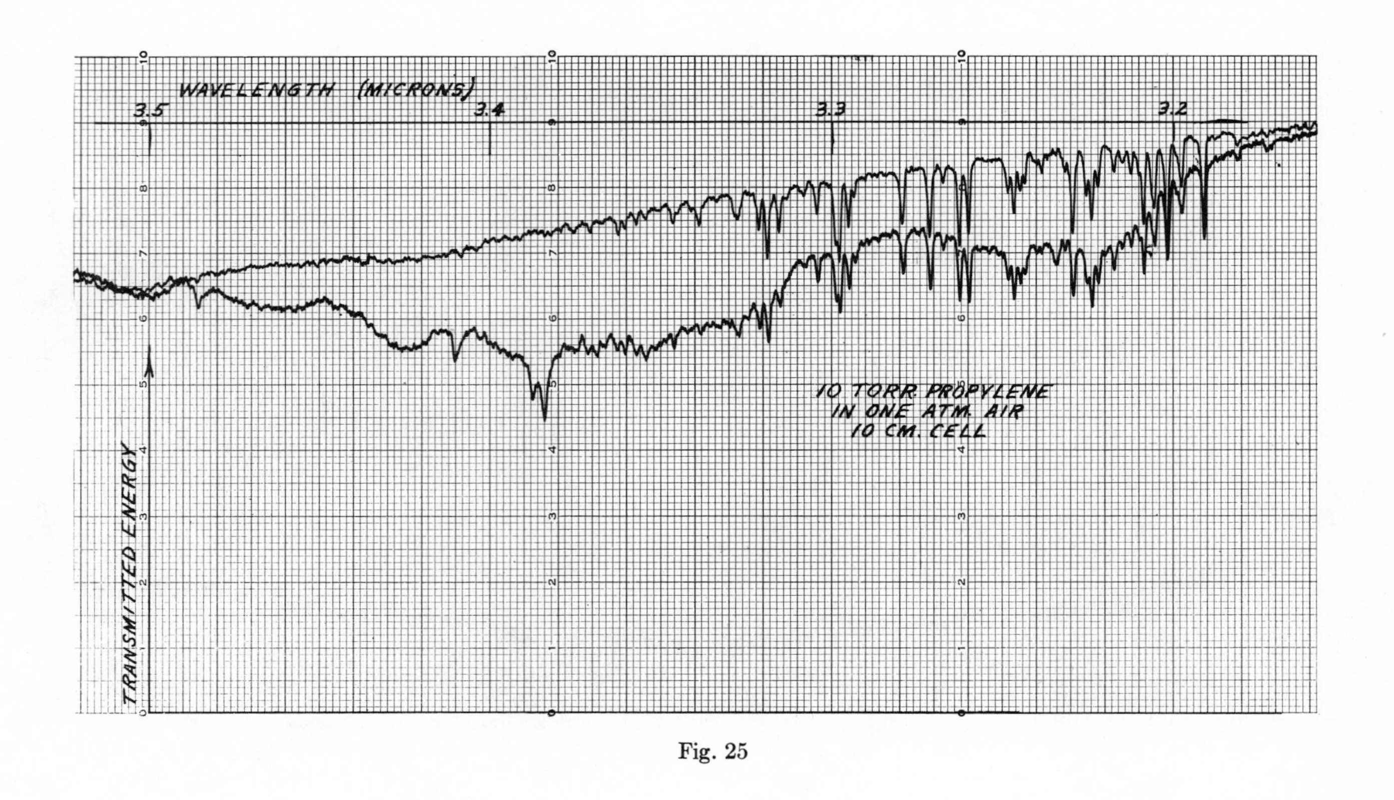

WAVELENGTH (MICRONS)
3.5
3.4
3.3
3.2
10 TORR PROPYLENE
IN ONE ATM. AIR
10 CM. CELL
TRANSMITTED ENERGY

Fig. 25

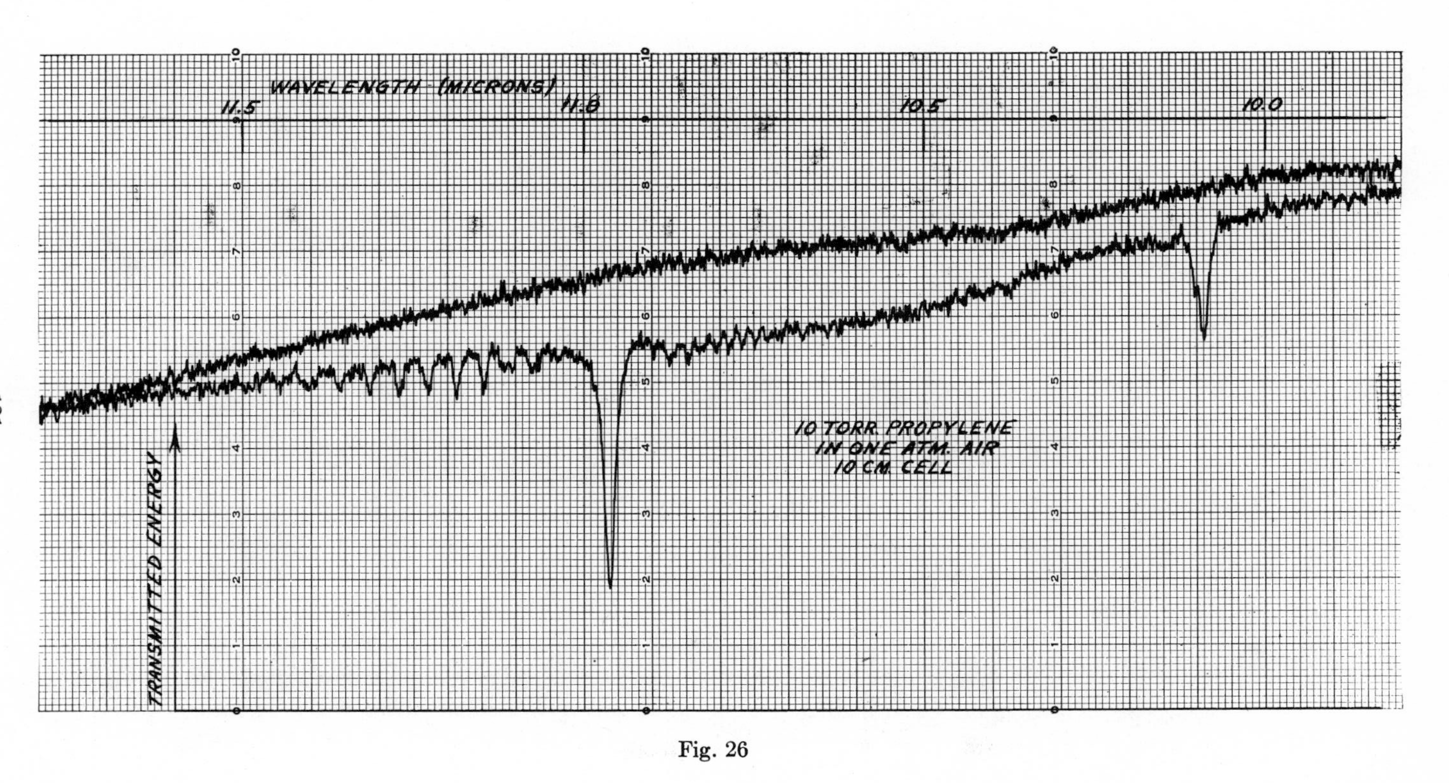

WAVELENGTH (MICRONS)
11.5
11.0
10.5
10.0
10 TORR. PROPYLENE
IN ONE ATM. AIR
10 CM. CELL
TRANSMITTED ENERGY

Fig. 26

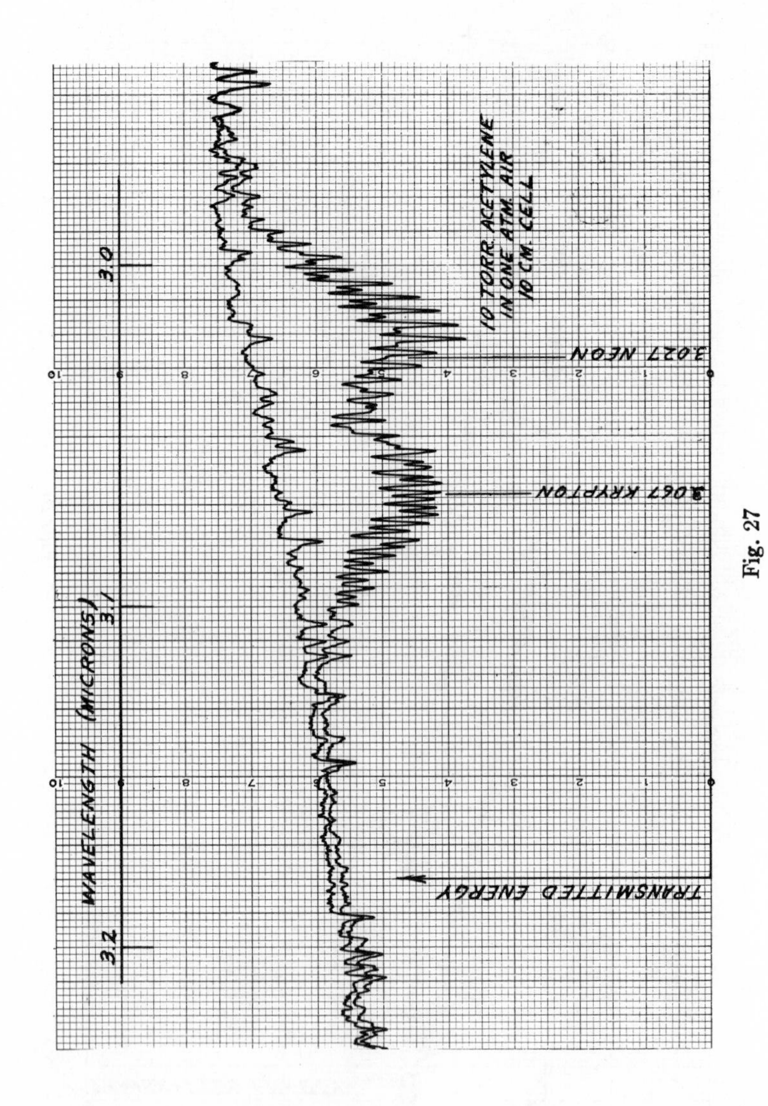

WAVELENGTH (MICRONS)
3.0
3.1
3.2
TRANSMITTED ENERGY
3.027 NEON
3.067 KRYPTON
10 TORR. ACETYLENE
IN ONE ATM. AIR
10 CM. CELL

Fig. 27

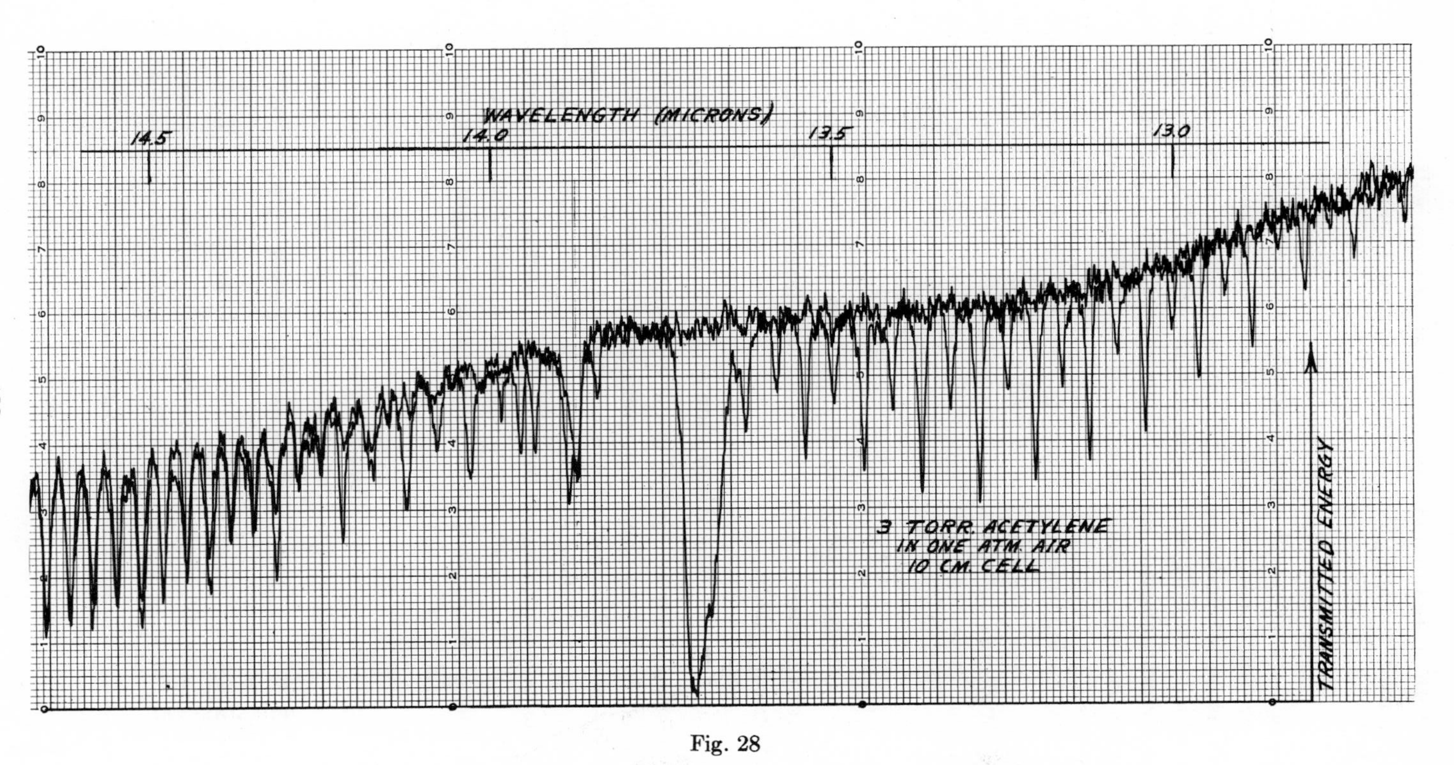
WAVELENGTH (MICRONS)
14.5
14.0
13.5
13.0
3 TORR. ACETYLENE
IN ONE ATM. AIR
10 CM. CELL
TRANSMITTED ENERGY

Fig. 28

atm^{-1}. With a 100-m light path, 1% absorption will be produced by the Q-branch if the acetylene is present in air at 1 pphm.

18. *Peroxyacetyl Nitrate*

Although peroxyacetyl nitrate (PAN) has not been studied in the present program, its spectrum is included because of the importance of PAN as an air pollutant. The spectrum has a number of distinctive bands which not only led to the discovery of PAN, but were useful also in determining its molecular structure. The location of a band near 5.4 μ is most unusual in infrared spectra and is probably the best characteristic indication of the organic peroxy nitrates. Peroxyacetyl nitrate can be differentiated from its homologues by means of the band at 8.6 μ. The carbonyl band at 5.75 μ was a key to the understanding of the structural formulas of the compounds. The PAN molecule is one of the largest of pollutant molecules and cannot be expected to exhibit any fine structure in its spectrum except at extremely low pressures. A review of the organic peroxy nitrates has been published recently by Stephens (24). The spectrum of Figure 29 is reproduced from the paper by Stephens, Darley, Taylor, and Scott (25). The value of 2.8 torrs for the PAN pressure has been calculated from Stephens' revised absorption coefficients.

C. The Absorption of Laser Radiation

The measurement of air pollutant concentrations by the attenuation of infrared laser lines has been considered in several papers in recent years (26–28). The method depends upon finding coincidences between laser lines and pollutant absorption lines. The selectivity and sensitivity of the technique derive from the very narrow spectral width of a laser line. When a coincidence between a laser line and a pollutant line is found, the absorption coefficient can be remarkably high. Furthermore, double coincidences are very unlikely. Thus a laser line which

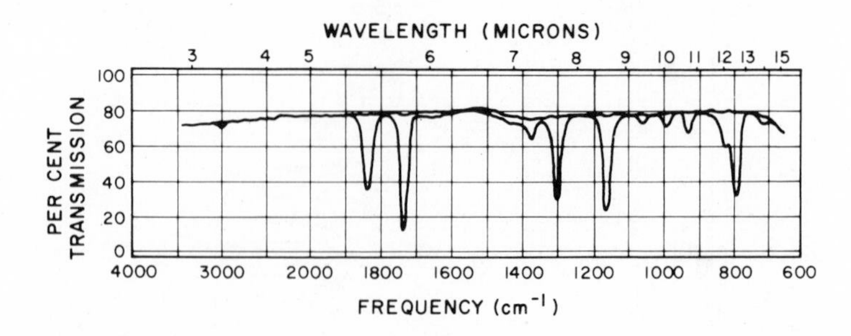

Fig. 29

happens to fall on an absorption line of one pollutant is not likely to also fall on the absorption lines of other pollutants. However, there are exceptions to this principle, as in the case of the C–H bands of organic materials near 3.4 μ.

In order to determine the selectivity and sensitivity associated with a chosen laser line, laboratory measurements of absorption must be made. When unresolved fine structure is present in the laboratory recording of the spectrum, the absorption coefficient at a laser frequency cannot be read from the spectrum but must be measured by passing the laser radiation through a known amount of the gas. This direct measurement of the laser absorption has been necessary in most of the individual cases discussed here. In the larger nonsymmetric molecules, however, there is very little fine structure in the spectrum at atmospheric pressure; one is justified in estimating the absorption coefficients directly from the spectrum.

1. *Absorption Coefficients*

Absorption coefficients for chosen laser lines have been measured by passing the laser radiation through a sample of the gas in a 10-cm absorption cell and then into a spectrophotometer. The laser line intensity was measured with an empty cell, with gas in the cell, and with various pressures of added inert gas, using carbon dioxide and iodine lasers. The carbon dioxide laser was tuned from line to line by means of a diffraction grating; the iodine laser was used without tuning.

The results of the absorption measurements at atmospheric pressure are presented in Table II. The lines measured are shown on the spectra of the preceding section, which includes numerous laser lines not measured but which are listed in Patel's table (29). The absorption coefficients listed apply to the attenuation expression $\ln I_0/I = \alpha cl$, where I_0 is the incident laser intensity, I is the transmitted laser intensity, c is the partial pressure of the absorber in atmospheres, and l is the path length in centimeters. The limit of detection depends on the path length and the ability of the instrument to discriminate small changes in intensity. If the instrument is able to discriminate a 1% difference between I_0 and I, for example, and if the absorption coefficient is 20 $atm^{-1}\ cm^{-1}$, then a 500-m path will allow the detection of a pollutant concentration of 1 pphm in air.

2. *Effects of Pressure Reduction on the Spectra*

At constant temperature and at pressures above about 0.05 atm the absorption coefficient, α, at wave number, ν, will depend on pressure,

TABLE II

Absorption Coefficients for Laser Lines

Compound	Laser line		Absorption coefficient, α, at 1 atm, cm^{-1} atm^{-1}	Comment
Methane	Ne	3.3913	15.0	Ref. 30
	I	3.431	less than 0.5	Figure 5
Ethane	Ne	3.3913	4.4	Figure 7
	I	3.431	3.8	
Propane	Ne	3.3913	8.9	Figure 8
	I	3.431	4.7	
Butane	Ne	3.3913	13.1	Figure 9
	I	3.431	6.3	
Pentane	Ne	3.3913	15.4	Figure 10
	I	3.431	9.0	
Hexane	Ne	3.3913	17.9	Figure 11
	I	3.431	11.5	
Acetaldehyde	Ne	3.3913	0.44	Figure 13
	I	3.431	0.39	
	Xe	3.5070	1.6	
	Xe	3.6849	2.1	
Nitric oxide	I	5.497	1.2	Measured by absorption of laser radiation.
Sulfur dioxide	Ne	7.427	15.9	Figure 19.
Ozone	CO_2	9.483	10.8	Ref. (23).
Ethylene	CO_2	10.231	5.0	Measured by absorption of laser radiation.
		10.529	36.0	
		10.650	4.5	
Ammonia	CO_2	10.231	15.4	Measured by absorption of laser radiation.
		10.717	32.0	
		10.739	15.8	
		10.762	4.8	
Methanol	CO_2	9.655	17.0	Figure 24.

P, approximately as follows, where C_1 and C_2 are constants characteristic of the absorbing gas.

$$\alpha_\nu = \frac{C_1 P}{(\nu - \nu_0)^2 + C_2 P^2}$$

At the center of the line ($\nu = \nu_0$), α will be inversely proportional to pressure. In the wings of the line α will be directly proportional to pressure. Near the steeply rising sides of the lines α will vary with total pressure in a more complicated way, and as pressure is changed, α may go through a maximum.

The change of line profile with pressure is illustrated in Figure 30, which was plotted from the formula above. A half-width of 0.1 cm^{-1} at the half-intensity points was chosen as being typical of absorption lines of gases at normal temperature and pressure.

The area under the plotted spectral line depends on the transition probability of the vibration-rotation change and therefore remains constant as long as the number of molecules in the light beam remains constant.

The extent to which an absorption band changes with pressure reduction depends on the spacing of the individual absorption lines. When there are hundreds of lines per wave number, as may be the case with some of the larger pollutant molecules, the absorption band will not develop a fine line structure until pressure is reduced to a few hundredths of an atmosphere. When there are about 10 lines per wave number, as is the case with some of the lighter pollutant molecules, the spectrum

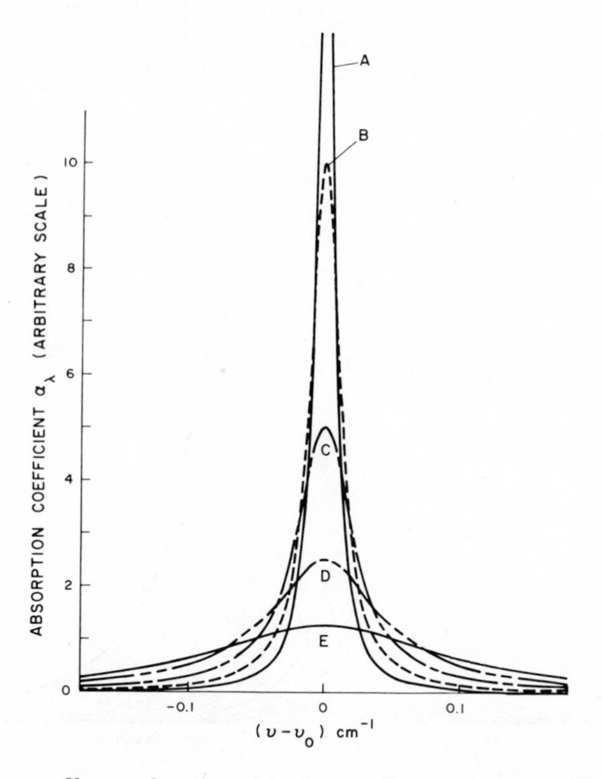

Fig. 30. Line profile as a function of pressure. Curves A through E correspond to pressures of 48, 95, 190, 380, and 760 torrs.

will be continuous at one atmosphere pressure but will break into individual lines with only a fourfold reduction in pressure. This is illustrated in Figures 31 and 32, which show 7 lines spaced 0.1 wave number apart. In the atmospheric pressure case there is a single absorption feature almost a wave number in width. The second case (Fig. 32) shows the same 7 lines with pressure reduced to one-fourth and the light path made 4 times longer. The integrated absorption seen by an instrument of ordinary resolution would still be the same after pressure reduction, but the true spectrum would have acquired the indicated fine structure which can be used to advantage by some instrument designs.

It is probably true that the spectrum of every gaseous air pollutant can thus be broken into a fine line structure by pressure reduction. The method will fail only when the lines are closer together than their Doppler width, which in the infrared vibration-rotation spectrum is on the order of one-thousandth of a wave number.

The principal practical application of the pressure-reduction method of creating fine struuture is to distinguish between molecules with overlapping spectra. It is shown in Section VI that the negative filter

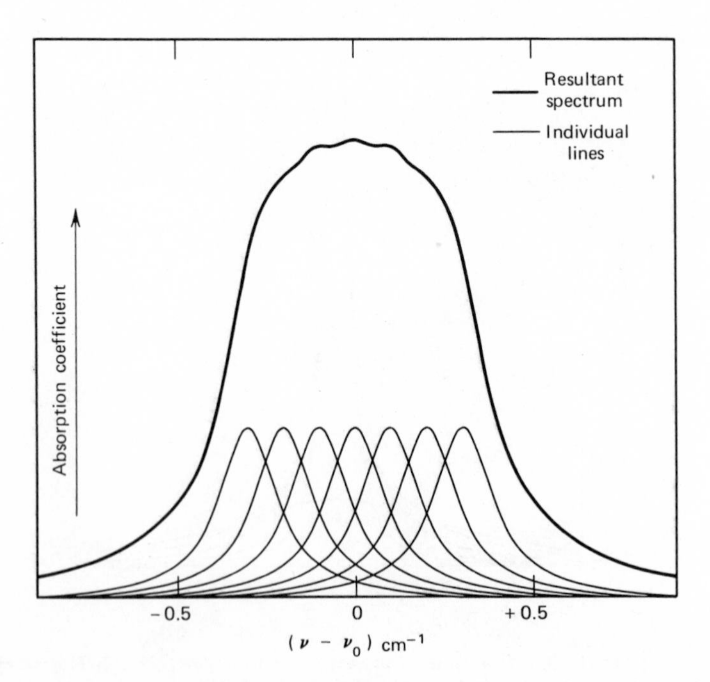

Fig. 31. Spectral lines at atmospheric pressure, with their resultant absorption band.

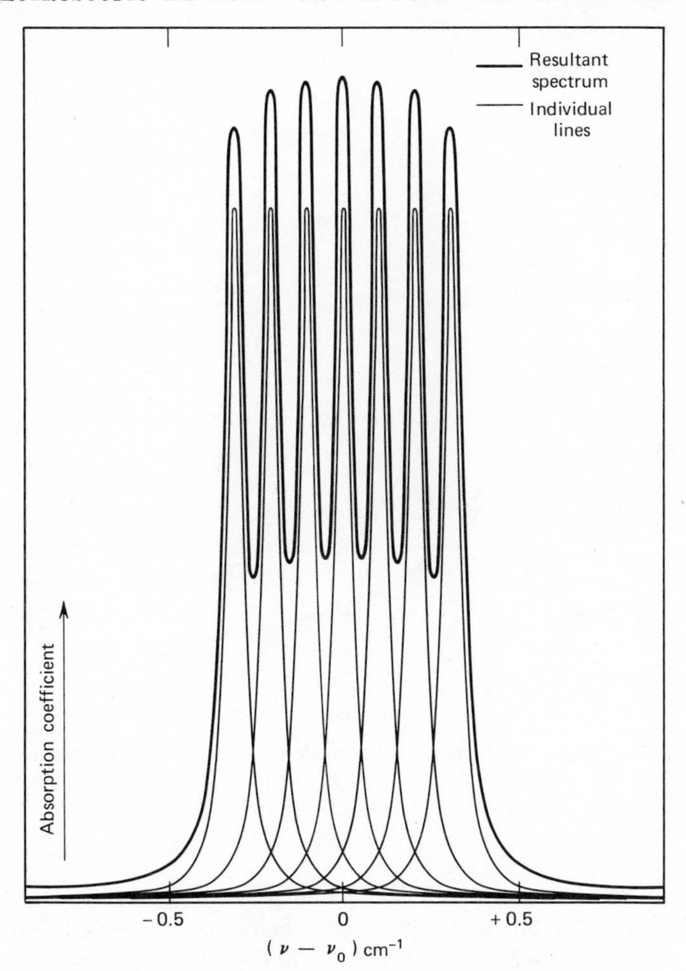

Fig. 32. Spectral lines at one-fourth atmospheric pressure, with their resultant absorption band.

type of nondispersive analyzer has remarkable discriminating properties when observing the overlapping fine spectra of molecular gases at low pressures. In addition, lasers and other narrow band sources can be tuned to individual absorption lines and can measure one low pressure gas while looking right through others.

It is important to consider the effect of line narrowing on the absorption coefficient exhibited toward a single laser line. This is illustrated in Figure 33 in which is plotted the absorption coefficient as a function of pressure for cases where the laser line is at various distances from

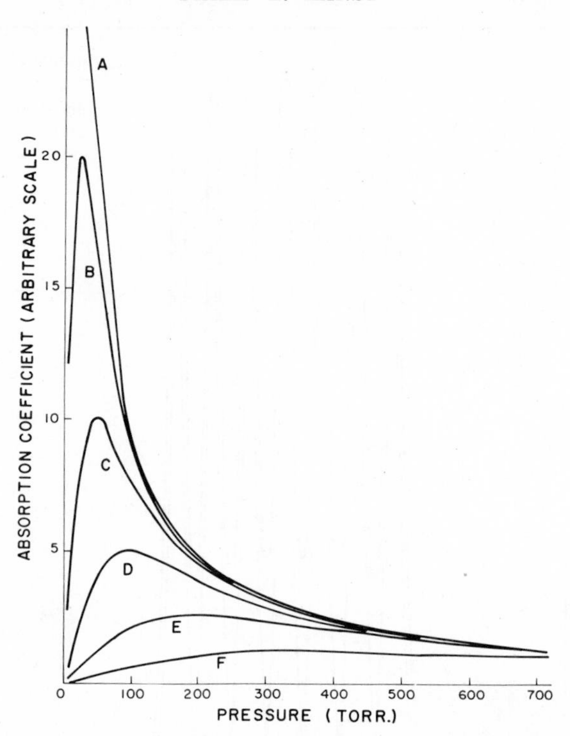

Fig. 33. Absorption coefficient as a function of pressure and distance from line center. Distances from line center $(\nu - \nu_0)$ in cm^{-1} for individual curves are as follows: A—0.0016, B—0.0032, C—0.0065, D—0.0125, E—0.0250, and F—0.050.

the absorption line center. The plots were generated from the line shapes of Figure 30 and show the behavior described previously. Presented in Figures 34 and 35 are examples of the absorption of laser lines by the pollutants ethylene and ammonia. These cases qualitatively exhibit the pressure versus absorption characteristics predicted by Figure 33. The data are discussed further in the following.

3. *Individual Cases of Laser Radiation Absorption*

a. Methane

The coincidence of the methane line and the neon laser line at 3.3913 μ was recognized shortly after the discovery of the laser. The use of the absorption of the laser radiation as a measure of methane pollution has been proposed by a number of writers including, for example, R. K. Long (27). Values of the absorption coefficients of methane and

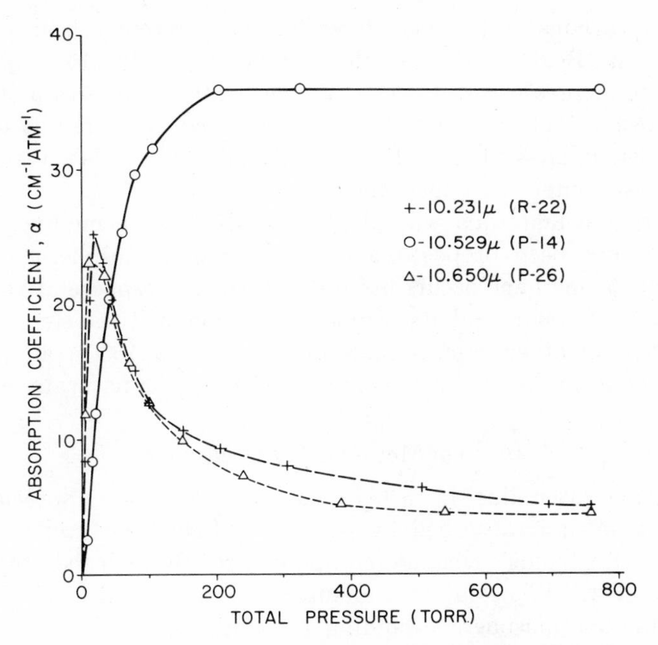

Fig. 34. Ethylene absorption coefficients as a function of total pressure.

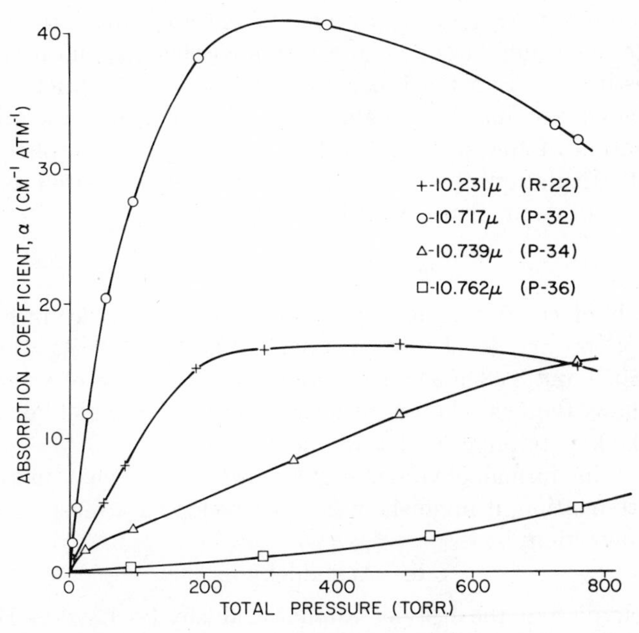

Fig. 35. Ammonia absorption coefficients as a function of total pressure.

other hydrocarbons at the neon laser line have recently been published by Jaynes and Beam (30), the value for methane being 15 (cm^{-1} atm^{-1}) at 1 atm total pressure and about 50 (cm^{-1} atm^{-1}) at pressures of 0.04 atm and lower. When these two values are fitted to the family of curves in Figure 31, it is estimated that the laser line is displaced about 0.01 cm^{-1} from the center of the methane line.

Most organic materials will absorb the 3.3913-μ neon line, and the attenuation has been proposed as a monitor of atmospheric pollution. Unfortunately, methane occurs naturally in the atmosphere at concentrations of 1 to 2 ppm, and its strong absorption will interfere with the measurement of other organic molecules. With a 100-m air path, for example, the neon line will be attenuated 25% by the naturally occurring methane.

b. NonMethane Hydrocarbons

The iodine laser line at 3.431 μ is not absorbed appreciably by methane, while it is absorbed by heavier hydrocarbons and other C–H containing compounds. The iodine line will see through the atmospheric methane and will be useful as a monitor for the general level of organic pollution in the atmosphere, excluding methane.

At atmospheric pressure ethane and the heavier paraffins have essentially continuous spectra in the vicinity of the neon and iodine laser lines. Therefore it is unnecessary to perform an actual attenuation experiment to determine laser line absorption coefficients, and it is sufficient to make estimates from the laboratory spectra. Absorption coefficients thus estimated are listed in Table II. Simultaneous measurement of the attenuation of the neon and iodine lines by an atmospheric sample will permit calculation of the methane concentration as well as the concentration of all nonmethane organic pollutants.

c. Formaldehyde

The result of comparing laser lines with formaldehyde lines is somewhat disappointing, as shown in Figure 12. The strong xenon laser lines at 3.5070 and 3.6849 μ, for example, fall in regions of weak absorption. Perhaps the best choices of laser lines for formaldehyde measurement are the krypton lines at 3.4873 and 3.4885 μ.

Although the formaldehyde fine structure is somewhat inappropriate for laser attenuation, it might be used for developing a sensitive gas-filter method of detection.

d. Acetaldehyde

Acetaldehyde and the heavier aliphatic aldehydes have C–H absorption extending to 3.7 μ and beyond, as shown in Figure 13. Fine structure does not appear in the C–H band at atmospheric pressure. The strong xenon laser line at 3.6849 μ will be absorbed by acetaldehyde and its heavier homologues, but will not be absorbed to an appreciable extent by other air pollutants. The absorption of the 3.6849 μ line will therefore serve as a quantitative and selective indicator of the presence of aldehydes in the air.

e. Deuterium Oxide

Several laser lines appear that might be used to monitor deuterium oxide, especially the krypton lines at 3.4885 and 3.4873 μ, as shown in Figure 16. It is not possible to determine from the figure how close these laser lines are to the D_2O lines. The absorption coefficients should be measured experimentally.

f. Carbon Monoxide

Although carbon monoxide has been measured by nondispersive infrared techniques for many years, there may be cases when a laser absorption method of measurement will be preferred. Two candidate laser lines are marked on Figure 17. The iodine line at 4.8619 μ is one possibility (28), but not a good choice as the neighboring CO line is rather weak and is approximately 0.8 wave numbers removed. The xenon line at 4.6097 μ falls within 0.44 wave numbers of the center of one of the strongest CO lines. Although the absorptivity of CO with respect to this xenon line has not been measured, it is expected to be high and to provide a basis for monitoring the CO concentration in ambient clean air and in polluted air.

g. Nitric Oxide

Although the nitric oxide fundamental band falls in a spectral region where there is also absorption by water vapor, some nitric oxide lines do seem to fall in the narrow transmission "windows" between water lines. If a laser line can be found which is coincident with one of the transmitted NO lines, the attenuation of the laser line could be used as a measure of NO, even in humid air. Two candidate lines, the 5.403-μ neon and the 5.497-μ iodine, are indicated in Figure 18. The absorption coefficient at the 5.497-μ iodine line was measured to be approximately 1.2 (cm^{-1} atm^{-1}). No measurement was made on the 5.403-μ neon line.

h. Sulfur Dioxide

The sulfur dioxide band centered at 7.35 μ is very strong and does have a neon laser line located within it at 7.427 μ. The attenuation of this laser line will be a sensitive measure of the presence of SO_2

in an air sample. Interference by water vapor may not be serious as the laser line is favorably situated between water lines. The extent of interference should be determined experimentally. If it is more serious than anticipated, then the possible beneficial effects of reducing sample pressure should be investigated.

i. Ozone

A number of strong CO_2 laser lines fall within the principal infrared absorption band of ozone as shown in Figure 20. The spectral region covered by this band is relatively free from water vapor absorption and absorption by other pollutants. A selective and sensitive ozone measurement should be obtained if the two neighboring laser lines, P-24 and P-26, are both passed through an air sample. Since the absorption coefficient is very low for P-24 and relatively high for P-26, a small amount of ozone in the path will change the intensity ratio of the two transmitted lines by an observable amount. Laboratory experiments should be performed to determine whether water vapor or other possible interfering compounds will change the intensity ratio.

j. Ethylene

The 10.5-μ band of ethylene encompasses many CO_2 laser lines. The absorption of 21 of these lines was measured in the laboratory, and it was found that 3 were absorbed strongly. The absorption coefficients of these three were studied as a function of pressure and are plotted in Figure 34.

The P-14 line, falling near the center of the Q-branch, shows the highest absorption coefficient that remains relatively constant at pressures between 760 and 200 torrs but which decreases as total pressure is lowered further. The pressure variation is in accord with the location of the laser line just off the peak of the Q-branch. The absorption is mainly due to the overlapping wings of many lines, so that a pressure variation similar to curves E and F in Figure 33 is obtained.

The strong absorption of the R-22 and P-26 CO_2 lines is somewhat surprising, since strong lines do not appear at these wavelengths in the ethylene spectrum. However, the pressure variation of the absorption coefficients explains the absorption and also provides an excellent example of the type of curves calculated for Figure 33. Both cases appear to involve a near coincidence of the laser line with an isolated ethylene line which was not resolved in the spectrum. The shape of the plots in Figure 34 indicates that each laser line must fall within about 0.003 cm^{-1} of the center of the absorbing ethylene line.

k. Ammonia

The absorptivity of ammonia at 21 CO_2 laser lines has been measured. Four lines were found to be strongly absorbed at atmospheric pressure. The pressure variation of the absorption of these four lines was studied and is presented in Figure 35. The strongest absorption is at the P-32 line, which is near the center of a cluster. The next strongest absorption is rather surprisingly at the R-22 CO_2 line where no ammonia line shows in the spectrum. This case is similar to the absorption of R-22 and P-26 in ethylene and must likewise be due to the presence of an unresolved single line which is nearly coincident with the laser line. The pressure variation of absorption places the laser line about 0.02 cm^{-1} from the center of the absorbing ammonia line. The P-34 and P-36 CO_2 lines also show appreciable absorption coefficients in ammonia. Each of these behaves with pressure as though the absorption were due primarily to the wings of ammonia lines.

l. Methanol

Methanol is a case similar to ozone, ethylene, and ammonia in that there are neighboring laser lines toward which methanol exhibits wide variations in absorption coefficients. If both the P-30 and P-32 lines shown in Figure 24 are passed through an air sample simultaneously, the change in their intensity ratio will be a sensitive and selective measure of the presence of methanol.

4. *Increasing Absorptivity by Zeeman Shift*

It has been shown in the spectra and in the previous discussions that usually one can find a laser line within a pollutant absorption band. Unfortunately, only a few cases of a close coincidence of the laser line and an absorption line are found in the spectrum. There are many more near coincidences in which the absorption coefficient could be made much larger by a slight shift of the laser frequency. Actually the lines of the atomic gas lasers can move and have moved the required small amounts by means of a magnetic field to give the effect known as the Zeeman effect.

Line shifts of several tenths of a wave number have been achieved using modest magnetic fields easily achieved by a simple solenoidal magnet on the laser (67). This amount of shift will be sufficient to create many new coincidences between laser lines and pollutant absorption lines. A case now under investigation at the laboratory of the Air Pollution Control Office is to move 3.5070 and 3.6849-μ lines of the xenon laser in order to create coincidences with specific aldehyde absorption lines.

The Zeeman effect results in a line splitting rather than just a line shift. A splitting of atomic energy states results from the coupling of the magnetic moment of the transition electron with the external magnetic field. The effect of the energy level splitting is to produce several closely spaced spectral lines in place of the original single line. The new lines are symmetrically distributed about the position of the original line. Three spectral lines are usually seen when looking across the magnetic field, and two are seen along the direction of the field.

A laser discharge in a solenoidal magnet emits in the direction of the field, so normally there are two components to the emission—one line at slightly higher frequency than the original and the other at slightly lower frequency. If both components are used together in an absorption experiment, the splitting is merely equivalent to a slight broadening of the original laser line, and the benefits derived from the splitting are minimal. If the two lines can be separated, the benefits are much greater because one line will move onto the peak of the pollutant absorption line and the other away from the peak. This creates a significant difference in absorption of the two components that can serve as a most sensitive and selective indicator of the presence of the chosen pollutant in an air sample.

Since the splitting in a practically achievable magnetic field is about a tenth of a wave number, it is very difficult to separate the two components with a spectrophotometer or other dispersive instrument. The two components are different in polarization, however, which does allow their separation.

One component is right circularly polarized and the other is left circularly polarized. These circular polarizations can be converted to two plane polarizations at right angles to each other by passing the radiation through a birefringent quarter-wave plate. These right-angle plane polarizations can then be separated by another birefringent element—an analyzer. In fact, if the analyzer is rotated continuously the two components will be passed alternately.

If the two components are equal in intensity, there will be no intensity alternation in the light passed by the rotating polarizer; but if one has been absorbed by the air sample to a greater extent than the other, there will be an intensity alternation at the rotation frequency. Such an intensity alternation can be seen by a detection system with tuned electronics even if it is only a very small percentage of the total intensity. Therefore, it should be possible to extend the limits of detect-

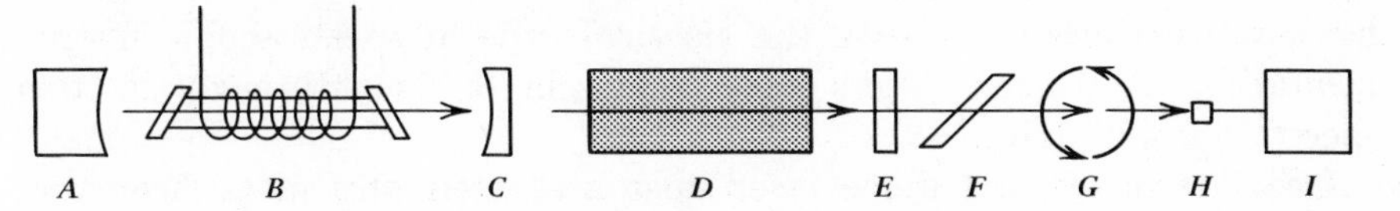

Fig. 36. Zeeman-split laser system for pollutant measurement in differential absorption mode. *A*, wavelength selection module; *B*, discharge tube with solenoid; *C*, laser output mirror; *D*, absorption cell; *E*, quarter-wave plate; *F*, adjustable transmitting plate near Brewster's angle; *G*, rotating analyzer; *H*, detector; *I*, tuned electronics.

ability of the system by using high quality detection electronics and long integration times.

Another important feature of this differential absorption system is that it should minimize the problem of laser noise. In a one-component absorption system, fluctuations in source intensity must be distinguished from the intensity changes due to the absorption. Although a reference intensity may be obtained by a beam splitter or by a sample-in, sample-out method, it is better to use a reference line in the spectrum. Such a reference line is precisely what is provided by the Zeeman splitting. The component which moves away from the peak of the absorption line is in effect a built-in reference line whose intensity fluctuation at the source will follow the intensity fluctuations of the line to be absorbed. Electronics which take the intensity ratio will therefore be immune to the laser fluctuations. A schematic diagram of the experimental system being built is shown in Figure 36. The equipment train consists of (*a*) a wavelength selector module which may be a rotatable diffraction grating, a Fabry–Perot etalon, or a gas-filled filter; (*b*) the laser discharge tube with magnetic field coil wrapped around it; (*c*) the laser output mirror; (*d*) the absorption cell, which may be a multiple-pass cell of very long path; (*e*) the quarter-wave plate; (*f*) a rotatable transmitting plate near Brewster's angle which is adjusted to equalize the two components when there is no gas in the absorption cell; (*g*) the rotating analyzer crystal; (*h*) the detector, which may be a thermocouple, a bolometer, a pyroelectric cell, or a photoconductive cell; and (*i*) the tuned electronics for observing the ratio of the intensity of the two components.

D. Infinite Resolution Spectroscopy

There is a significant amount of characteristic fine structure available in the infrared spectrum of a polyatomic molecular gas at low pressure. Although this has been recognized for many years, the fine structure has not been accessible with the classical type of spectroscopic instrumentation. However, lasers may represent a breakthrough in this spectral resolution barrier.

Gas laser lines and diode laser lines are essentially monochromatic, their spectral widths being far smaller than the width of the narrowest gas absorption line. Thus if some method could be found to continuously shift the laser wavelength, one would have the equivalent of a scanning monochromator of infinite resolving power.

Several techniques of shifting the laser emission have been studied in recent years. One is the Zeeman-effect tuning just discussed. This method is capable of displacing the emission a few tenths of a wave number from its unperturbed position. Although this amount of shift is sufficient for certain practical applications, it is not enough for spectroscopic studies.

Another tuning technique is the use of a rotatable diffraction grating in a dye laser (Section V). There are dyes that emit in the ultraviolet, visible, or near infrared. A single dye can have its emission shifted through hundreds of wave numbers. Thus one has continuous tuning available from about 3000 Å to 10,000 Å wavelength. The disadvantages of the dye laser for pollution studies are mainly (*1*) the dyes do not emit a narrow enough spectral line for truly high resolution absorption spectroscopy, and (*2*) a precise setting of the emission wavelength is difficult to achieve and to maintain. A further disadvantage for pollution detection is that most pollutants do not exhibit characteristic spectral fine structure in the regions of dye laser emission.

Another laser tuning method is to use the so-called parametric amplification effect. In a crystal a fundamental laser frequency can be divided into two new frequencies. The new frequencies must be equal to the original frequency, but their relative values can be changed by rotating the crystal. This type of laser tuning can be applied in the infrared and gives a narrow laser line, but to date it has the disadvantage of requiring very intense bursts of laser radiation and carefully selected and controlled crystals.

Still another laser tuning technique is under investigation by Hinkley and co-workers at the M.I.T. Lincoln Laboratories (31,32). This method appears to be the one most likely to be applied to pollutant detection in the near future.

The laser source used by Hinkley is the lead-tin-telluride diode which is discussed in Section V. Increasing the current through the diode increases the temperature of the emitting heterojunction, which in turn moves the laser resonance frequency. As will be shown in Figure 59, a continuous tuning of several tenths of a wave number is possible. This

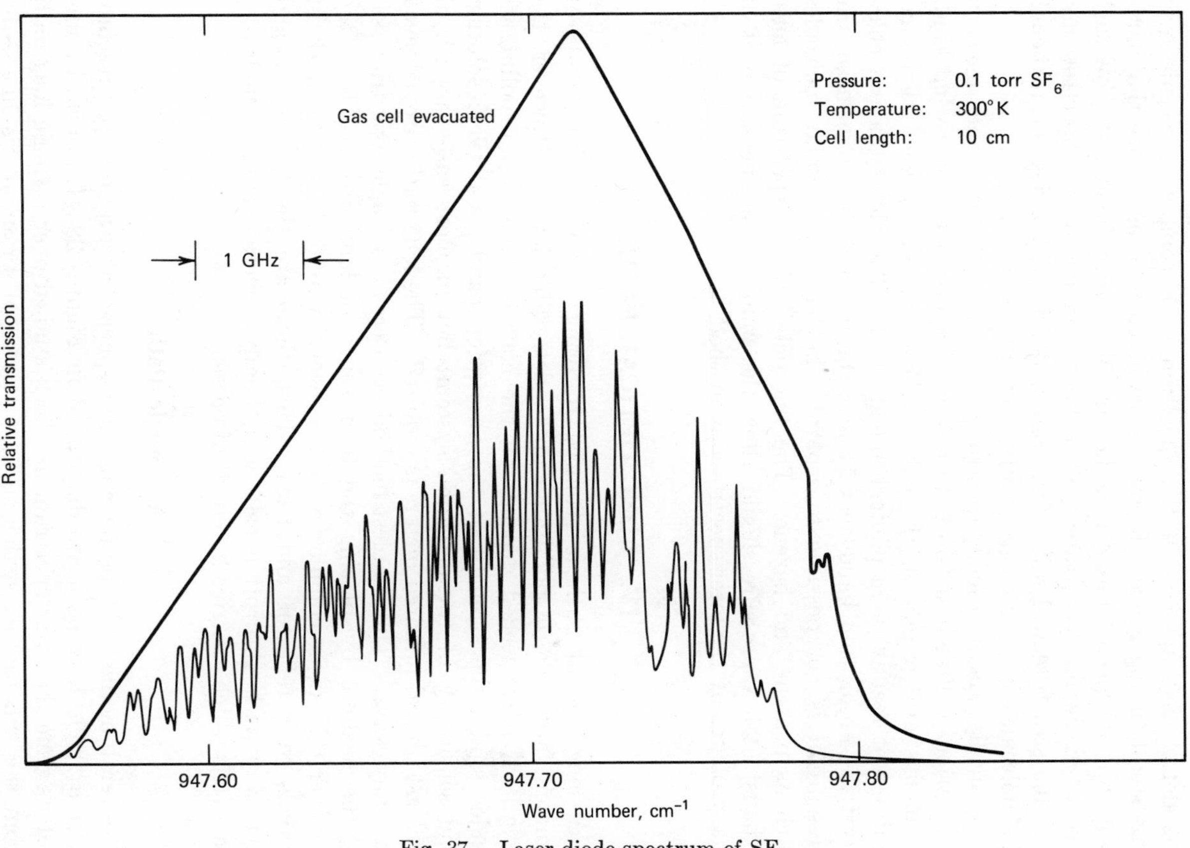

Fig. 37. Laser diode spectrum of SF_6.

does not seem like a very wide tuning range, but Figure 37, the spectrum of sulfur hexafluoride taken from Hinkley's recent paper (32), illustrates the detail that appears in the spectrum of a complex molecule over this small tuning range. It can be predicted with assurance that large molecules of specific interest in the air pollution problem will show such characteristic fine structure. As is known in microwave spectroscopy, the fine structure will differ from one molecule to the next even though the molecules may have very similar molecular structures.

Typically, laser diode spectroscopy could identify and measure the various olefins in auto exhaust. It is important to know what kinds of olefins are present in the exhaust because each type of olefin has a different reactivity in photochemical smog. The olefins have a characteristic absorption band near 11 μ in the infrared, but at atmospheric pressure it is not possible to separate the bands of similar molecules such as pentene and hexene. The fine detail in the spectrum of auto exhaust at low pressure should allow the simultaneous measurement of many olefins by the laser spectroscopy method.

IV. LONG OPTICAL PATHS

A long optical path is required to bring the absorption bands of ambient atmospheric pollutants up to detectable levels. However, the length of path needed depends on the concentration of the pollutants under study, the strength of the absorption bands, and the resolution and sensitivity of the detection system; but in most cases a path of at least several hundred meters is required. The long paths are achieved in many ways. The simplest method is to create a single long traversal of the medium by placing the light source a long distance from the receiver. Next in complexity is the two-way path in which the light travels to a reflector and back. Then there are the triple-path and four-way path absorption cells and, finally, several types of multiple-pass cells which can give hundreds of passes.

A. Single Path

Several methods of performing atmospheric absorption spectroscopy over a single long path are diagramed in Figures 38–41. Since Fraunhofer's time the solar spectrum has been studied over a single long path from the sun to the earth's surface. Some of Fraunhofer's lines—the telluric lines—were due to gases in the earth's atmosphere and therefore represent one of the earliest applications of long-path absorption spectroscopy. The study of the atmosphere by means of the solar spectrum

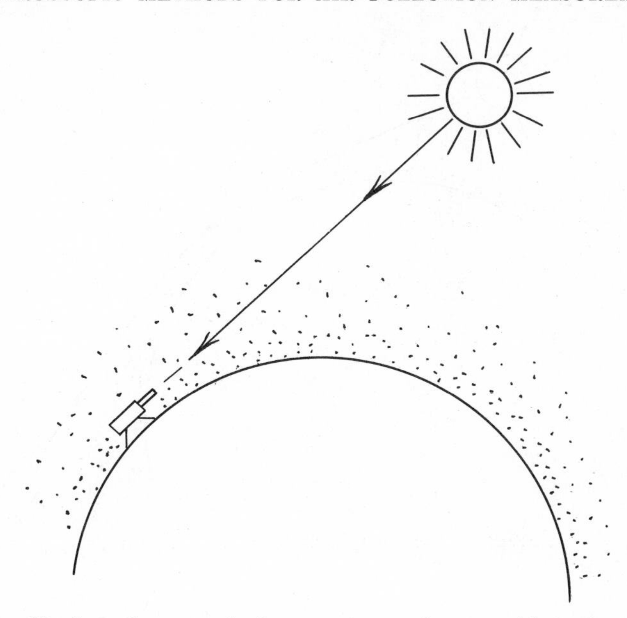

Fig. 38. Single-path atmospheric spectroscopy from earth's surface to sun.

is still being utilized at many locations in the world with useful new data being produced as a result of improvements in detection systems and increases in spectral resolution. The type of optical system used for such work was diagramed in Figure 1.

The work of the Royal Belgian Observatory (Section II) in producing the photometric atlas of the infrared solar spectrum is probably the most outstanding example of what can be achieved by looking at the sun.

A more recently developed technique of looking through the earth's atmosphere at the sun may be called atmospheric-limb spectroscopy, Figure 39. Murcray and co-workers at the University of Denver have been reporting some impressive absorption spectra taken from a high-altitude balloon looking at grazing incidence through the upper atmosphere to the sun. Most notable was their discovery of nitric acid vapor in the upper atmosphere (33). The optical path through the atmosphere from the sun to a high-altitude receiver can be very long—hundreds of miles—with a very large total air mass being traversed. However, the exact air mass traversed is somewhat uncertain. The sun's rays are refracted through the upper atmosphere to the receiver, and it is difficult to calculate their exact path because of uncertainties in the

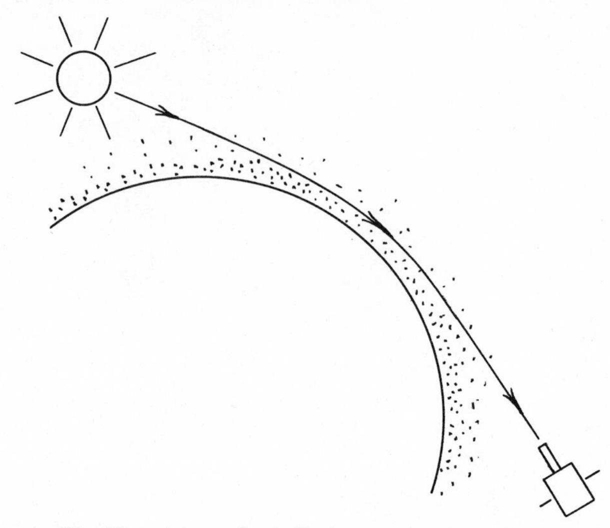

Fig. 39. Atmospheric limb spectroscopy.

temperature, density, and other physical characteristics of the upper atmosphere.

Atmospheric limb work is not unique in producing a path through a large air mass. Such a path is also traveled by the rays of the setting sun as seen by an observer on the earth's surface. The significant advantage of the atmospheric limb technique for the study of the upper atmosphere is that the rays do not pass through the water vapor which is concentrated in the lower atmosphere. This is the reason that Murcray and co-workers were able to see nitric acid bands which cannot be seen from the earth's surface.

Looking downward to the earth's thermal emission, Figure 40, is another example of a single-pass absorption experiment. The thermal radiation is a continuum in which absorption lines due to molecules in the atmosphere will appear. For a typical earth temperature of 290°K, the peak emission intensity will occur at about 10-μ wavelength. This happens to fall in a region of the spectrum in which many characteristic absorption bands of atmospheric pollutants can be seen. One important condition that must be satisfied for the asborption bands to be seen is that the air sample must be cooler than the radiating source. This is not always true. Generally the high-altitude sensing system will be looking down through regions of various temperatures, and the interpretation of the spectra will be a complicated matter, as discussed in detail in Section V.

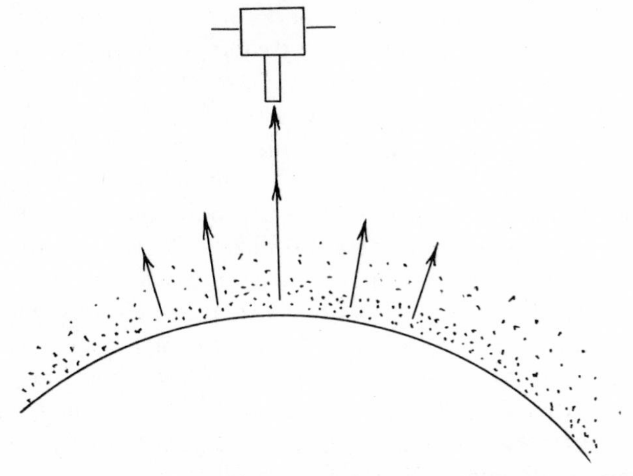

Fig. 40. Observing thermal emission through the atmosphere.

An instrument which observes thermal emission spectra is being developed and flight-tested for NASA at the Convair Corporation in San Diego. This instrument, discussed in Section V, registers the presence of the pollutant absorption lines in the spectrum by looking through filters containing the gases to be measured.

To pass light across a straight horizontal path from source to receiver is the usual way of performing absorption spectroscopy. For the study of pure gases, commercial spectrophotometers use cells a few centimeters in length. For the study of weak absorption lines, such as the lines of atmospheric impurities, the source and receiver are moved farther apart, as illustrated in Figure 41.

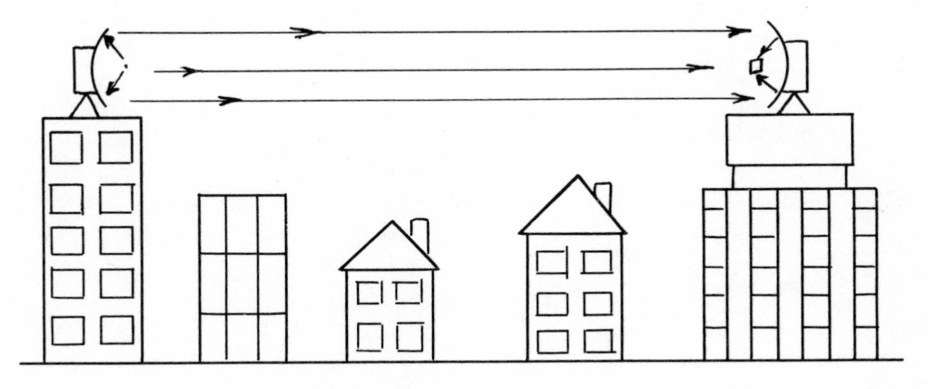

Fig. 41. Horizontal single path.

Limitations on horizontal single-pass experiments arise from the spreading of the optical beam as it travels through the atmosphere. When incoherent sources such as glowers, lamps, and carbon arcs are used, the ability to collimate the beam is restricted by the finite dimensions of the source. When laser beams are used, the collimation is limited by atmospheric turbulence. The turbulence is a much lesser limitation, and there have been some truly impressive accomplishments in passing laser beams through the atmosphere. Television pictures have been sent many miles through the atmosphere on a laser beam containing just a few milliwatts of power. A 2-W argon laser beam was projected all the way to the moon and seen by the cameras on the Surveyor spacecraft (34). That beam carried less power than is emitted by the dial light on a kitchen appliance. There is little doubt that with simple equipment a low-power laser beam can be transmitted and received over long paths through the polluted lower atmosphere.

B. Double Path

Three double-path experiments are illustrated in Figures 42 and 43. Figure 42 is a schematic diagram of experiments in which the reflected sunlight is observed by sensors outside the atmosphere. The atmosphere studied in this type of experiment has in the past been a planetary atmosphere, and the platform for the sensing system has been the earth. This has meant that the radiation not only passed twice

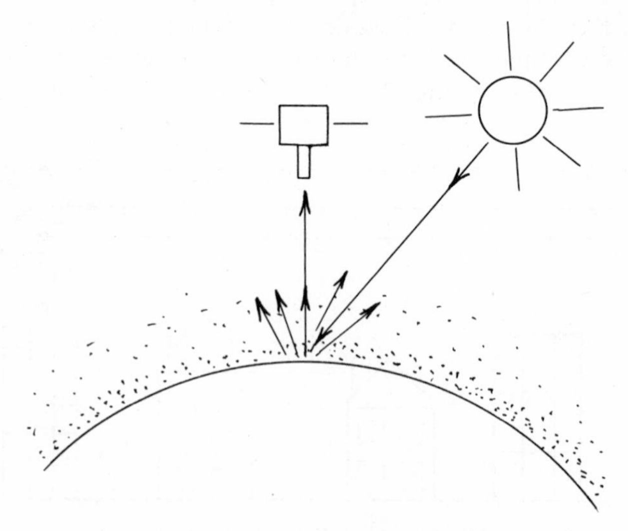

Fig. 42. Double path through the atmosphere.

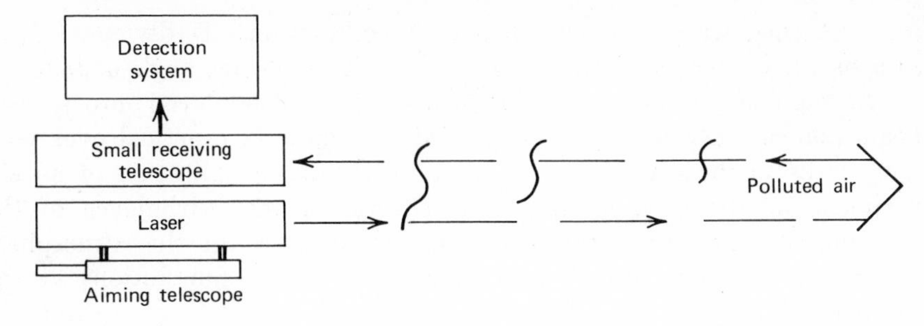

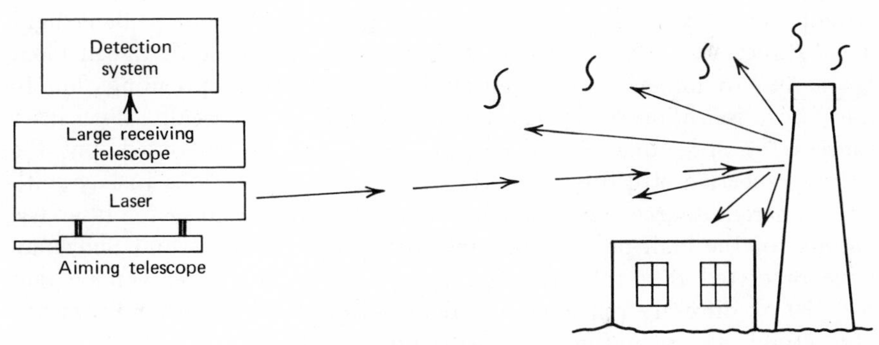

Fig. 43. Double laser paths.

through the planetary atmosphere, but also passed through the earth's atmosphere, which is a major source of interference. This interference now is being avoided by placing the telescope above the earth's atmosphere.

Orbiting sensing systems are pointed both ways—out to the planets and down to earth. In the case of Venus, one sees not the surface but the clouds. In the case of earth, the surface is seen between the clouds, providing that the wavelength is not too short. At short wavelengths in the ultraviolet, one will see the atmosphere rather than the earth's surface because of the strong Rayleigh scattering of the sunlight.

The upper drawing in Figure 43 shows a retroreflecting optical system. The retroreflectors used in laser experiments, such as the ones on the moon, are almost always cube corners. They consist of three mutually perpendicular intersecting plane mirrors. A ray of light incident on a cube corner is returned in the direction from which it came, regardless of the initial angle of incidence. If collimated light falls on the corner, it is returned with the same degree of collimation. If divergent light falls on the corner, it continues to diverge with the same angle as before.

The high degree of collimation which can be introduced into a laser beam makes the experiment diagramed in Figure 42 practical over very long paths in the atmosphere. Probably the basic limitation of an experiment involving lasers and retroreflectors is the turbulence of the atmosphere. The refractive index inhomogeneities in the atmosphere spread the beam as it travels and cause intensity fluctuations at the receiver.

The bottom sketch of Figure 43 shows a beam of light being reflected from a natural target. This approach is very appealing in its simplicity, but it presents serious difficulties because of the weakness of the return signal. The diffuse reflection at the target spreads the incident light in all directions. With incoherent sources the spreading of the incident beam due to imperfect collimation and the further spreading due to the diffuse reflection render the technique quite impractical. With laser sources the spreading of the incident beam can be restricted, but the diffuse reflection still dissipates the energy in all directions just as with an incoherent source. Nevertheless, there is some promise for laser use because of the high power available from lasers. Jacobs and Snowman have predicted that the radiation from a 10-W CO_2 laser can be sent out 300 m, diffusely reflected, and detected at the transmitter by means of a receiving system of modest design (26).

C. Triple and Quadruple Paths

A triple-pass cell with large optical aperture was used for the study of atmospheric gases many years ago by Pfund (35). This type of cell has two spherical mirrors facing each other at a separation equal to the mirror radius of curvature. Each mirror has a central hole. Light from the source is passed through one of the holes to the opposite mirror, which it fills. The beam is then reflected in a state of collimation to the other mirror that focuses it out the opposite hole to the receiver. The cell is diagramed in Figure 44. A Pfund cell with 1-m diameter mirrors and a 30-m radius of curvature is in use by Professor John Strong at the University of Massachusetts in Amherst. An example of a four-pass cell is the Perkin-Elmer 1-m absorption cell, diagramed in Figure 44. Light enters the cell from the side, experiences four traversals, and then is passed out the other side. There are three reflections from spherical mirrors and two from plane mirrors. The optical aperture is large, but the path of 1 m is not really very long. There are no mirror adjustments in the cell; it can only be used at the fixed path of four passes.

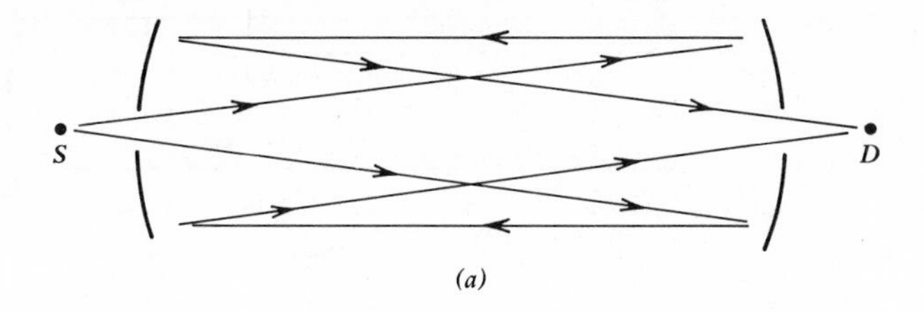

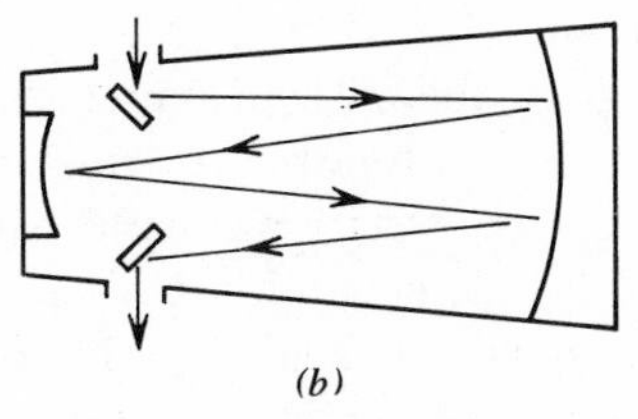

Fig. 44. Three-pass and four-pass cells. (*a*) Pjund cell. (*b*) One-meter cell.

D. Many-Pass Cells

1. *The White Cell*

The three-mirror multiple-pass cell known as the "White" cell (9) has been of great value to spectroscopy for more than 25 years; its applications still are being extended. The principal virtue of the White cell is that it reimages the source after each double pass of the radiation up and down the cell. This reimaging keeps all the energy that is collected on the entrance pass enclosed within the mirror system until the exit pass. Therefore, reductions in beam intensity result only from absorption by the mirror surfaces and the gas between the mirrors. The optical aperture of the White cell is the same as the optical aperture involved in the focusing of the input beam by the first mirror it encounters, hence the title of White's original paper, "Long Optical Paths of Large Aperture." The performance of the White cell has been analyzed and commented upon by a number of previous authors (27,36). The performance is not analyzed in detail here, but we describe the path of the light through the cell and offer some comments on the limitations of the cell's performance. This discussion is the basis of a subsequent explanation of a new eight-mirror long-path cell devised by P. L. Hanst and V. H. Early, and further extension of the multiple-pass technique to cells with any even number of mirrors at each end.

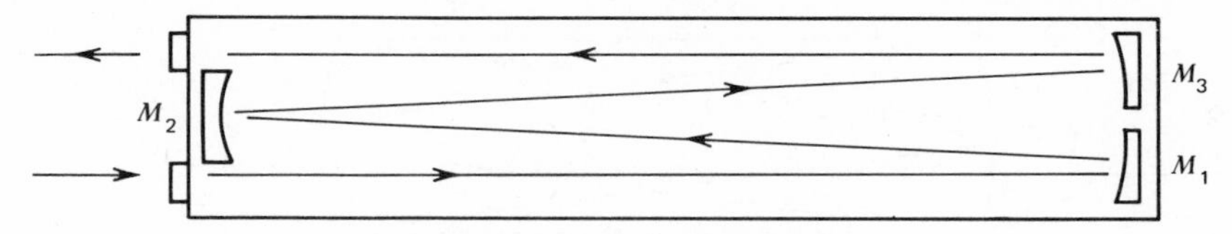

Fig. 45. White cell (4n pass).

The operation of the White cell is described by following the light from the source into the cell and through at least four passes, with the aid of Figures 45 and 46. The light from the source initially is focused into a real image in the entrance aperture of the cell. In Figure 46 this is designated the zeroth image. It must fall above the horizontal center line of the mirror, M_2. After passing through the zeroth image, the beam diverges and is collected by mirror M_1. The mirror M_1 is a spherical mirror situated two-focal lengths from the image, so that it refocuses the image, inverted, on M_2, which is another spherical mirror of the same focal length. The mirror M_1 is aimed so that this first image which in Figure 46 is marked 2 (2 passes), and falls below the horizontal center line of M_2. Mirror M_2 is now aimed so that the reflected diverging beam falls entirely on spherical mirror, M_3. The mirror M_3 is then aimed to form another image (marked 4) above the center line of M_2 alongside the zeroth image. If the image formed by M_3 is placed in the exit aperture of M_2, the total number of passes through the cell is the minimum of four. If the image formed by M_3 falls on M_2 symmetrically opposite the first image (numbered 2), the beam will be returned to M_1 at the required small angle with the input beam

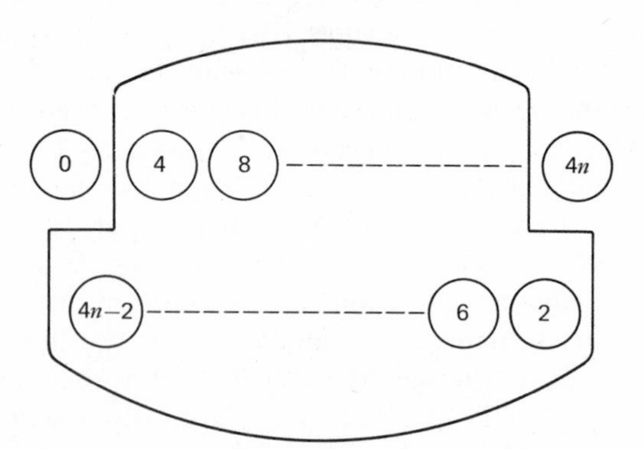

Fig. 46. Placement of images on M_2. (Image number equals number of passes through the cell.)

so that all the energy is again collected by M_1, and there will be at least four more passes through the system.

The number of images in the rows depends on the placement of the first image on the lower part of the mirror. If it falls exactly on the vertical center line, no more than four passes are possible. The farther to the right it falls, the greater the number of passes allowed. When the images are given only even numbers as in the Figure, the number of passes is equal to the number of the exit image. In practice, the number of passes is determined by counting the number of images in the bottom row and multiplying by four.

The maximum possible number of passes is obtained when the input image just grazes the edge of M_2, and the images on M_2 are placed so close together that they touch. (It is impossible for them to overlap.) Therefore, the size of the zeroth image and the width of M_2 are the basic geometric limitations on path length. In order to have as much energy as possible going through the system, an enlarged image of the source is usually projected into the entrance aperture, and the exit image is demagnified when the transmitted light is fed into the detection system. A trade off therefore exists between the maximum possible path length and the source magnification. Another path length limitation in practice is the sensitivity of the lateral adjustment of the mirrors. At very long paths, the images are packed very closely together, and a minute change in lateral adjustment, resulting from temperature variations, for example, can change the path length substantially. Even at shorter paths, changes in lateral mirror adjustments can create problems, because they move the exit beam away from its proper exit aperture.

The only major source of energy loss in the system is the absorption at the mirror surfaces. If R is the reflectivity of the mirrors, the fraction of the input energy which will be lost at n reflections will be $(1 - R^n)$. Some sample calculations show the magnitude of this loss. The infrared reflectivity of a gold surface will be about 98%. With 50 reflections, about 63% of the energy will be absorbed at the surface and 37% transmitted. With 150 reflections about 95% of the energy will be absorbed and 5% transmitted. In going from 50 to 150 passes the path is tripled, but the transmitted energy is reduced sevenfold. If the light source is a laser, the excessive loss might be acceptable; but if it is a thermal source, it would not be acceptable. Through the use of multilayer coatings, mirrors with reflectivities higher than 99% are becoming available for use in most regions of the spectrum, including the infrared. These new mirrors considerably enhance the applicability of the many-pass cell to atmospheric problems.

2. *The Eight-Mirror Long-Path Cell*

It has been stated that the close packing of the source images on the mirror M_2 of the White cell is a limitation. There are two aspects to this limitation: (1) at long paths the allowed width of the source image, hence the energy throughput of the system, is restricted by the width of the mirror, M_2; and (*2*) at long paths the lateral adjustment of the mirrors becomes too delicate for reliable cell performance.

The eight-mirror long-path cell produces four rows of images; each image again corresponding to two passes. For a chosen mirror width and path length, the tendency toward lateral misalignments in the eight-mirror cell is therefore only half as great as in the White cell, and the allowable image width is twice as great.

The eight mirrors are all spherical mirrors; they all have the same focal length; and the distance traveled by the beam between mirrors is two focal lengths. The four mirrors at the in-focus end of the cell are rectangular, and the other four at the out-of-focus end can be square or round. All eight mirrors have individual adjustments. The operation of the system is diagramed in Figure 47 with an illustration of the basic unit of eight passes. Figure 48 is a photograph of the in-focus end of the eight-mirror cell with a multiple-passed helium–neon laser beam made visible by smoke. A description of the alignment of the mirrors will also serve to trace the path of the rays in the system. (*1*) The zeroth image is focused in the entrance aperture. (*2*) The divergent entrance beam is collected by the first mirror and imaged on the right side of the second mirror, which is the bottom rectangle at the in-focus end of the cell. (*3*) The divergent return beam is directed to the third mirror. (*4*) The third mirror collects the light and directs it to the left side of the fourth mirror. (*5*) From the fourth mirror the beam is directed, diverging to the fifth mirror. (*6*) The fifth mirror sends a converging beam to the sixth mirror. (*7*) The sixth mirror sends a diverging beam to the seventh mirror. (*8*) From the seventh mirror the converging beam finally goes to the eighth mirror where it forms an image alongside the zeroth or input image. (*9*) When the eighth mirror is aimed at the first mirror, the light beam on its ninth pass travels along a path which forms a very small angle with the path of the input beam. Therefore, further sets of eight passes take place, and new sets of four images are formed alongside the initial set of four images, as shown in Figure 47. As in the White cell, the total number of passes depends on the separation of the images on the mirrors. The path length can be controlled by changing the position of the image on the eighth mirror. The total number of passes can

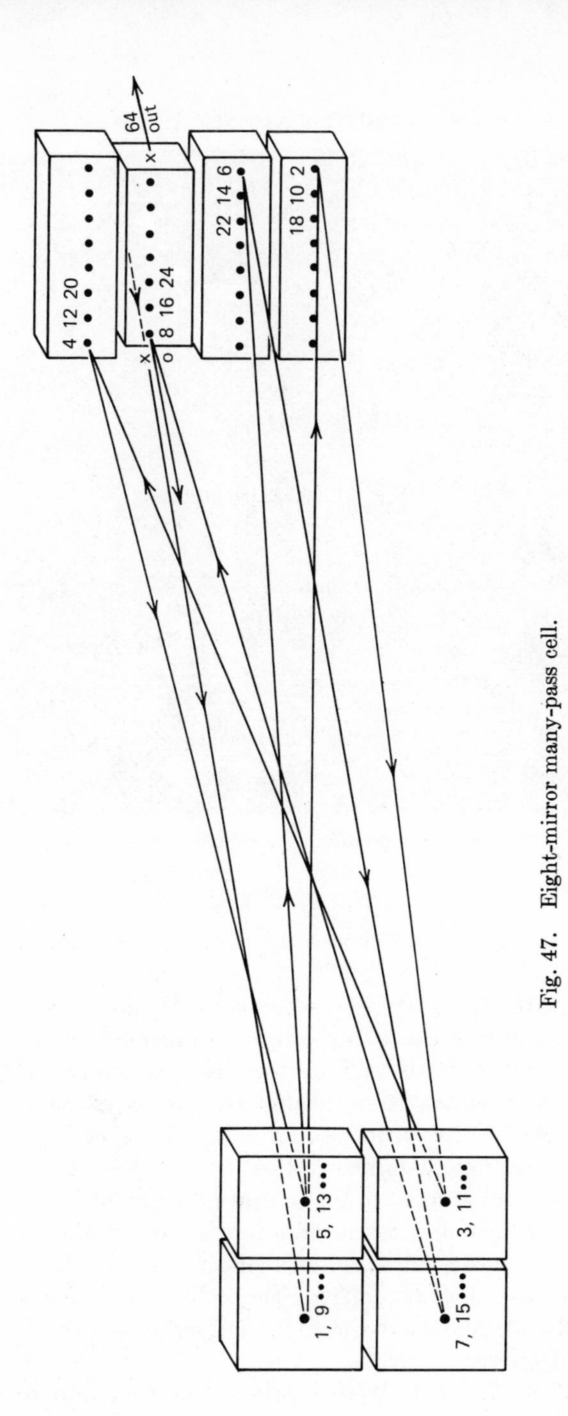

Fig. 47. Eight-mirror many-pass cell.

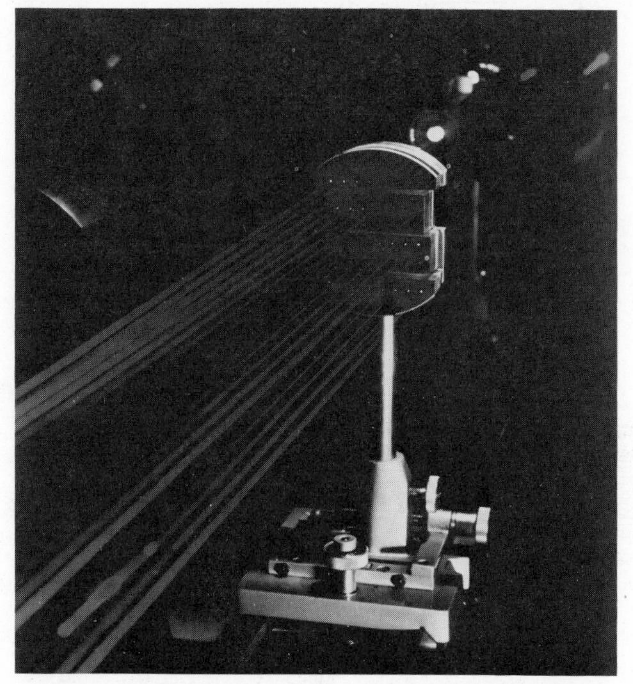

Fig. 48. Multiple passing of a helium–neon laser beam.

be changed in units of eight passes by turning one screw behind mirror number seven.

When used with incoherent sources, the eight-mirror cell has a disadvantage, compared with the White cell, in requiring twice as much mirror surface at the out-of-focus end of the cell. At very long paths this disadvantage is, in effect, compensated by the extra amount of mirror surface available at the in-focus end. If this is considered to be an equal trade off, the eight-mirror cell still has the advantage of being less sensitive to misalignment in the lateral adjustment.

When a narrow pencil of laser radiation is passed through a long-path cell, there is no consideration of mirror-size trade offs, because small mirrors will suffice at both ends of the cell. In that case the lateral adjustment advantage which the eight-mirror cell has over the White cell should be decisive.

A method of changing the path length in the eight-mirror cell without changing any mirror adjustments is illustrated in Figure 49. The bottom mirror at the in-focus end of the long-path cell receives the beam after the second pass and every subsequent set of eight passes. Small mirrors

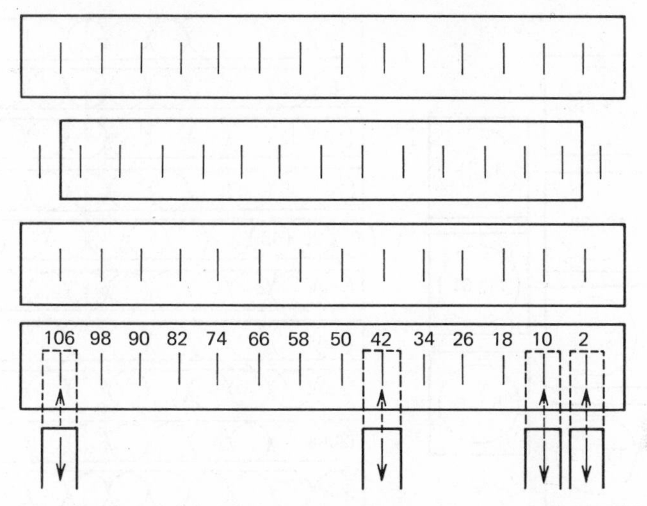

Fig. 49. Method of coupling the light beam out of the cell.

whose surfaces are at a 45° angle with the direction of the beams are placed below the row of images, as illustrated. When any one of the small mirrors is raised into the light path, it couples out the energy after the indicated number of passes. In practice, one would not require all possible path-length increments, but would probably be satisfied with changes of path length in multiples of three or four. Four coupling-out mirrors would suffice. These are shown in the figure at passes 2, 10, 42, and 106. The coupling-out mirrors could be raised and lowered in a vacuum by magnets. Once the eight multiple-passing mirrors were lined up, the alignment of each array could be made permanent by casting the array in epoxy cement.

3. *Larger Mirror Arrays*

It appears to be a simple matter to extend the principle of the eight-mirror long-path cell to larger arrays of mirrors. The total number of mirrors must be a multiple of four—there must be an even number of mirrors at each end of the cell. In Figure 50 the two ends of a possible 20-mirror cell are diagramed, with numbered images. The alignment basically would be the same as in the eight-mirror cell. The beam would be passed from one mirror to the next until finally the twentieth reflection (the tenth-focused image) would occur alongside the zeroth or input image. When the beam on the twenty-first pass is returned to the first mirror at a slight angle with the input beam, multiple passing will take place. Adjacent images on the bottom mirror

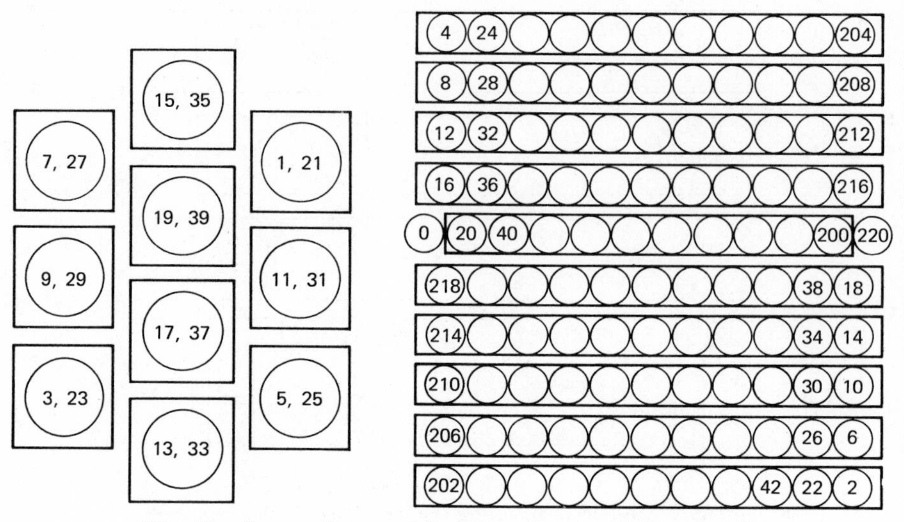

Fig. 50. Image placement in a 20-mirror multiple-pass cell.

will have path increments of 20 passes. As in the case of the eight-mirror cell, coupling out can be done by means of small mirrors placed in front of the bottom rectangular mirror at the in-focus end of the cell. Again the aligned mirror system could be cast in epoxy. It is not expected that the 20-mirror system would be practical for use with incoherent sources, but it definitely would be practical for use with lasers.

The many-pass cells described here must be compared for laser applications with the many-pass cells described by Herriott and Schulte (37). Their cell will be referred to as the nearly confocal cell, and the cell described in the present section will be called the multimirror cell. The nearly confocal cell is in fact a three-mirror system requiring two main spherical mirrors separated at slightly less than their radius of curvature and an additional small perturbing mirror. When a laser beam is passed between the two main mirrors, it is reflected back and forth: the locus of intersections of the beam with each of the mirrors is an ellipse. The perturbing mirror is placed at the last spot of one of the ellipses from which it returns the beam in a direction slightly different from the direction of the input beam, thus creating many new pairs of ellipses each of which is somewhat displaced from the first pair. The system can be adjusted so that an array of many spots appears on each mirror and there are hundreds of passes.

The basic limitation of the nearly confocal cell is the energy losses at the reflecting surfaces, as it is in the case of the multimirror cell. In this regard the systems are equivalent. However, in regard to convenience of operation, it appears that the multimirror cell may have some advantages. Although a laboratory comparison between the two systems has not been made, it seems worthwhile to state both advantages and disadvantages of some of the aspects of the multimirror cell.

Possible advantages are the following: (*1*) Coupling out the beam of the multimirror cell has been shown to be a relatively simple matter of raising a small mirror at the desired position in the bottom row of images. (*2*) Separation of the various beams in the multimirror cell is excellent; if very small spots, hence very clean separation, are desired, the input laser beam can be focused at the entrance aperture. (*3*) Lining up the mirrors of the multimirror cell is surprisingly easy. The experimental model used in the present study had ordinary 10-32 adjusting screws which presented no difficulty. (*4*) The number of passes of the multimirror cell is not a sensitive function of the direction of the input beam. As long as the beam hits somewhere on the first collecting mirror, the cell will give the intended number of passes. (*5*) The larger the number of mirrors, the less sensitive the system is to misalignment of one of the mirrors.

Following are two possible disadvantages, with comments: (*1*) The multimirror cell has many mirror adjustments, each of which may be a possible source of misalignment. This disadvantage might be nullified by the expedient of cementing together each array of aligned mirrors by means of epoxy cement as mentioned previously; the only adjustments would then be the aiming of each array as if it were a single mirror. (*2*) The multimirror cell is complicated. Although this complexity cannot be denied, the expense of constructing the cell is minimized by cutting each of the two arrays of mirrors from one large mirror. In constructing the cell used in the present work, two telescope mirrors were cemented to predrilled aluminum plates, and then each was cut into the appropriate set of four smaller mirrors.

V. RADIATION SOURCES

A. Man-Made Thermal Sources

The choice of a radiation source for an absorption spectroscopy experiment involves a compromise among several factors. The highest possible emission intensity in the spectral region under study is desired. At a fixed spectral resolution a brighter source permits a longer light path through the sample; at a fixed path length a brighter source permits a higher spectral resolution. Other factors in the choice of a light source are the need for a continuous spectrum and the convenience and reliability with which the source can be operated. For work in the infrared the source chosen is generally a heated solid. The spectral emission characteristics of four frequently used sources are plotted in Figure 51. Note that the vertical scale has been compressed by plotting the logarithm of the emission intensity. An informative article giving addi-

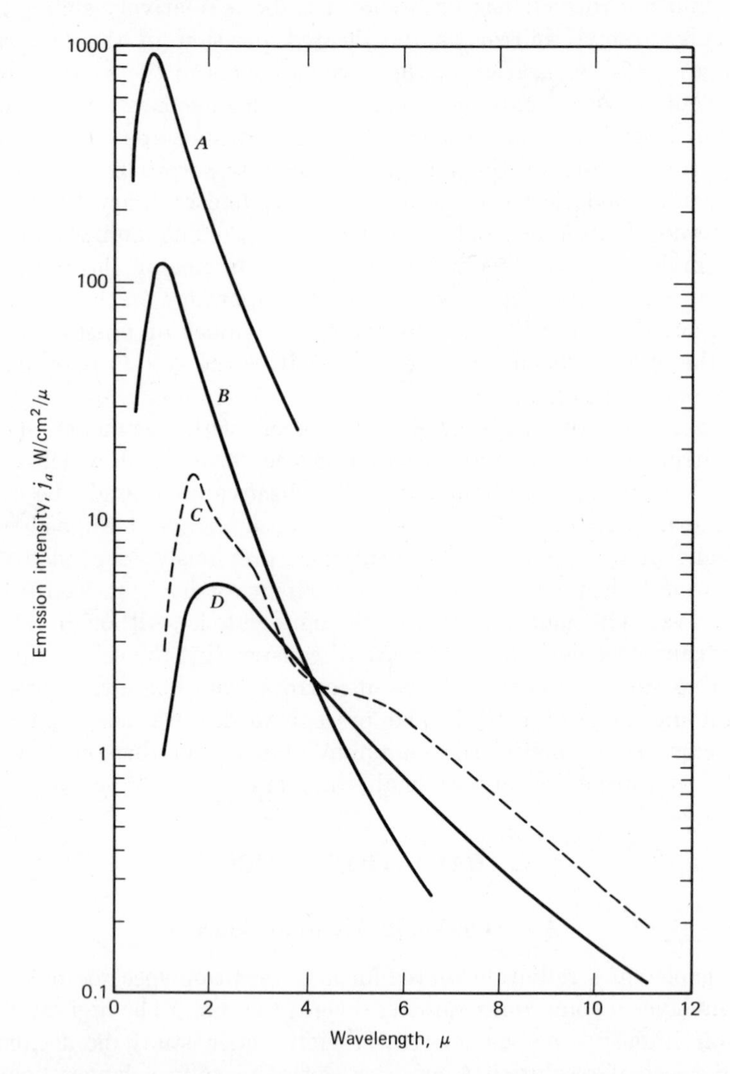

Fig. 51. Spectral emission characteristics of four sources of infrared continuum. *A*, carbon arc at 3950°K; *B*, tungsten filament at 3000°K; *C*, Nernst glower at 2000°K; *D*, globar at 1400°K.

tional details on radiant sources is published in the *German Encyclopedia of Physics* (38). The spectral brightness, amount of radiation per unit surface area per unit wavelength interval, is given by Planck's law:

$$J_\lambda = \frac{\epsilon_\lambda C_1}{\lambda^5}\left(\exp\frac{C_2}{\lambda T} - 1\right)^{-1}.$$

The two source parameters determining brightness are therefore emissivity, ϵ_λ, and temperature, T. At the shorter wavelengths, source temperature is the predominant factor. For example, in the ultraviolet, electrically heated gases of low emissivity are used as sources as they have extremely high temperatures. At longer wavelengths the emissivity factor can become predominant, shown by the long wavelength crossover in Figure 51.

The carbon arc is definitely the brightest of the four sources at all wavelengths of interest and has a high temperature and a uniformly high emissivity. Unfortunately the carbon arc is not convenient to use in the laboratory, because the electrodes are consumed and an electrode feed mechanism is needed.

The tungsten lamp is the best of the sources illustrated if one is working at wavelengths shorter than 2.5 μ. At these wavelengths the high temperature of the tungsten filament counteracts its relatively low emissivity. Furthermore, the lamp is a convenient source to use. Although the glass envelope of the lamp absorbs the infrared at wavelengths of 2.6 μ and longer, the useful range of the lamp can be extended to longer wavelengths by use of quartz or sapphire windows. Since the emissivity of the metal decreases with longer wavelengths, the Nernst glower surpasses the tungsten lamp in brightness at about four μ.

The Nernst glower, a resistance–heated rod composed of refractory oxides, has wide applications in infrared spectrophotometers. It is a convenient source to use, it operates at high temperatures in air, and it has a good emissivity characteristic in the infrared. The usual Nernst glower is a relatively narrow source, the rod being only about 1.5 mm in diameter. When the detecting instrument has a large entrance pupil, the larger globar may be preferred.

The globar is a silicon carbide rod heated by an electric current. It is somewhat less convenient to use than the Nernst glower and more subject to burn-out. The globar has a uniformly high emissivity and is found in larger sizes, the usual rod being about 6 mm in diameter.

B. The Sun as a Source of Radiation

Direct sunlight is so intense that it is an easy matter to gather enough sunlight to record a spectrum. In most experiments in which one looks at the sun, a small collecting mirror or lens is sufficient. In fact, too large a collector may damage the detecting instrument through overheating. Considerable solar energy can be focused into an instrument because the solar surface is inherently extremely bright. The sun radiates like a black body at 5800° K and delivers approximately 1400 W of radiant energy per square meter of surface at the top of the earth's atmosphere. Some energy is removed by scattering and by absorption in the atmosphere, so that when sunlight reaches the earth's surface it has been reduced in intensity by an appreciable amount, depending upon the atmospheric conditions and the path of the rays through the atmosphere. For an approximate calculation, assume that one-half of the direct intensity is lost on passing through the atmosphere. Then a mirror of 1-m^2 area will collect and focus 700 W of energy. An $f/1$ mirror would focus this into a region about 1 cm^2 in area. Such a collecting system is a solar furnace and is capable of melting almost any material.

When the sun-illuminated earth is viewed from an airplane or a spacecraft, the inherent brightness of the illuminated surface is very much smaller than the brightness of the sun's surface, but the spectral distribution, as modified by the atmosphere, is basically the same. The brightness is simply the product of the solar irradiation intensity and the diffuse reflectivity of the surface. The irradiation intensity is the sum of the direct intensity and the intensity of the sunlight scattered by the atmosphere. Assume, for example, that the irradiation totals 1000 W/m^2, which is 0.1 W/cm^2; of this, only part is reflected. The total reflected intensity seen by an instrument pointed toward the surface is therefore quite small and is about equal to the total amount radiated by the surface. Since the spectral distributions of the solar radiation and the earth radiation are quite different, an instrument looking toward the earth will see mainly solar radiation when operating in the visible and mainly earth radiation when operating in the longer infrared region. This is illustrated in Figure 52. At about 3-μ wavelength the instrument will see an equal amount of radiation from each source.

If the detection system is tuned to the ultraviolet, it will see mainly Rayleigh scattered radiation from the atmosphere. The shorter the wavelength, the higher the contribution of Rayleigh scattering, as given by the inverse fourth-power law. This scattered ultraviolet light is a limiting factor in the development of instruments that remotely sense pollutants by observing reflected sunlight. When clouds appear in the field of view of an instrument looking down from high altitudes, the cloud reflection predominates, and it is difficult to see the earth at any wavelength shorter than microwaves.

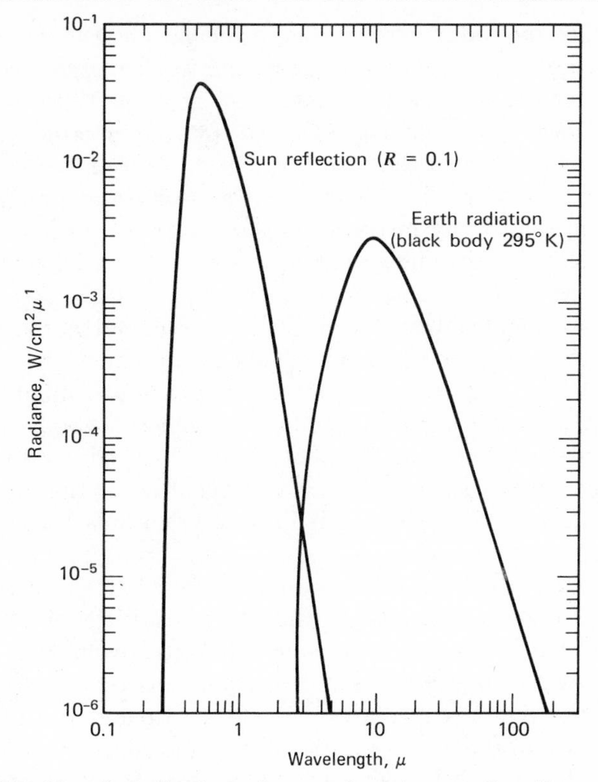

Fig. 52. Contributions of sun and earth to earth radiance (11).

C. The Earth as a Source of Radiation

Direct measurement of the pollutants in the lower atmosphere requires instruments which look downward, as well as instruments which look up to the sun. If the downward-looking instrument works with the infrared vibration–rotation spectrum, the earth will be the chief source of continuum radiation for absorption experiments.

As shown in Figure 52, a typical earth temperature sets the maximum emission intensity at about 10-μ wavelength. This maximum fortunately falls in the center of a principal transmission region of the atmosphere where many useful absorption bands of pollutants fall, as shown in Figure 4.

The spectral region between 3 and 5 μ also should be available for use by downward-looking instrumentation if at all possible, because of the many distinctive molecular bands it contains and its relative freedom from interference by water vapor. The principal obstacle to the use of the 3 to 5-μ region is that the earth radiation in that region is approximately two orders of magnitude weaker than at 10 μ. This weaker radiation is at least partially compensated by the greater detectivity of detectors that operate in the 3 to 5-μ region, such as the lead sulfide and indium antimonide cells. Therefore, it appears to be practical to do remote pollution sensing by instruments that operate between 3 and 5 μ. In Section VI.E, one instrument that has been developed for this region is discussed.

The downward-looking instrument that observes earth radiation can operate both by day and by night. This is an important advantage over instruments that operate only during the hours of sunlight, because it is important to study nocturnal pollution problems. Some important aspects of night-time pollution that should be observed are the buildup of photolyzable species such as NO_2, the stagnation of the air at night, and the possible furtive release of industrial pollutants under cover of darkness.

It would appear from the discussion thus far that the use of earth radiation has so many advantages that there is little question as to the applicability of earth-oriented remote sensing systems to pollutant measurement. Unfortunately, one major problem is that the atmosphere not only absorbs but also radiates. If the earth and the atmosphere are about at the same temperature, radiation and absorption are nearly balanced, and there is little structure to be seen in the spectrum, no matter how high the pollutant concentration.

To visualize the physical principle controlling whether an absorption spectrum will appear, one should recall that when a sample is in thermodynamic equilibrium, the absorptivity and emissivity are equal at any selected wavelength. If $\epsilon_{e(\lambda)}$ and T_e are the emissivity and temperature of the radiating earth, respectively, the radiant intensity will be given by Planck's law:

$$J_{e(\lambda)} = \frac{\epsilon_{e(\lambda)} C_1}{\lambda^5} \left(\exp \frac{C_2}{\lambda T_e} - 1 \right)^{-1}$$

If $\alpha_{a(\lambda)}$ is the absorptivity of the atmosphere, the amount of energy absorbed will be

$$A_{a(\lambda)} = \alpha_{a(\lambda)} J_{a(\lambda)}$$

The amount emitted by the atmosphere having emissivity $\epsilon_{a(\lambda)}$ and temperature T_a will be

$$E_{a(\lambda)} = \frac{\epsilon_{a(\lambda)} C_1}{\lambda^5} \left(\exp \frac{C_2}{\lambda T_a} - 1 \right)^{-1}$$

The ratio of the amount absorbed to the amount emitted will be

$$\frac{A_{a(\lambda)}}{E_{a(\lambda)}} = \frac{\alpha_{a(\lambda)}}{\epsilon_{a(\lambda)}} \epsilon_{e(\lambda)} \frac{\exp \dfrac{C_2}{\lambda T_a} - 1}{\exp \dfrac{C_2}{\lambda T_e} - 1}$$

Since at thermal equilibrium $\alpha_{a(\lambda)}/e_{a(\lambda)} = 1$, the extent of absorption (or emission) depends on the earth emissivity, $\epsilon_{e(\lambda)}$, the two temperatures, T_e and T_a, and the wavelength, λ. The earth emissivity will always be less than unity, so that equal earth temperature and atmospheric temperature would actually lead to emission of the pollutant lines. For the purpose of the following calculation of temperature effects, however, earth emissivity will be assumed to be unity.

On the short wavelength side of the black-body emission curve (e.g., at 3 μ for the radiating earth of Figure 52), the absorption will be near its maximum when the atmosphere is only slightly cooler than the earth. On the long wavelength side of the radiation peak, larger temperature differences are required to bring out the absorption. As the atmospheric temperature is lowered, the absorption approaches a limit, $\alpha_{\max}$. The ratio of actual absorption, α_a, to maximum absorption, $\alpha_{\max}$, is plotted in Figure 53 as a function of wavelength and atmospheric temperature, assuming an earth black-body temperature of 300°K. The figure shows, for example, that when the air is 10°K cooler than

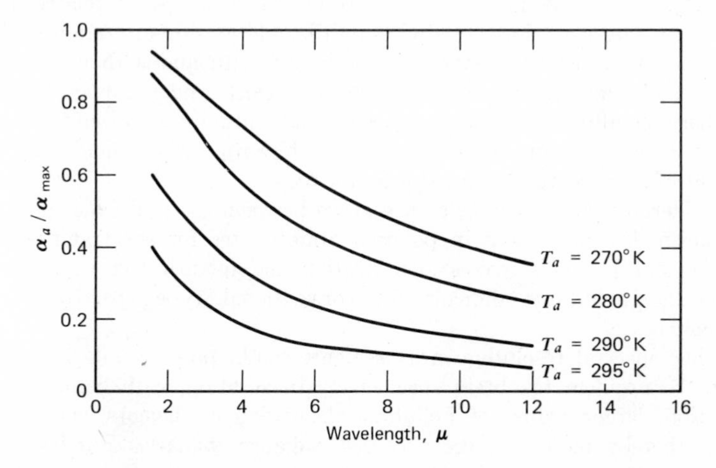

Fig. 53. Ratio of actual absorption, α_a, to maximum possible absorption, $\alpha_{\max}$, for atmospheres at temperatures, T_a, slightly cooler than the radiating earth at 300°K.

the earth, an absorption band at 3.4 μ will have 40% of its maximum possible intensity, and a band at 11 μ will have 14% of its maximum intensity.

In practice, the determination of absorptivity is extremely complicated. Actually the earth has an emissivity less than unity which varies with wavelength and with type of terrain. The atmosphere is not isothermal and is not necessarily cooler than the surface. In order to estimate properly the absorption (if any), one must have a knowledge of the earth surface temperature distribution, the earth emissivity characteristics, the gas absorption coefficients, the air temperature distribution with altitude, and the pollution distribution with altitude. Most of these factors will change as functions of time of day, season of the year, and atmospheric meteorological conditions. Calculations of the absorption signal expected under typical conditions are in ref. 11. Quoted below are conclusions set forth in ref. 11.

> From these calculations the following conclusions can be drawn with respect to the remote measurement of air pollutants in the infrared region.
>
> The difference between the earth's surface temperature and the effective temperature of the pollutants is the principal variable that affects the sensitivity of the instrument to changes in radiance with and without the existence of air pollution.
>
> Since the atmospheric temperature changes with altitude, a knowledge of this temperature profile and of pollutant concentration profile shape is necessary to make quantitative measurements. Methods of obtaining this knowledge either directly or from a priori considerations should be investigated.
>
> In general, the pollutant concentration profile shape is related to the temperature profile. Further studies on this relationship are required.
>
> Significant, measurable changes in radiance, with and without even relatively small amounts of air pollution, will exist under a variety of atmospheric conditions. However, these changes are less than linearly related to increasing pollutant concentrations. This appears to hold true independent of spectral resolution, at least for $\Delta\lambda \geq 0.1$ μ.
>
> Considerable spectral interference exists between the individual pollutants that normally are present in polluted atmospheres for spectral resolutions of about 0.1 μ. This makes identification and quantitative measurements of specific pollutants difficult for conventional type spectrometers or radiometers.
>
> Higher spectral resolution measurements would possibly alleviate the interference problem to obtain specificity. In addition, with higher spectral resolution, larger values of pollutant absorption coefficients would be obtained, thereby increasing the observed radiance changes. Further studies are required in this general area.

D. Pollutants as Radiation Sources

In the preceding discussions it is mentioned that infrared instruments which look to the earth from high altitudes may under some conditions see pollutant emission lines rather than absorption lines. Such emission lines will always be seen when the pollutants are at a higher temperature than the brightness temperature of the optical background. Smoke from a chimney will emit the characteristic vibration–rotation bands of its molecular constituents. It will also emit continuum because of the hot particulates, but in a smoke of low opacity the effective emissivity of the particulates will be small, and the molecular emission bands will be observable. Radiometers of fixed bandpass and of the spectral scanning type are available for the observation of the optical emission from such effluents. The problem is not so much in recording the emission as it is in interpreting the record. When using a remote sensing radiometer it is difficult to separate the factors of temperature, emissivity, concentration, and spatial distribution of emitters. Discussions of these important problems may be found in the infrared literature.

The pollutant emission may also be used as a source in instrument systems for absorption spectroscopy. In this case, the pollutant gives an emission spectrum which is closely matched to the absorption spectrum and therefore discriminates against possible interferences. The absorption coefficient which the pollutant under consideration shows when exposed to its own radiation will be high, while the absorption coefficient shown by all other pollutants will be low. Instead of the laser-line-absorption-line coincidences previously discussed, this will be a case of emission-line—absorption-line coincidences.

One case involves placing the emitting gas in a heated enclosure with a transmitting window. Then the emitting gas has an intensity distribution and spectral line widths characteristic of its own temperature. Since the absorbing gas is at a lower temperature, the spectral match is not perfect, and the absorption coefficient averaged over the band will vary with the temperature of the emitter.

Another case causes the gas to fluoresce in the infrared. In this case there may be a very close match between the spectral distributions of emission and absorption because the rotational and translational temperatures of the emitter may be near room temperature even though its vibrational temperature is very high. Such a fluorescence has been obtained from the carbon monoxide molecule by the ARKON Company, Berkeley, Calif., and is being applied in a carbon monoxide detection system. Infrared fluorescence is a relatively new phenomenon and has not yet been demonstrated for many compounds. It certainly appears to be worthy of further investigation.

E. Lasers

The laser is a key development which has revitalized optical science in recent years. Nonlinear effects in optics, for example, are not observable with the intensities available from thermal sources of radiation but are readily observable with high intensity lasers. The subject of holography was understood and proved in the prelaser days, but it did not become practical until the laser supplied the necessary degree of optical coherence. Optical communications is another new subject in which lasers are promoting rapid development.

Lasers surpass the limitations of incoherent sources in many ways. They can deliver higher power; they can produce shorter pulses; they have narrower spectral lines; they can be better collimated and focused. It is natural that lasers are leading to new and better methods of air pollution measurement.

There are four principal classes of lasers, all of which are being applied to atmospheric problems. These classes, distinguished by the physical state of the lasing medium, are *(1)* gases, *(2)* liquids, *(3)* semiconducting crystals, and *(4)* dielectric crystals and glasses. Each type of laser has the following basic features of laser design. *(1)* An emitting medium is enclosed within reflectors that constitute a resonant optical cavity. *(2)* There is a means of exciting the lasing medium to a population inversion. *(3)* There is a means of allowing the stimulated emission to escape from the optical cavity.

Aside from the basic laser characteristics, the four types of lasers are very much different. Consider size, for example. Gas lasers are usually large, some being 1 m long or more. Semiconductor lasers are very small, some being only 1 mm long. Liquid lasers and dielectric crystal lasers are intermediate in size, being perhaps a few centimeters in length. The spectral, temporal, and intensity characteristics of lasers also cover a wide range. Nearly all emission wavelengths are now available, ranging from 0.27 μ in the ultraviolet to 337 μ in the far infrared. The duration of the laser action can vary from less than 1 nsec (10^{-9} sec) to continuous operation. Pulse repetition rates can be one per second, one million per second, or anything in between. Power levels can be greater than 1 GW (10^{9} W) or as small as 1 μW (10^{-6} W).

Many books on lasers have been published in recent years, and it is not appropriate to attempt to review the properties of lasers here in any detail. Several of the better books are listed in the bibliography (see refs. 39,40, and 41). It will be useful, however, to discuss some aspects of laser operation which have application to pollution sensing instrumentation.

1. *Gas Lasers*

Of the many gases that have been shown to produce laser action, the most promising for application to pollution detection appear to be CO_2, the rare gases, and iodine. There are several reasons that these lasers appear practical. It is important, for example, that they can be operated continuously on a sealed-off sample of gas. Furthermore, these gases offer many possible laser lines. The N_2O and CO lasers also offer many useful laser wavelengths, but they cannot be operated sealed off. N_2O and CO lasers require a continuous supply of fresh laser gas that limits the possibilities of long-term operation.

Carbon dioxide can supply any one of hundreds of different lines between about 9 and 12-μ wavelength. A different family of laser lines is available from each isotopic species of molecule. Emission lines available from the common isotope $^{12}C^{16}O_2$ are marked on spectra in Section III. Lines available from various isotopic species are shown in Figure 54 which is reproduced from a paper by Jacobs, Morgan, Snowman, and Ware (43). The amount of isotopic gas required for laser operation is very small; thus cost would not be an obstacle.

The iodine laser reported by Kim *et al.* (44) emits three useful lines at 3.4, 4.8, and 5.5 μ which fall respectively, within the CH, CO, and NO infrared bands. These also are indicated in spectra of Section III.

The rare gases emit hundreds of lines in the infrared with wide variations of wavelengths and intensities. An extensive table of these lines has been presented in a recent article by Patel (41).

To select a desired laser wavelength, one must introduce into the optical cavity a source of loss which varies with wavelength. A prism between the laser mirrors is a simple example. It refracts from the optical path all the emission wavelengths except for a narrow band centered about the desired lasing wavelength. Substituting a diffraction grating for one of the laser mirrors is another way of introducing wavelength selective loss. As the grating is turned, the laser can be tuned over a wide frequency range. If the wavelength selection method is to permit the maintenance of a good laser power output, no loss should be introduced at the desired lasing wavelength. A method of wavelength control used in NASA work is to put into the optical cavity a sample of absorbing gas, altering the gain curve without absorbing appreciably at the desired wavelength (45).

The physical basis for the shifting of the CO_2 laser emission by absorbing gases is illustrated in Figure 55. The tendency of the laser gas

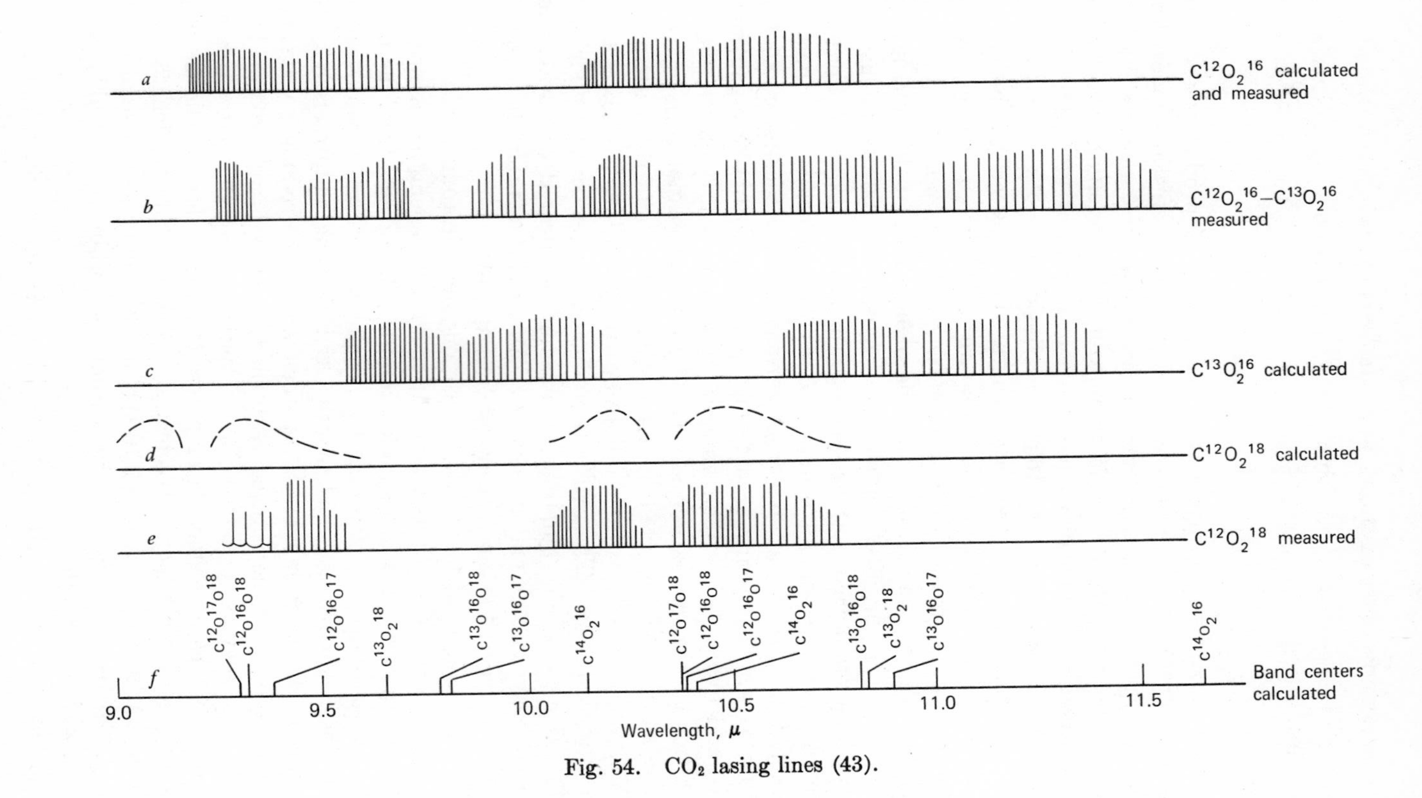

Fig. 54. CO_2 lasing lines (43).

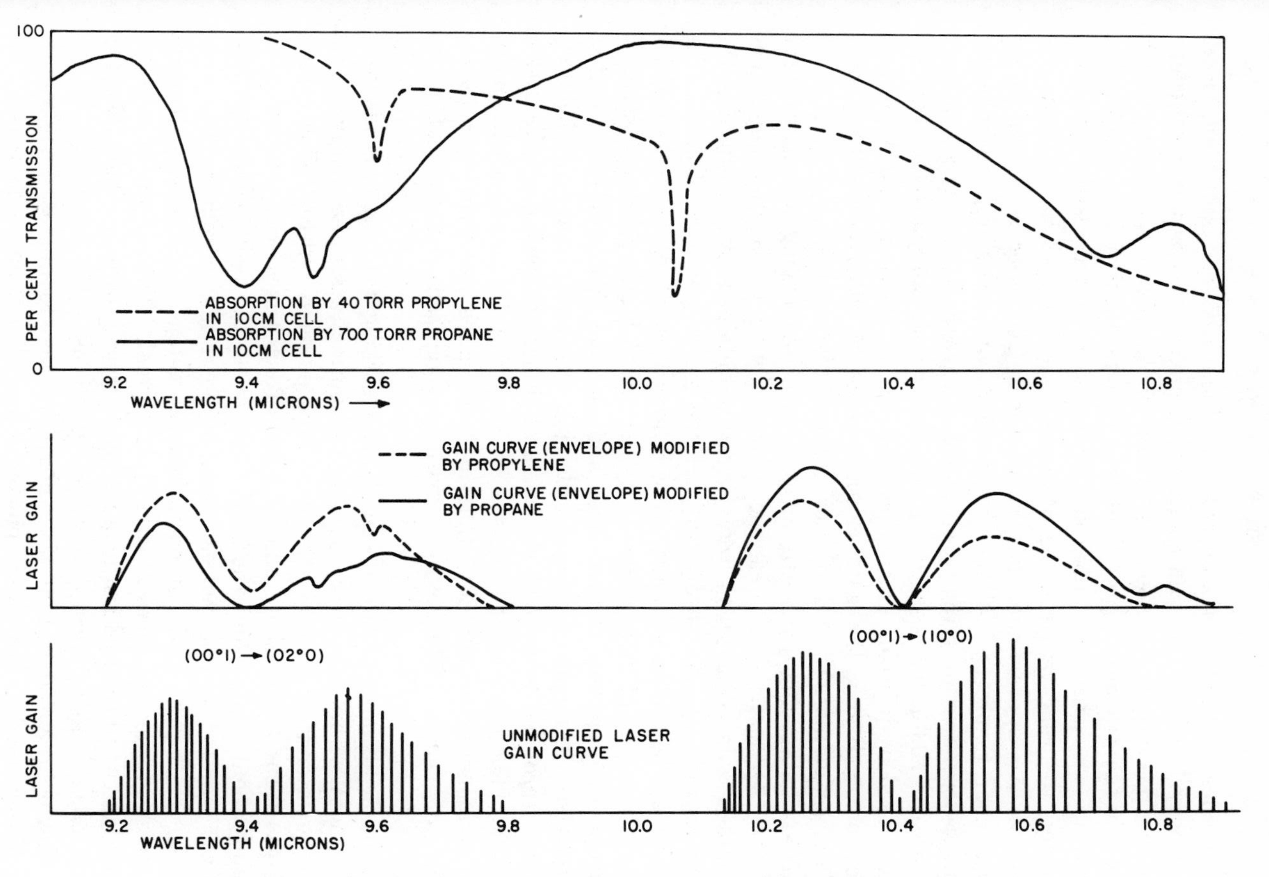

Fig. 55. Laser gain curves and absorption spectra of wavelength control gases.

to produce stimulated emission of radiation, called the laser gain, varies with wavelength. The gain characteristics of the lasing medium depend on the population distributions in the upper and lower energy levels and on the transition probabilities. The gain curve is similar to an absorption spectrum but does not follow its intensity distribution completely, the relative strengths of the 9.4- and 10.4-μ bands being reversed. An approximate gain curve for a CO_2 laser is shown at the bottom of Figure 55. If the cavity losses are not significantly wavelength-dependent over the spectral region concerned, the lines of highest gain near 10.6-μ wavelength are the ones which are emitted. The loss introduced by a prism, grating, or filter within the cavity can be visualized as modifying the gain characteristics so that the emitting lines of highest gain are shifted to a new region of the spectrum.

Many absorbing gases are available for modifying the gain curve of the CO_2 laser. They can be used singly or in combinations. The modified gain at a given wavelength is obtained by multiplying the original gain by the absorptivity of the gas at that wavelength. The gas should be chosen so that there will be as little absorptivity as possible at the new line of highest gain. At the top of Figure 55, the absorption spectra of propylene (dotted line) and propane (solid line) are plotted. Propylene is a good choice for driving the lasing down to the 9.2-μ region; propane is a good choice for holding the lasing to the 10.2-μ region. The modified gain curves are shown in the middle of the figure. It is found experimentally, that in these cases the laser runs at the new wavelength of highest gain without significant reduction of output power.

The inclusion of a second sample of absorbing gas in a separate compartment within the laser converts the normally steady energy output of the laser into a rapid stream of high-powered pulses. This sample of gas turns the laser on and off some 50,000 times/sec; the "on" time being about 1 μsec and the "off" time about 20 μsec. Since the total power emitted in the pulsed mode of operation is roughly the same as in the steady mode of operation, the intensity of radiation during a pulse is some 20 times greater than normal. The output of a pulsed CO_2 laser is illustrated in the oscilloscope trace shown in Figure 56. Details on the pulsing recently have been published (46).

For pollutant detection it is not necessary to operate the laser in a pulsed mode. However, pulsing does offer some operational advantages, such as permitting a better discrimination of the laser radiation from interfering environmental radiation and permitting a time code distinction between the laser line to be absorbed and its neighboring reference line.

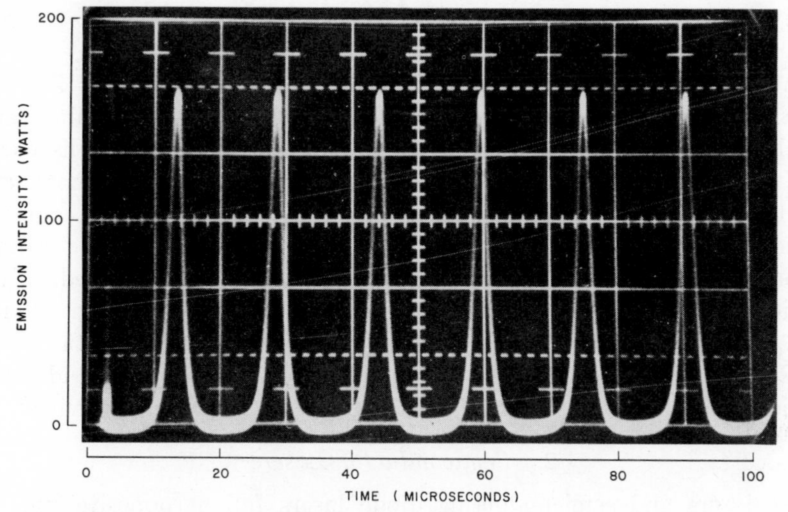

Fig. 56. Oscilloscope trace of the output of a pulsed CO_2 laser.

A portable CO_2 laser that has been built for atmospheric studies is sketched in Figure 57. The construction is all glass except for the windows and the electrodes. The laser has ground glass ball joints for alignment and glass stopcocks for admitting gases. The laser tube is surrounded by a sizable gas reservoir to allow long-term operation on one gas filling. Two gas compartments are built into the laser—one for the wavelength shifting gas and one for the gas which causes repetitive pulsing. If pulsing is not desired, the second gas compartment can be removed.

The iodine and rare gas lasers can be operated in the same apparatus as the CO_2 laser. Windows, electrodes, glass tubing, mirrors, and power supply all can be the same. Only the filling gas is different. Iodine vapor is supplied by a small amount of crystalline iodine placed in the discharge tube. Pressure of a few torrs of helium is also required

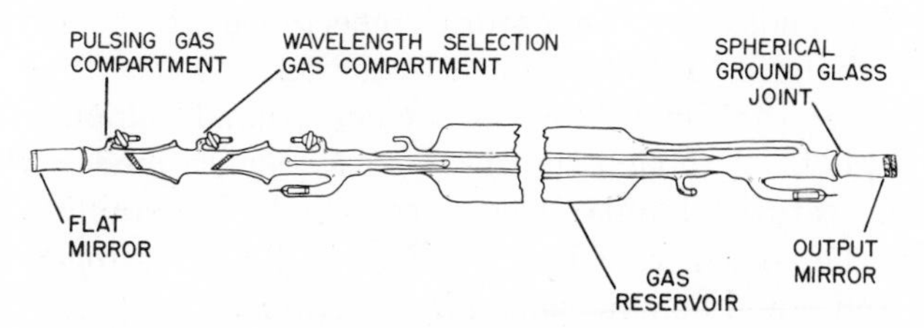

Fig. 57. Laser for pollutant detection—operable on CO_2, rare gases, or iodine.

to give maximum power output. The rare gas lasers operate at gas pressures usually less than 1 torr (40).

The atomic gas lasers have the virtue of being susceptible to Zeeman effect tuning. This class of laser includes the rare gas, iodine, and metallic vapor lasers. With a magnet coil surrounding the discharge tube, the emission lines generally are split into two components, one of which moves to higher frequency and the other to lower frequency. These two components can be separated because of their different polarizations, and the differences in their absorption can be observed. This application of the Zeeman effect is discussed in some detail in Section III. If initial expectations are borne out in the developmental work, the Zeeman-effect splitting of the emission lines will extend considerably the application of the atomic gas lasers to pollution measurement.

2. *Semiconductor Lasers*

Gas lasers and semiconducting diode lasers fall at opposite ends of the range of laser sizes, the one being about a thousand times longer than the other. Nevertheless, the two types of lasers have common properties applicable to the pollution measurement problem. They both are electronically pumped, for example, and they both emit a wide range of wavelengths in the infrared. The diode laser is a single semiconducting crystal which has been processed in such a way that one side of the crystal is an n-type semiconductor in which electrons carry the current and the other side is a p-type semiconductor in which holes carry the current. The electrons are in a higher energy state than the holes, and at the interface, or p-n junction, recombination of electrons and holes takes place with the emission of light. When the electrons are driven across the junction by an external field, a strong population inversion is set up, and conditions are conducive to an efficient laser action.

A schematic diagram of a diode laser is given in Figure 58. Typically, the diode is 1 or 2 mm long and a few tenths of a millimeter wide and high, as indicated by the relative size of the connecting leads in the drawing. The ends of the crystal are cleaved parallel to form the Fabry-Perot resonant cavity. There may be a reflecting coating deposited on the ends, but the natural reflectance at the crystal faces can give enough optical feedback for laser action.

The cooling of the diode lasers is an operational complication which must be accepted. Although the shorter wavelength diodes will operate at room temperature, liquid nitrogen cooling is preferable. For diode operation in the longer wavelength infrared, liquid nitrogen cooling is mandatory, and liquid helium cooling is preferable.

Since the lasing takes place in a volume at the p-n junction that

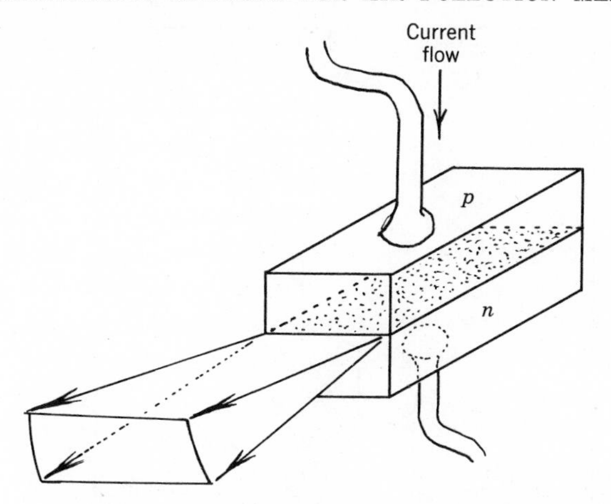

Fig. 58. Diode laser.

is only a few microns deep, diffraction effects spread the emerging laser beam into a fan. The fan is shown in Figure 58 with a typical spread of 15 deg in one direction and 5 deg in the other. If the emission is to be passed through a long-path cell, the beam should be collimated by an appropriately chosen mirror which may be cylindrical, spherical, or toroidal.

Of special interest in air pollution studies are the lead-tin-tellurium diodes recently developed at the M.I.T. Lincoln Laboratories (47,31). Laser action at any wavelength between 6 and 32 μ can be obtained from these diodes by proper choice of composition. The stoichiometric composition $Pb_{0.88}Sn_{0.12}Te$ emits at 10.6 μ, and the composition $Pb_{0.77}Sn_{0.23}Te$ emits at 28 μ. A given diode can be tuned in a discontinuous manner over a range of several wave numbers by changing the current through the crystal. Completely continuous tuning actually takes place for several tenths of a wave number, and then a jump occurs of perhaps half a wave number to the next continuous tuning range. This is illustrated in Figure 59. The continuous tuning is due to a change in the laser resonant frequency because the crystal temperature goes up with increasing current. The discontinuity results from a laser mode change. The most promising application of the diodes is in "infinite resolution infrared spectroscopy" (Section III).

3. *Dye Lasers*

The organic dye laser is a relative newcomer to the laser field, but it occupies a unique place by being tunable to any wavelength in the

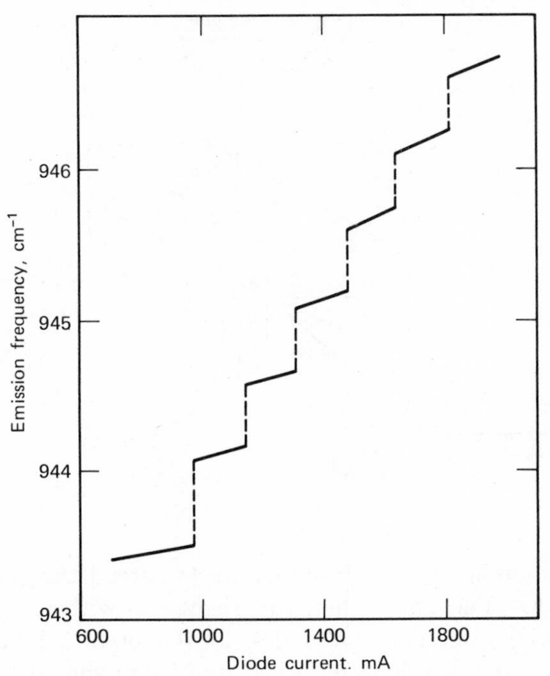

Fig. 59. Tuning of $Pb_{0.88}Sn_{0.12}Te$ diode laser (31).

visible and near ultraviolet. An additional advantage of the dyes is that the laser medium is inexpensive and unbreakable. This contrasts to the ruby with the risk of shattering a crystal worth thousands of dollars.

Furumoto and Ceccon have developed a technique of pumping the dye lasers with a cylindrical shell flashtube that allows production of megawatts of power in microsecond long laser pulses (48). Typical dyes used are Rhodamine and Coumarin. These are dissolved in alcohol, and the solution is placed inside a cell that is surrounded by the xenon-filled flash lamp. When a high-voltage capacitor is suddenly discharged through the xenon lamp, the emission intensity is so great that the absorbing ground electronic state of the dye becomes depleted. A population inversion then exists, and laser emission can take place. With broad-band reflectors at the ends of the laser cavity, the wavelengths of laser emission may cover a span from 200 to 300 Å. This broad-band emission is characteristic of the dye lasers. It results from a high transition probability for laser emission over a broad spectral region. If one of the laser mirrors is replaced by a diffraction grating, the laser emission spectrum narrows to a fraction of an angstrom at the wavelength for which the grating is set. As the grating is turned, the dye laser is then tuned to new wavelengths. A coupling-out device suggested by Hard permits the energy to be removed from the dye laser cavity in a fixed direction regardless of the angular adjustment of the grating (49). The apparatus is drawn in Figure 60. With each dye having a tuning range of several hundred angstroms and there being many dyes from which to choose, laser action can be obtained at any wavelengths from the near ultraviolet to the infrared.

The most obvious application of the dye laser to atmospheric problems is in resonance scattering experiments. When the laser is tuned to an atomic resonance line such as the sodium D line at 5890 Å, a strong fluorescence may be observed from resonant material which is in the path of the laser pulse. The intensity of such resonance will be many times greater than the intensities of other types of scattering such as Rayleigh and Raman scattering. An application of the dye laser to the study of atmospheric sodium has been reported by Bowman, Gibson, and Sandford (50). Additional resonance scattering experiments are known to be in progress at this time.

4. *Optically Pumped Crystal Lasers*

The ruby laser, the first laser discovered, has been applied extensively in atmospheric probing. The backscattering of intense bursts of ruby

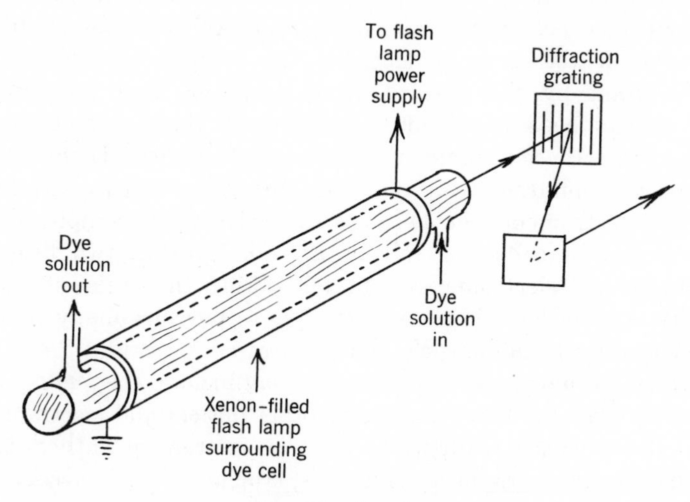

Fig. 60. Dye laser.

light has been used to measure dusts and haze layers in the atmosphere, as well as to measure the distance to the moon. Lasers which employ neodymium-doped crystals or glasses also are being applied in meteorological studies and may surpass the ruby in performance.

The crystal is excited to population inversion by very intense flash lamps, usually xenon filled, which are placed with the crystal in an elliptical reflector. Pulses shorter than a nanosecond (10^{-9} sec) with power higher than a gigawatt (10^9 W) can be obtained.

Although very useful in meteorological studies, optically pumped crystal lasers do not yet have any extensive application to the detection of gaseous air pollutants. However, there is optimism that the laser pulses will detect gaseous pollutants by means of Raman scattering. The ruby laser has been used in Raman experiments in which the ruby light was doubled in frequency to the ultraviolet by means of a nonlinear crystal (51). As the ultraviolet pulse propagated up through the atmosphere, Raman-shifted backscattered radiation due to nitrogen, oxygen, and water vapor was observed. Now that the technique of observing Raman backscatter has been demonstrated for the major constituents of the atmosphere, work is in progress to extend it to the minor constituents. The sensitivity of the method must be raised by several orders of magnitude before it becomes applicable to the ambient air pollution.

VI. DETECTION INSTRUMENTATION

At the detection end of the optical system, the information contained in the spectrum must be extracted in the most efficient manner to fulfill the objective of the measurement. If the objective is to monitor a specific pollutant by the absorption of a chosen laser line, then the detection system need be no more complicated than a combination of collecting mirror, thermocouple, amplifier, and recorder. If the objective is to monitor a pollutant by comparing absorption at two wavelengths, the detecting instrumentation becomes somewhat more complicated, and might consist of a collecting mirror, a beam splitter, two filters, two thermocouples, an electronic system for taking the ratio of voltages, and, finally, a recorder. If the objective is to measure many pollutants simultaneously by examining the entire infrared spectrum, then the detection system becomes very complex and may include choppers, gratings, slits, mirrors, beam splitters, and many other optical and electronic components. The efficiency with which such an instrument gathers and examines the available radiation may be crucial to the success of the measurements.

Factors to be considered in evaluating the performance of spectroscopic instruments include the following. (*1*) *Discriminating power.* Can the instrument measure a chosen pollutant in gas mixtures containing many other pollutants? (*2*) *Information content of output.* With what accuracy and degree of assurance does the instrument provide concentration measurements and on how many different pollutants? (*3*) *Speed.* How long does it take to make the chosen measurements? (*4*) *Versatility.* Can speed be exchanged for greater information content? Can continuous monitoring of a single pollutant be substituted for periodic measurement of many pollutants? (*5*) *Reliability.* Is the instrument subject to misalignment because of thermal or mechanical stresses? Are there moving parts that may deteriorate? Are there optical surfaces that are subject to deterioration because of reactions with pollutants, water vapor, or oxygen in the atmosphere? (*6*) *Size, cost, and convenience.* How big is the instrument? How much does it cost? Does it require frequent maintenance? Is the output easily interpreted?

If an instrument rates high in one of the six factors, it may necessarily rate low in another. For example, an instrument whose output has a high information content may not be convenient for operation in a routine manner and may require careful data interpretation. Discriminating power, information content, and speed are interdependent factors. One may be substituted in favor of another. To a large extent these three factors derive from the instrumental properties of throughput, resolution, and multiplexing.

Throughput is a measure of the degree to which the instrument conserves energy. A radiometer has very high throughput, since nearly all of the collected energy is focused onto the sensing element. A scanning spectrophotometer of the conventional type has a low throughput when used for high resolution work because of the narrow entrance slit which blocks out most of the incident light.

Resolution is the ability of the instrument to recognize small details in the spectrum. The greater the resolution, the higher the information content of the output. In spectrophotometers resolution generally can be exchanged for scanning speed. In order to record all the information in the spectrum of a gas at atmospheric pressure, an instrument should have resolving power of 0.1 cm^{-1} or better; that is it should be able to distinguish intensity differences between points in the spectrum only 0.1 cm^{-1} apart. This degree of resolution is achieved only in laboratory research instruments.

Multiplexing is the simultaneous observation of multiple spectral resolution elements. At a fixed resolution, multiplexing permits faster data acquisition. A radiometer observes a wide region of the spectrum at all times, but it is not a multiplexing instrument because it has only one element of spectral resolution. However, a photographic spectrograph is a multiplexing instrument because many elements of spectral resolution are simultaneously recorded.

A. Radiometers

A radiometer measures intensity of electromagnetic radiation in some spectral interval—frequently a wide spectral interval. Radiometers are used mainly for remote temperature measurements. They give a brightness temperature, which requires a knowledge of emissivity in order to be converted to a true temperature. Basically, a radiometer consists of a collection mirror, a filter, a chopper, a detector, and a calibration source. For operation in several spectral regions, various filters may be used. A schematic diagram of a filter-wheel radiometer is shown in Figure 61. As the wheel rotates, sequential measurements in the various spectral regions are obtained. Details on the design and performance of radiometric systems may be found in several of the excellent books available on infrared technology (52,53).

One of the most noteworthy remote sensing experiments to date was the filter-wheel radiometer experiment performed on the Mariner spacecraft which flew past Mars in July 1969. This experiment employed a wedged interference filter which yielded a spectral resolution ($\lambda/\Delta\lambda$) between 100 and 200 over the spectral region 1.9 to 14.3 μ. A spectral scan took 10 sec and every twelfth spectrum was recorded through a

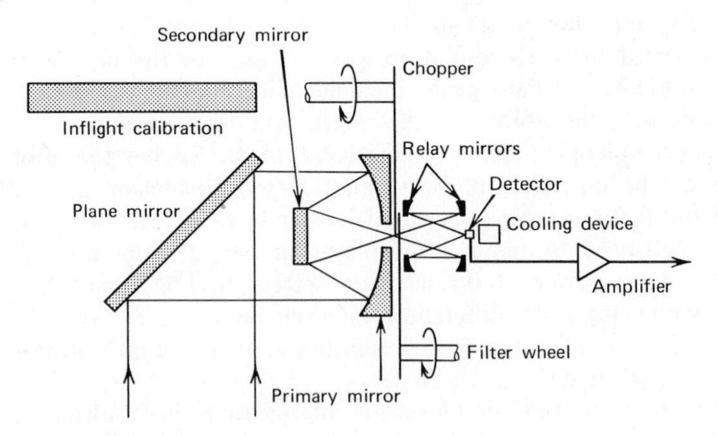

Fig. 61. Schematic diagram of a filter-wheel radiometer.

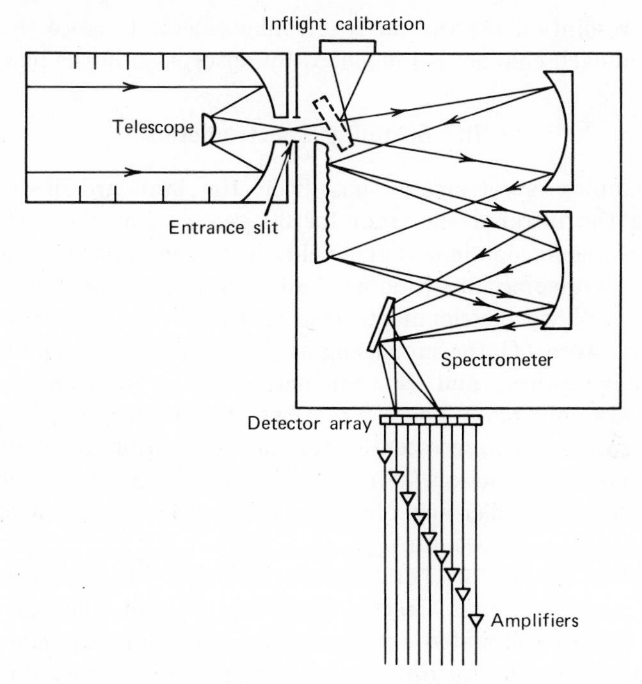

Fig. 62. Schematic diagram of a polychromator radiometer.

polystyrene film for calibration. Lead selenide and mercury-doped germanium detectors were used. Results of the experiment reported by Herr and Pimentel have included the detection of solid carbon dioxide in the upper atmosphere of the planet (68,69).

A second form of the multichannel radiometer is the polychromator-radiometer diagramed in Figure 62. This system uses a dispersing element and a bank of detectors in the focal plane. The obvious advantage is that all channels are observed simultaneously, and the integration time for each channel is increased over that of sequential observation. The multiplexing property of the polychromator-radiometer leads to a higher signal-to-noise ratio in remote sensing applications than the filter-wheel radiometer, but a heavy penalty is paid in instrument complexity and weight. The operational polychromator-radiometer called "Satellite Infrared Spectrometer (SIRS)" used in the TIROS and NIMBUS spacecraft weighed approximately 90 lb, which is unusually heavy for a space-flight application (54). When a multichannel radiometer is used to view a source of known emissive properties, it serves as a low resolution absorption spectrophotometer because the intensity observed in each channel is influenced by absorption in the light path.

B. Scanning Spectrometers

The scanning spectrometer has been the standard instrument for measuring the infrared spectrum for 30 years. For the first 20 years, prism instruments dominated the field; in the last 10 years grating instruments have come into wide use and have largely displaced the prism instruments. Two developments leading to the use of infrared grating instruments were *(1)* the increasing availability of high quality gratings at reasonable prices, and *(2)* the perfecting of long-wavelength pass filters for the infrared. These filters, which are based on the new semiconducting materials, are essential for the elimination of the overlapping higher orders of diffraction. Grating-filter combinations are now readily available for high efficiency operation in the first order in any chosen part of the infrared spectrum.

Most of the new infrared grating instruments are of the Czerny-Turner type. In the single-pass Czerny-Turner instrument, the light from the entrance slit is collimated by the first mirror and directed onto the grating, after which the diffracted light goes to a second mirror and onto the exit slit. The scan is achieved by rotating the grating. In the double-pass instrument the diffracted light is not immediately focused onto the exit slit but is sent through a chopper to a mirror and back to the collimator. The chopped light goes to the grating for a second diffraction and the diffracted chopped light is then focused onto the exit slit. The double-passed instrument has a higher degree of scattered light rejection and higher resolution than the single-passed instrument. These instruments are easy to operate and can be built from readily available components. A schematic diagram of a scanning single-pass spectrometer as it might be devised for flight applications is shown in Figure 63.

When there is ample source energy available, the scanning spectrometer seems to be the best choice for a spectroscopic experiment. In the recent outstanding work of Murcray *et al.* a 0.50-m scanning Czerny-Turner grating spectrometer was carried aloft on a balloon for observation of the solar spectrum (55). Resolution on the order of one-half wave number was obtained.

In applications where only a limited amount of source energy is available, the scanning spectrometer is usually not the best choice because the entrance slit gives it a low throughput, and the exit slit allows the detector to see only one resolution element at a time. Such low energy

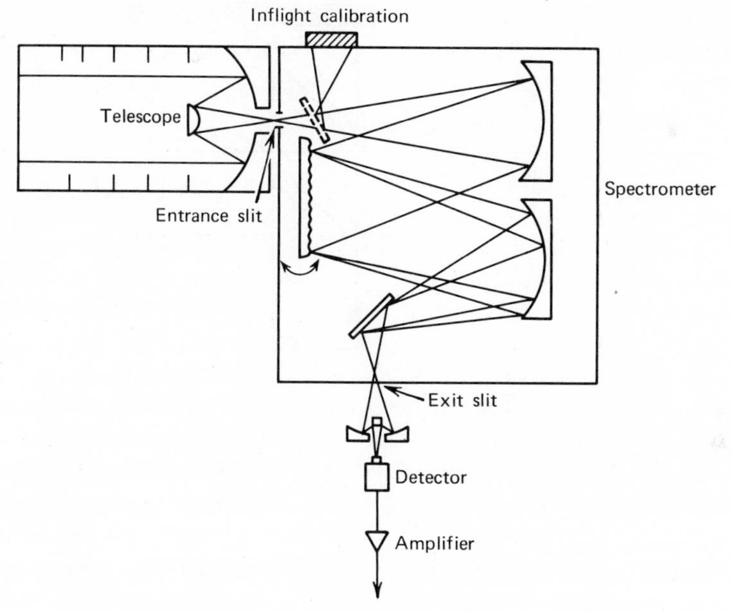

Fig. 63. Schematic diagram of a scanning spectrometer.

applications include long-path spectroscopy using artificial sources and remote sensing applications that do not use the sunlight directly. For these applications the several types of multiplexing instruments offer significant advantages.

C. Interferometers

The Michelson interferometer spectrometer is probably the most versatile of the multiplexing instruments. It can deliver as much spectral information as the scanning grating spectrophotometer, but in a much shorter length of time. The greater speed results from a larger throughput and a high degree of multiplexing.

Figure 64 is a schematic diagram of the instrument, showing source, collimating and focusing lenses, beam splitter, mirrors, detector, and a compensator plate to make the optical paths of the reflected and transmitted beams equivalent. During the travel of the movable mirror, the two beams interfere with one another at the detector and produce fluctuations in intensity which are characteristic of the spectrum of the incident radiation. The detector output is actually the Fourier transform of the spectrum and is called the interferogram. At any mirror

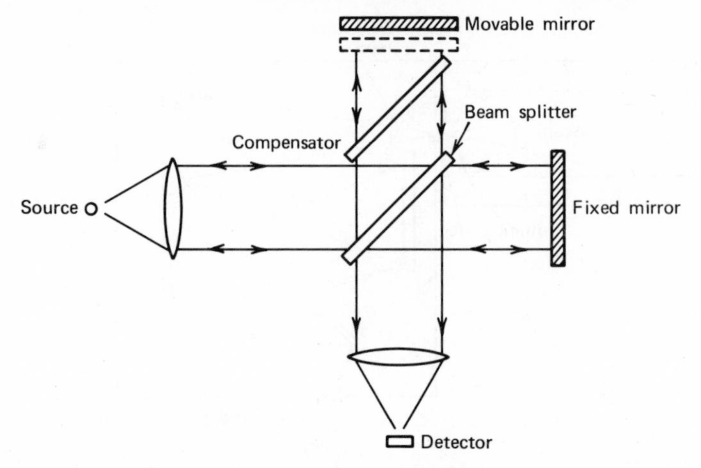

Fig. 64. Schematic diagram of a Michelson interferometer.

displacement, x, the observed intensity $I(x)$ will be given by the expression

$$I(x) = B_\nu \cos(2\pi\nu x)$$

where ν is the wave number and B_ν is the spectrum. The presence of the function B_ν in the expression for intensity at any mirror displacement, X, is indicative of the multiplexing property of the interferometer spectrometer; all frequencies contribute to the signal at all times. If the source is monochromatic, then B_ν is zero at all but one value of ν, and intensity as a function of mirror displacement is a pure cosine function. For sources which are not monochromatic the intensity at any given mirror displacement is given by the integral over the whole spectrum.

$$I(x) = \int_{\nu_1}^{\nu_2} B_\nu \cos(2\pi\nu x)\, d\nu$$

The intensity variation, $I(x)$ as a function of mirror displacement, x, is the interferogram. It contains each spectral resolution element $B_{\nu i}$ coded according to its individual cosine function, $\cos(2\pi\nu_i x)$. The whole spectrum, B_ν, is obtained by taking the Fourier transform of the interferogram by digital or analog computation:

$$B_\nu = \int_0^{x_{max}} I(x) \cos(2\pi\nu x)\, dx$$

The resolving power of an interferometer depends upon the number of sampling points and upon the accuracy with which the sampling points are measured. The maximum possible resolving power depends upon the travel distance of the movable mirror. This maximum resolving power, R_ν, at some point in the spectrum is the wave number, ν, divided by the wave number width of the resolution element, $\delta\nu$. This is equal to the number of half-wavelengths which fit into the maximum path difference, x_{max}.

$$R_\nu = \frac{\nu}{\delta\nu} = \frac{x_{max}}{(\lambda/2)}$$

This limitation on resolving power is analogous to the grating width limitation on the resolving power of a grating spectrophotometer. A 4-in. travel in an interferometer, for example, gives the same theoretical resolution limit as a 4-in.-wide diffraction grating in a spectrophotometer. In both cases the amount of information actually extracted from the spectrum depends on the number of sampling points.

The extent of the throughput advantage of the interferometer has been a somewhat controversial subject among spectroscopists. There definitely appears to be a throughput advantage, but it may not be as great as was originally claimed. Throughput is a function of the instrumental resolution and the instrument size. It is stated by Vanasse (56) that in considering instruments of comparable size and comparable transmission factors, a Michelson interferometer has a throughput energy gain of 100 over a grating spectrophotometer. However, he also indicates that a spectrophotometer is generally a larger instrument than is an interferometer and the spectrophotometer is likely to have a larger transmission factor. It may be that the typical grating spectrophotometer and the typical Michelson interferometer will have roughly equal throughputs under conditions of equal resolution. If that is the case, the interferometer still has the advantage of smaller size.

The multiplexing advantage of the interferometer results in a higher detector signal-to-noise ratio at a given instrumental scanning speed. Conversely, for a given signal-to-noise ratio the spectrum can be scanned faster. The signal-to-noise gain over an instrument which observes one resolution element at a time is equal to the square root of the number of resolution elements which are simultaneously observed. If the spectrum between 500 cm^{-1} and 1500 cm^{-1} is to be covered with a resolution of 0.1 cm^{-1}, it is necessary to record 10,000 resolution elements, and the theoretically possible speed gain of the interferometer over a spectrophotometer is a factor of 100.

The interferometer's performance is limited mainly by the mirror control system. The movable mirror must remain perpendicular to the beam and must move with a high degree of precision and reproducibility. One successful method of controlling the movement of the mirror is to introduce a laser beam into the interferometer and to monitor the moving cosine wave which is the interferogram of the laser line.

The data manipulation required to obtain the spectrum from the interferogram is also a limitation, although not a fundamental one. Spectroscopists are used to seeing the results of their measurements traced out in real time. The extra computational steps involved in obtaining the Fourier transform are an inconvenience as well as an extra expense. Interferometers now are available with built-in computers, which eliminate the delay in obtaining the spectrum, but considerably increase the price of the instrument.

D. Optical Correlation Instruments—Dispersive Type

The optical correlation instruments achieve multiplexing by the use of filters whose transmission characteristics match the spectrum of the compound to be detected. When the matched filter is placed in the focal surface of a spectrometer, the instrument is an optical correlation instrument of the dispersive type, also called a dispersive matched-filter spectrometer. The filter may be compared to a comb with irregular teeth in which the spaces between the teeth match the spectral lines. The filter searches the spectrum for those features characteristic of the gas to be detected. If the filter matches the desired spectrum very closely, the selectivity of the instrument will be high; the desired spectrum will be selected from a complex mixture of spectra, and the remainder will be rejected. The more structure there is in the spectrum being sought, the more effective the matched filter will be. These instruments normally are designed to detect a single compound, so the information content of their output must be classified as low. However, their degree of detection sensitivity and selectivity is high, even when working with weakly radiating sources.

A dispersive optical correlation spectrometer for use with a balloon-borne telescope has been described by Strong (57). The matched filter was designed for water vapor detection in planetary spectra. It consisted of 21 slots cut in a metal plate at positions corresponding to 21 water vapor lines in the 1.13-μ water band. Barringer and Schock have described a matched-filter correlation spectrometer for remote detection of iodine vapor (58). Correlation instruments also are described by Williams and Kolitz (59). An SO_2 measuring instrument of the dispersive optical correlation type is commercially available (60). In these instruments the matched filters are photographic reproductions of the spectra of the compounds of interest.

The correlation, or lack of correlation, between the spectrum and the filter is determined by oscillating the spectrum on the filter while all the light transmitted by the filter is being focused on a single detector. If correlation exists, the transmitted light will fluctuate in intensity at the oscillation frequency. The amount of fluctuation will depend on the strength of the correlated absorption lines. Alternatively, the spectrum can remain stationary and the matched filter can be oscillated.

The matched filter gives a higher signal-to-noise level than does a single-exit slit because of the multiplexing. Nevertheless, in many applications additional signal gain is needed. It may be sought at the input end of the spectrometer by using multiple-entrance slits. The multiple-entrance slits produce overlapping multiple spectra. The filter must be matched to the overlapping spectra for the throughput gain of the multiple slits to be realized and for the selectivity of the method to remain high. A multiple-slit matched-filter spectrometer is diagramed in Figure 65.

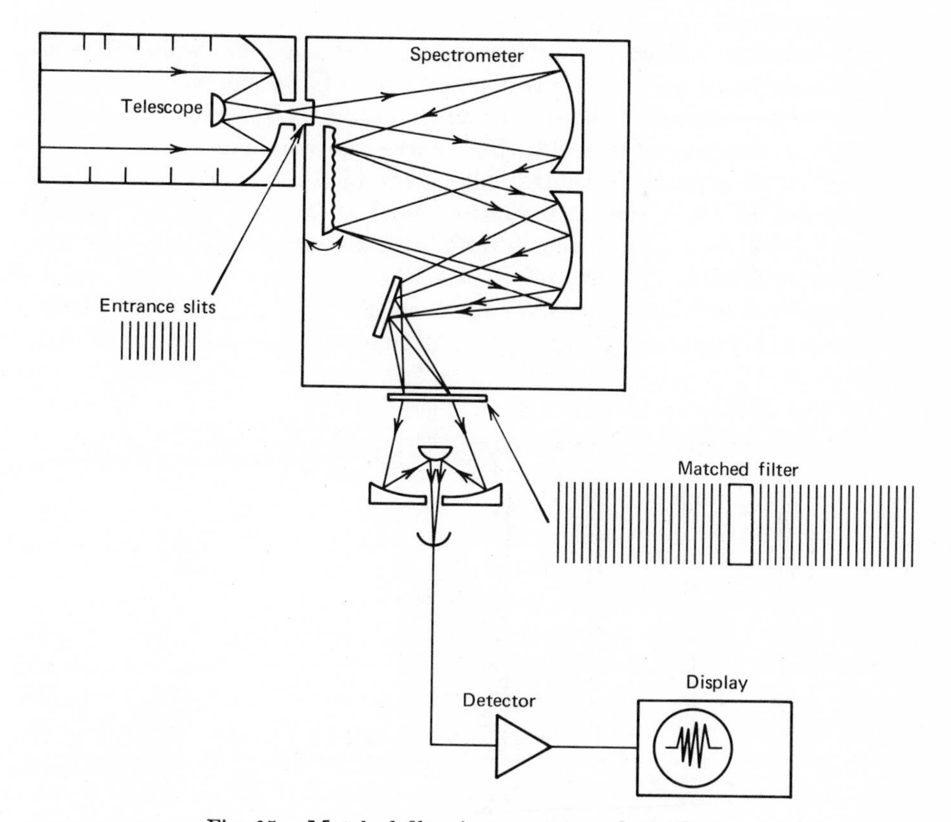

Fig. 65. Matched-filter instrument configuration.

E. Optical Correlation Instruments—Nondispersive Type

The nondispersive type of optical correlation instruments have been known for many years, indicated by an early publication of A. H. Pfund (61). Designed with the same objective as the dispersive type of optical correlation instruments, they monitor the presence of a single compound in a gas mixture. Nondispersive instruments are used widely in gas monitoring in plants and laboratories and now are gaining recognition for use in remote sensing of gases in the atmospheres of the earth and planets (62,11).

The key component of the nondispersive analyzer is a filter whose absorption spectrum closely resembles the spectrum of the compound to be measured. This matched absorption filter searches the non-dispersed incident radiation for the presence of spectral features which correlate with the filter's own absorption characteristics. The filter usually consists of an absorption cell containing a sample of the gas to be measured.

The principle of spectrum matching is shown in Figures 66, 67, and 68. Figure 66 shows the absorption spectrum of two channels of a non-dispersive instrument. The left channel contains a gas filter cell which introduces absorption lines into the source continuum as shown. The right channel contains a neutral filter that has been adjusted to yield the same total absorption as the left channel, as indicated by the shaded area. If the detection system alternates between channels, it will see constant intensity. Figure 67 shows the absorption changes when matching spectral lines appear in the incident radiation. The intensity transmitted by the left channel has not changed appreciably, while the

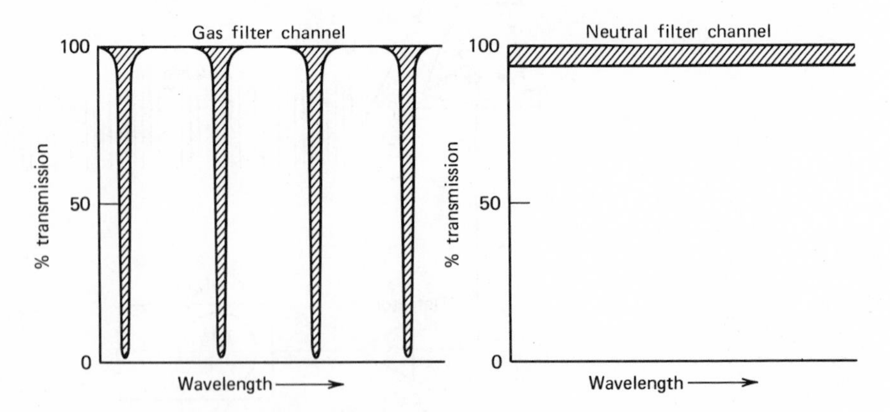

Fig. 66. Balanced absorption in two channels of a nondispersive analyzer.

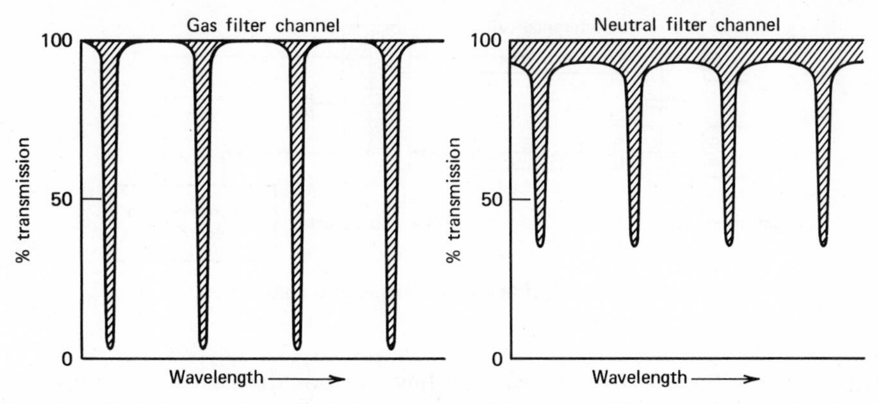

Fig. 67. Unbalance in nondispersive analyzer due to matching spectrum in incident light.

intensity transmitted by the right channel has decreased. The detection system now sees unequal intensities that appear as an alternating signal at the channel switching frequency. Figure 68 shows the intensity changes when the nonmatching spectral lines of a second compound appear in the spectrum. These lines reduce the energy by the same proportion on each side. Since the amount of alternating signal is not changed, the second compound is not seen by the instrument.

Commercial nondispersive infrared analyzers, known as NDIR instruments, have been on the market for many years. Design features and applications of the instruments were discussed by Hill and Powell in

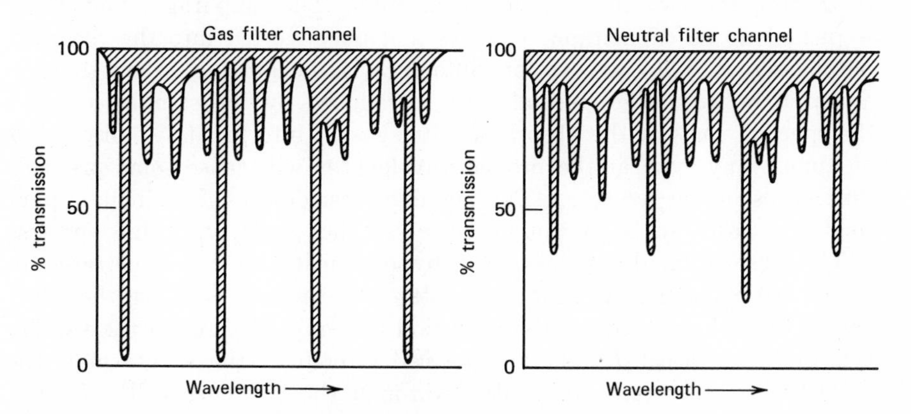

Fig. 68. Retention of state of unbalance in nondispersive analyzer when nonmatching spectrum appears in incident light.

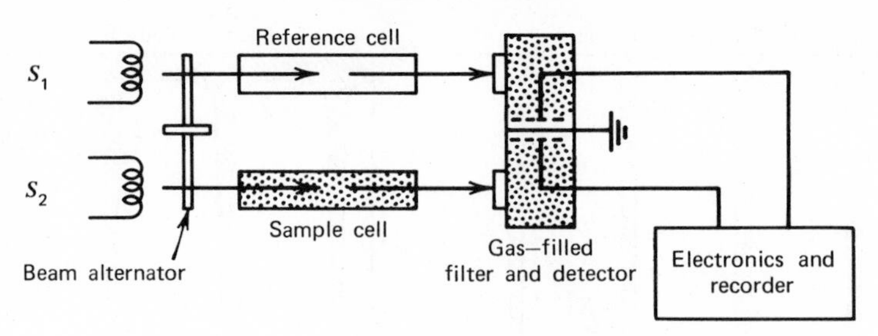

Fig. 69. Luft-type nondispersive analyzer.

their recent book (70). However, they did not discuss in any appreciable detail the reasons for the deficiencies in performance of certain types of NDIR systems, nor did they indicate the directions which future instrument improvements might follow.

The most widely used NDIR systems follow the design described in 1943 by Luft (71). The optical system is outlined in Figure 69. There are two separate sources of radiation with one beam passing through a reference cell and the other through the sample cell. A two-compartment gas-filled detector receives the two beams. The filling is a pure sample of the gaseous species to be measured in the air sample. This gaseous absorption is supposed to give the instrument a degree of selectivity. The gas in the detector can be considered to be the filter gas. As the beam alternator passes first one beam and then the other through the cells and into the receiver compartments, absorption differences in the cells lead to absorption differences in the filter gas. These differences cause pressure differentials that vibrate the thin diaphragm separating the two receiver compartments. The diaphragm movement is picked up as capacitance changes and is converted into the recorded output signal. The absorption difference is the greatest when there is a match between the spectral lines introduced by the sample cell and the spectral lines of the absorbing filter gas. Unfortunately, absorption of almost any type at the proper wavelengths will cause some pressure differences in the receiver. The possible interferences thus include particulates, gases with continuous spectra, and gases with line spectra.

The absorption differences in a hypothetical case are illustrated in Figure 70. The left column shows the absorptivity plot for the filter gas, A, and the spectra of the radiation transmitted through the sample cell in three cases: I_1; the sample cell is empty (the spectrum is the same as the spectrum transmitted through the reference cell); I_2; the sample cell contains the gas to be measured; and I_3; the sample cell

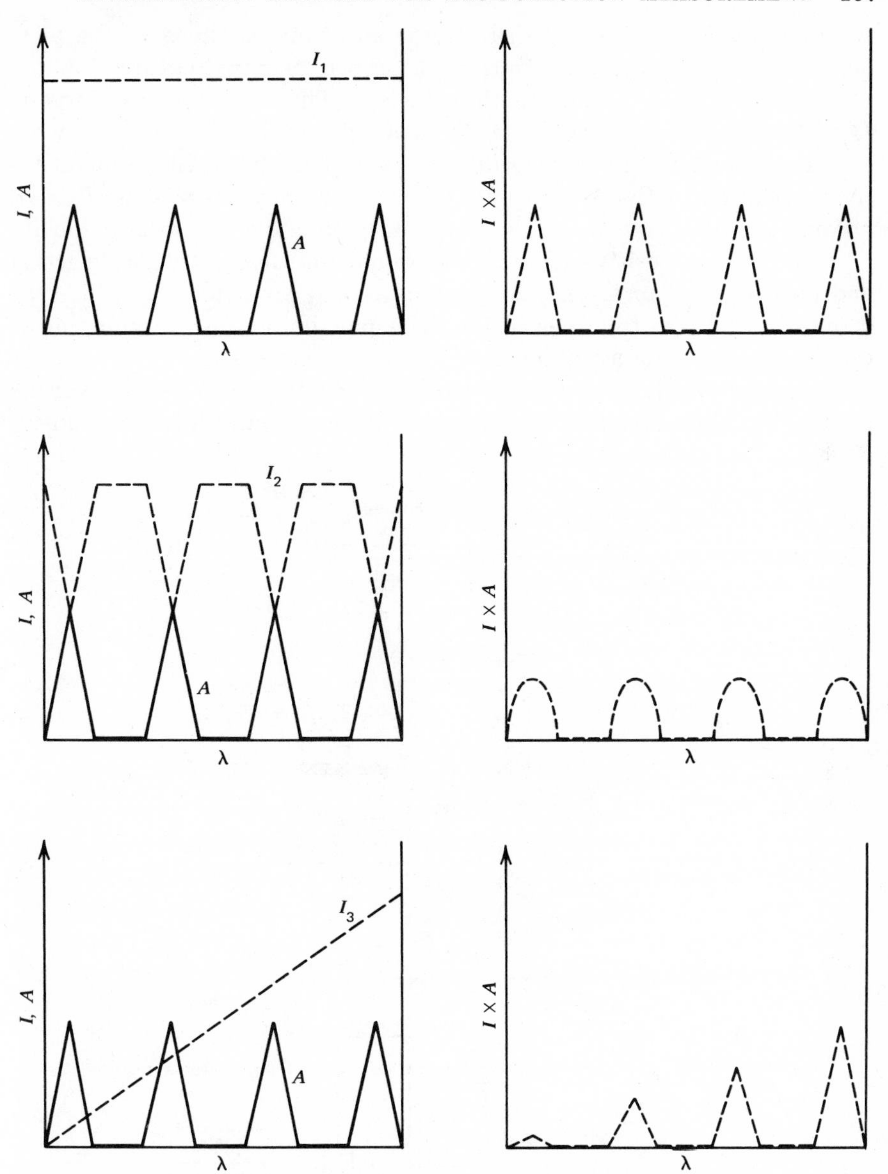

Fig. 70. Operation of Luft-type infrared analyzer.

contains another gas which distorts the spectrum as shown. The plots on the right show the amounts of absorbed energy indicated by the products of I and A for the three cases. The $I_1 \times A$ is the largest; $I_2 \times A$ is smaller and would result in a substantial unbalance between the sample side and reference side of the cell pair, thus giving a quantitative indication of the presence of the gas to be measured. The $I_3 \times A$, however, is also less than $I_1 \times A$, showing that the gas responsible for I_3 would give a positive reading and would interfere with the detection and measurement of the gas to be measured by spectrum I_2. This type of interference has been one of the principal limitations on the application of commercially available nondispersive analyzers.

The poor selectivity of the system described above is due to the crude nature of the correlation operation it performs. The instrument

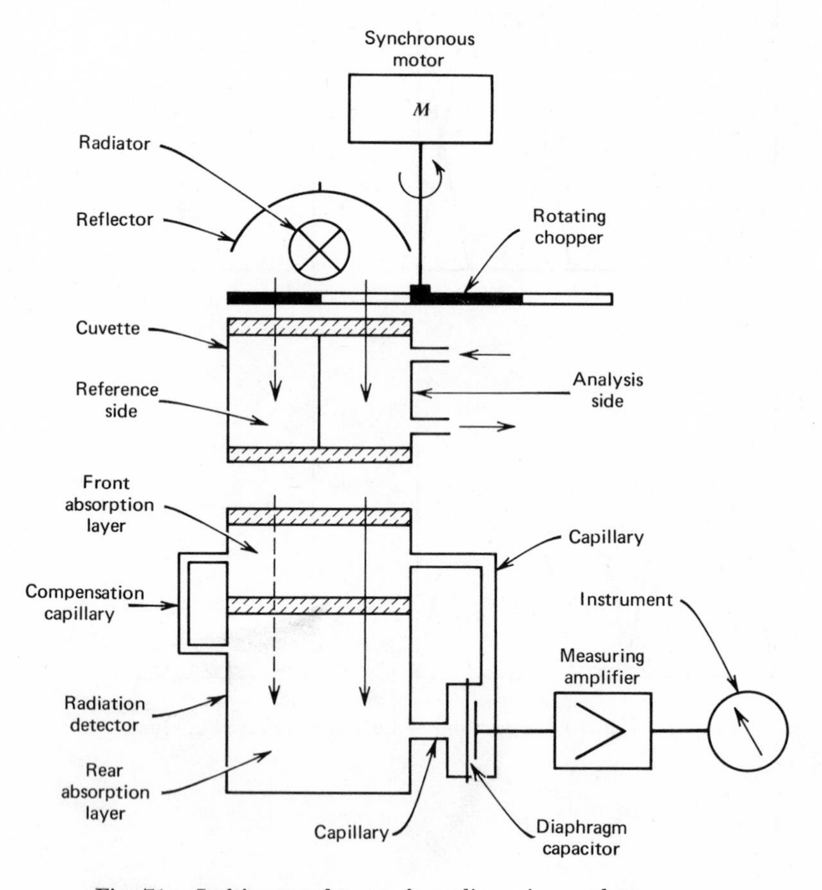

Fig. 71. Luft's second type of nondispersive analyzer.

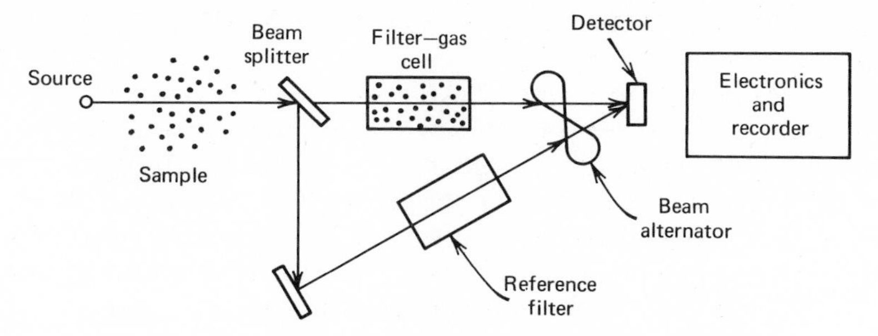

Fig. 72. Cross-correlation type of nondispersive analyzer.

basically operates on only one channel, because the signal produced by the reference channel is always the same.

A substantial improvement in the pneumatic detector was described by Luft, Kesseler, and Zörner in 1967 (72). This improvement is included in the instruments marketed by the H. Maihak Company of Hamburg, Germany. The improvement derives from a reorientation of the two gas-filled absorption compartments. They are now placed in series, instead of side-by-side. This is illustrated in Figure 71. In this arrangement the front gas compartment absorbs radiation at the center of the lines, and the rear compartment absorbs radiation in the wings of the lines. With an incident continuous spectrum the two amounts of absorption can be made equal by adjusting the receiver gas pressure. If matching absorption lines then appear in the incident continuous radiation, absorption differences and pressure differences will occur in the receiver. If a nonmatching spectrum were to appear in the incident light, the absorption change would be roughly the same in each compartment of the receiver, and the nonmatching spectrum would not be seen.

A true cross-correlation type of nondispersive analyzer is diagramed in Figure 72. This system is similar to the so-called negative filter NDIR instrument marketed by Leeds & Northrup Company. Although the system is diagramed with two filters, it could have many. The radiation source may be an artificial source or a natural source such as the sun or the radiating earth. The sample may be dispersed in the open atmosphere, or it may be contained in short- or long-path cells in the laboratory. A closely related detection system called a cross-correlating spectrometer has been described by Goody (62). He proposes an instrument for the study of the planetary atmospheres which uses a pressure-modulated gas absorption cell to perform the same basic cross-correlation operation which is performed by the cell pairs in the instrument described here.

In the instrument of Figure 72, the examination of the spectrum is performed by comparing the amount of radiation transmitted by the filter cell to the amount transmitted by a reference cell containing no filter gas. The transmissions of the two cells are balanced when the incident spectrum is known not to contain absorption or emission lines of the compound to be detected. When the matching spectral lines appear in the spectrum, a transmission difference results. For an appreciable transmission difference to exist, there must be a close spectral match, with the result that the system is much less subject to interference than the early Luft system previously discussed. This immunity to interference is illustrated in Figures 73 and 74. Figure 73 shows the adjustment of the absorption in the reference cell, A_2, to match the absorption in the filter cell, A_1, when a continuous spectrum, I_1 is incident. The amounts of energy absorbed are equal, as given by the areas under the plots of the products $I_1 \times A_1$ and $I_1 \times A_2$, (dotted lines). When the matching absorption, I_2, appears in the incident light, the amounts of signal reduction shown by the plots of $I_2 \times A_1$ and $I_2 \times A_2$

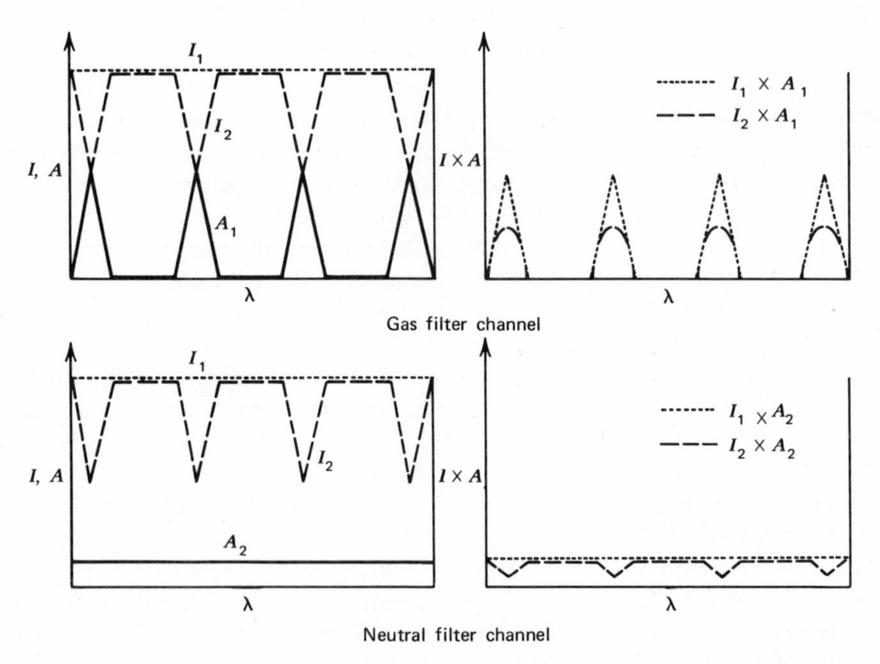

Fig. 73. Response of cross-correlation type of nondispersive analyzer to gas being measured.

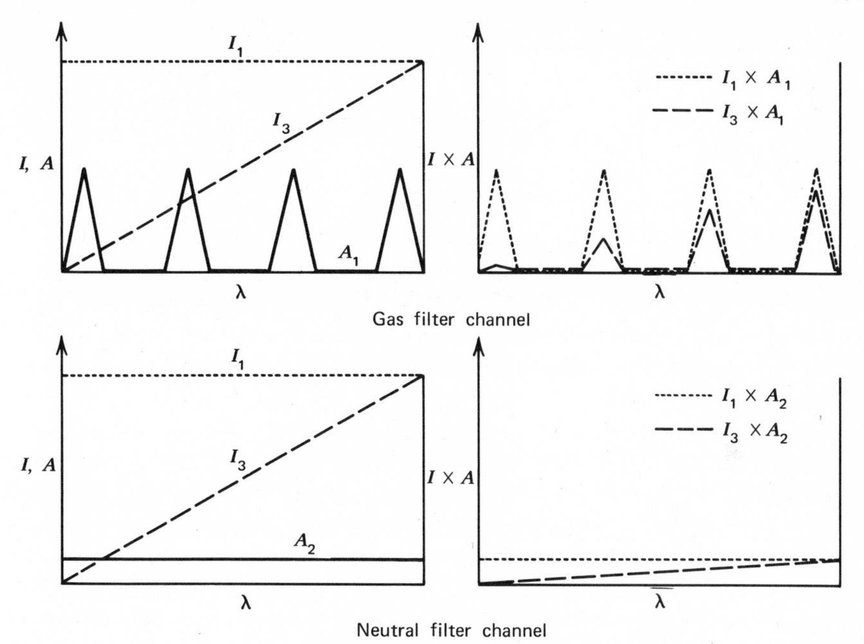

Fig. 74. Response of cross-correlation type of nondispersive analyzer to potential interference.

are different in the two channels. Thus there will be an unbalance and a positive reading. Figure 74 shows the spectra and products when the nonmatching spectrum, I_3, is incident. In this case the areas under the plots of products $I_3 \times A_1$ and $I_3 \times A_2$ are equal, showing that the nonmatching spectrum is not seen.

Laboratory measurements at the Convair Corporation have demonstrated the high degree of sensitivity and selectivity of the cross correlation type of nondispersive analyzer (11). The system derives its sensitivity and selectivity from a large throughput, a high degree of multiplexing, and a very high degree of spectral resolution. In throughput, the system is like a radiometer, there being very little restriction on the entrance aperture or on the acceptance angle. The multiplexing is high because the filter system performs the cross-correlation operation between the incident spectrum and the gas filter spectrum with essentially infinite spectral resolution and with all wavelengths impinging on the detector at all times. The resolution involved in examining the spectrum is limited only by the widths and shapes of the spectral lines of the sample gas and the filter gas.

The resolution involved in the gas filter cross-correlation operation

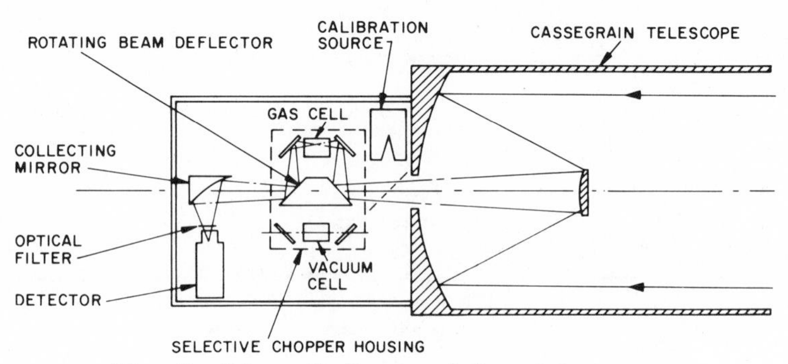

Fig. 5. Schematic diagram of Convair instrument.

can be appreciated by considering the following illustration. If the gas being examined is at atmospheric pressure, its lines will be approximately 0.1 cm^{-1} in width. For an optimum cross-correlation, the filter gas should have the same line widths as the sample gas. Typically, one might choose a bandpass filter to select an optimum spectral region 300 cm^{-1} wide, thus bringing nearly 3000 resolution elements into play. If it is possible to work at lower sample pressures, the filter gas pressure could also be lowered and a proportionally larger number of resolution elements would be brought to bear on the measurement. Line widths are proportional to gas pressure, down to a few torrs. If the sample and filter gas are both at 0.01 atm, a 300 cm^{-1} wide spectral band would include approximately 300,000 spectral resolution elements.

It has been suggested by Goody and others that the sensitivity of the gas-filter cross-correlation detection method can be increased by selecting specific source wavelengths. A basic form of wavelength selection has been mentioned previously. One uses a bandpass filter to select the spectral region where the compound to be measured has its strongest and most characteristic lines. Another method of wavelength selection would be to precede the nondispersive analyzer with a dispersive correlation system; select the desired regions of the spectrum by means of a suitable focal plane mask; and then recombine the selected regions of the spectrum before passing them into the nondispersive analyzer. A third method of selecting emission wavelengths would be to use as a source a sample of the gas to be detected that is excited to emission by heat, electric discharge, or radiation absorption. A method using the latter type of source would be a molecular analogue of atomic absorption spectroscopy.

A multiple-filter cross-correlation instrument has been designed and constructed at the Convair Corporation, under a NASA contract, for

Fig. 76. Photograph of Convair instrument.

use in the remote sensing of atmospheric trace gases. A schematic diagram of the instrument appears in Figure 75, and a photograph of the prototype in Figure 76. The telescope collects thermal radiation that has passed through the atmosphere from the earth. A rotating mirror passes the radiation alternately through four cells—one reference cell and three gas-filter cells. The transmitted radiation is then imaged on the detector, which is a nitrogen-cooled indium antimonide cell. The electronic system converts the four detected signals into ratios of transmittances. Flight-testing of the system from an airplane is currently under way. Carbon monoxide detection is being attempted initially, but nitrogen oxides, aldehydes, hydrocarbons, and many other pollutants are candidates for measurement with the future models of the instrument.

F. Using a Double-Beam Spectrophotometer as an Optical Correlation Instrument

A double-beam spectrophotometer can be operated as an optical correlation instrument. As such the spectrophotometer is being used in a nonscanning, infinite-resolution mode of operation. Experiments in the author's laboratory have demonstrated that when used as a correlation instrument, a spectrophotometer can make measurements that it cannot make in its scanning mode of operation.

The essential parts of an infrared correlation system are: (*1*) the source, (*2*) the sample compartment, (*3*) the filter channel, (*4*) the reference channel, (*5*) the bandpass filter, (*6*) the detection system. The parts of the spectrophotometer's system are used as follows. (*1*) The source is operated normally. (*2*) A sample compartment is placed in both beams; it may be the usual two absorption cells tied together or one absorption cell placed in front of either the source or the detector. (*3*) The filter channel is created by placing a sample of the filter gas in a short cell in the sample beam of the spectrophotometer. (*4*) The reference channel is created by adjusting the attenuating shutter in the reference beam of the spectrophotometer. (*5*) The monochromator serves as the bandpass filter. It may be used with maximum-width slits if desired, and will still serve as a sufficiently narrow bandpass filter. It should be set at the optimum operating wavelength indicated by the spectrum of the filter gas. This might be either the spectral region of strongest absorption or a point in the spectrum where absorption is moderate but overlapping with other spectra is at a minimum. (*6*) The detection and recording systems of the spectrophotometer are operated normally, with scale expansion if it is available.

In laboratory tests of this method a Perkin-Elmer 621 spectrophotometer was used with two 10-m gas absorption cells and with 20X-scale expansion. The filter cell was a 4-cm absorption cell with salt windows. It was positioned in the sample beam just where it exits from the source compartment into the transfer optics. Sensitizing gases were placed in this cell at pressures of a few torrs. The two 10-m cells were connected together and became in effect a single 10-m absorption cell in both beams. The experiment is diagramed in Figure 77.

The conduct of a measurement was as follows. (*1*) With the interconnected 10-m cells and the 4-cm cell evacuated, the monochromator was set at the desired wavelength, using wide slits. (*2*) Pressure of a few torrs of the sensitizer gas was then placed in the filter cell, the pen being driven down in proportion to the absorption. (*3*) The pen was then moved upward by closing the shutter in the reference beam; the most sensitive position was to set the pen near the bottom of the paper, using 20X-scale expansion. (*4*) The gas mixture to be measured was then allowed to flow into the interconnected 10-m cells. Through the correlation effect, the pen would be driven *upward* by an amount proportional to the concentration of the gas being measured.

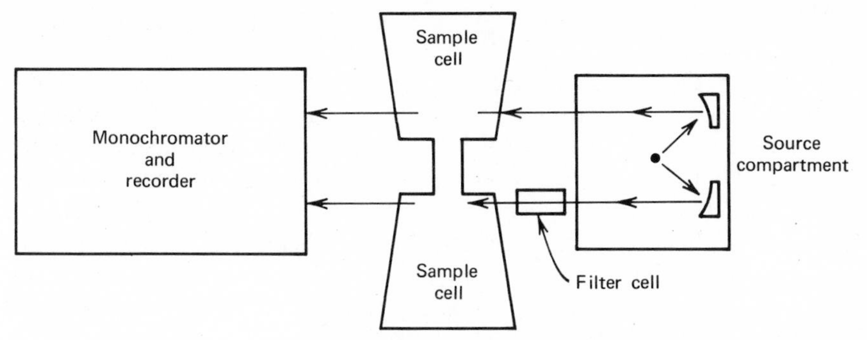

Fig. 77. Operation of a double-beam spectrophotometer as an optical correlation instrument.

Maximum discrimination against possible interferences was obtained at low sample pressures. This is exactly what was expected because at low pressure the spectra have narrow lines and a highly distinctive fine structure. The most selective spectral match (or cross-correlation) is obtained when the sample pressure and filter gas pressure are the same. However, in most cases the sample pressure could be raised significantly above the filter gas pressure, thus increasing the sensitivity of detection without losing the selectivity of the correlation.

The following measurements were made. (*1*) A few parts per million of nitric oxide in auto exhaust were measured with the monochromator set near 5.3 μ. (*2*) A nitrogen dioxide concentration of about 1 ppm in diluted auto exhaust was measured at about 6.2 μ. (*3*) A few parts per million of formaldehyde in auto exhaust was measured using the carbonyl band near 5.7 μ. (*4*) A few parts per million of formaldehyde in auto exhaust were measured using the C–H band near 3.6 μ.

Measurements (*1*), (*2*), and (*3*) are impossible to make when the spectrophotometer is used in the scanning mode of operation because of the interference of water vapor. Measurement (*4*) is not impossible but is difficult to make in the scanning mode because of interference from other organic materials in the sample. These results therefore may be considered a significant extension of the applicability of double-beam spectrophotometers to gas analysis problems. This extended capability derives from the unlimited resolution capacity of the cross-correlation mode of operation. Many analyses considered impossible because of spectral interferences now may be performed using the optical correlation mode of operation. Most of the stable air pollutants may be measured by this technique.

G. Infrared Filters

Filters for the infrared have been improved in recent years. Three standard types are available: short-wavelength pass, long-wavelength pass, and bandpass. These filters specifically are mentioned here because of the improvements that they have brought to high-resolution infrared spectroscopy and because of their anticipated contributions to nondispersive methods of pollution measurement. A review of filter development with references to earlier work is given in an article by Robert G. Greenler (63). A complete treatise on filters also has been published recently by H. A. Macleod (64).

The short-wavelength pass filters are not new and are a rather simple class of filters. For infrared purposes a set of short-wavelength pass filters can be assembled from common materials. For example, water is a short-wavelength pass filter with a cut-off near 1 μ; lucite has its cut-off near 2 μ, glass, at 2.7 μ, quartz, at 3.5 μ, lithium fluoride, at 7 μ, and calcium fluoride, at 10 μ.

The main objective of the short-wavelength pass filters in infrared spectroscopy has been to determine the level of scattered light in an instrument. Thus the early Perkin-Elmer single-beam spectrophotometers came with a lithium fluoride shutter that could be placed into the optical path periodically to determine what fraction of the detected light was short-wavelength scattered radiation.

The long-wavelength pass filters are much more important to infrared spectroscopy, but have taken much longer to be developed. They cannot be fabricated from common materials, but must be made from the extremely pure materials that have been perfected with the development of semiconductor technology. For example, silicon is a long-wavelength pass filter with a cut-on at 1 μ; it absorbs all radiation with wavelengths shorter than 1 μ and transmits a substantial fraction of all wavelengths longer than 1 μ. Germanium cuts on at 2 μ; indium arsenide, at 3.8 μ; and indium antimonide, at 7 μ.

A great contribution of the long-wavelength pass infrared filters has been the simplification of high-resolution grating spectroscopy in the infrared. For working with gratings at long wavelengths the difficulty in the past has been the elimination of the overlapping orders of diffraction. When a grating spectrometer is set at 10 μ in the first order, for example, it will also transmit 5 μ in the second order, 3.3 μ in the third order, 2.5 μ in the fourth, and down to whatever wavelength is transmitted by the optical components. If a filter such as indium antimonide is placed in an optical system adjusted for operation at 10 μ, it will cut off all orders of diffraction except the first. Now available are sets of infrared long-wavelength pass filters that permit working with gratings out to 40 μ.

The infrared bandpass filters are multilayer interference filters made from pure infrared transmitting materials of various indices of refraction. They now can be obtained commercially with bandpass widths of a fraction of a μ centered at any chosen wavelength out to 25 μ.

The main use of the infrared bandpass filter in pollution measurements would appear to be in connection with gas-filled filter cells used in nondispersive analysis. The bandpass filter can be used to select that region of the spectrum in which the nondispersive analyzer will operate with the highest selectivity and sensitivity.

H. Detectors

Detection is the basic operation of most optical systems. A radiometer contains a detector which is preceded by a collector and possibly a filter. An optical radar is a detection system, a collector, and a source. An absorption spectrophotometer consists of a detection system, a dispersing system, an absorption path, and a source. In all cases the performance of the system is improved when the performance of the detector is improved.

Detector development justifiably has occupied a prominent position in optics in recent years, with marked improvements. Detectivity steadily has been increased; response times have been shortened; spectral ranges have been widened; and detector cooling systems have been simplified.

Two main classes of detectors are found, thermal and photoelectric (65). The thermal detector is the more traditional type and is still extensively used. Thermal detectors, such as thermocouples and bolometers, have the important advantage of operating at room temperature. However, they do not have as high a detectivity as the photoelectric detectors, and their response times are much longer. A thermal detector is to be preferred in systems that have ample energy available and that operate at relatively low scan and modulation frequencies. Unfortunately, thermal detector development has not proceeded rapidly in recent years, and the cooled photoelectric detector frequently must be used to acquire desired system performance.

The new photoelectric infrared detectors, like the infrared filters and the diode lasers, resulted from the recent significant advances in semiconductor technology. In principle, an infrared photoelectric detector is the reverse of a diode laser. Instead of the electrons and holes creating photons, the photons create electrons and holes. The new infrared detectors are pure mixed crystals, but not necessarily the same mixed crystals as constitute the new infrared diode lasers.

The mercury-cadmium-telluride detectors are of interest at the present. These detectors are responsive from the visible to the far infrared. They can have response times considerably shorter than 1 μsec; they have detectivities that approach the theoretical limitation of the background noise (66). The mercury-cadmium-telluride detectors can be created in extremely thin layers either as single detectors or in multidetector arrays. They operate at liquid nitrogen temperatures or higher, which is probably their major advantage over the competing detectors, such as copper-doped germanium, requiring liquid helium cooling.

It is interesting to consider a system for atmospheric measurements that includes a mixed-crystal laser source, such as lead-tin-telluride, and a mixed-crystal detector such as mercury-cadmium-telluride. Both source and detector might be mounted together in the same cooler, and together they might bring to the system the maximum performance presently available.

Pyroelectric detectors are also of much interest at present (73)—detectors whose capacitance changes with temperature. The capacitance change results from a change in electric polarization of the ferroelectric detecting crystal. The pyroelectric detectors have a higher speed and a greater sensitivity than many other thermal detectors. They have the advantages of flat spectral response and room temperature operation.

I. Lasers in Remote Sensing

Lasers of various types are being applied in remote sensing, and applications of them are on the increase. It is difficult to say what the status of laser development is at the moment because laser technology is rapidly evolving. The applicability of lasers in remote sensing derives mainly from their ability to produce very brief but extremely powerful bursts of radiation. These pulses can be shorter than 1 nsec (10^{-9} sec) and more powerful than 1 GW (10^{9} W). The most powerful lasers are crystals which emit specific wavelengths, such as 1.06 μ. Dye lasers are not yet as powerful as the crystal lasers, but they have the ability to be tuned to any chosen wavelength over wide regions of the spectrum.

Remote probing with lasers generally depends upon scattering of the laser energy by the constituents of the atmosphere. Four principal types of scattering of optical pulses need to be considered: Rayleigh scattering, Mie scattering, resonance scattering, and Raman scattering. The scattering depletes the energy of the optical pulses and redistributes it in all directions with a definite intensity pattern that varies according to wavelength and type of scattering.

Rayleigh scattering happens when the dimensions of the scatterer are smaller than the wavelength of the light. The scattered light is concentrated along the line of flight of the pulse, with equal intensities in the forward and backward directions. Since the Rayleigh scatterers are small, they have high thermal velocities, causing the Rayleigh-scattered light to be Doppler broadened to a considerable extent. Since the intensity of Rayleigh scattering is a smooth function of wavelength, the scattered light does not identify the scatterer; thus its usefulness in atmospheric analysis is limited.

Mie scattering takes place when the dimensions of the particle are comparable to or larger than the wavelength of the light. The Mie-scattered light is concentrated in the forward direction, with a much smaller intensity backward. The relatively heavy Mie scatterers do not have high thermal velocities and hence do not broaden the laser line significantly through Doppler effect. The Mie scattering can be utilized to locate dusts and other aerosols in the atmosphere.

Resonance scattering occurs when the energy of the incident photon is the same as the energy difference between the ground state and an excited state of the scattering species. The resonance scattering is very strong and will have high intensity in both the forward and backward directions. Resonance scattering is most intense at low gas pressures. At atmospheric pressure, the resonant radiation is quenched by collisional deactivation. The line width of the scattered radiation will be determined by the pressure, temperature, and other properties of the scatterer. In a resonance-scattering experiment the scattering species must be known in advance, and the laser must be tuned to the resonant energy difference. The tunable dye laser offers promise of applicability to resonance scattering experiments. An outstanding example of resonance scattering was the glow observed from clouds of sodium and other substances released into the sunlit upper atmosphere.

Raman scattering differs from resonance scattering in that the energy of the incident photon does not have to coincide with the energy difference between electronic states of the scattering species. Thus the Raman scattering is much weaker than resonance scattering; but with the high incident intensities available from lasers, the Raman scattering is strong enough to be useful in many applications. The main distinction of Raman scattering is that the wavelength of the light is changed. The scatterer changes its energy state during the process of scattering with the increment of energy appearing in the scattered photon. These energy changes differ from one molecule to another and thus identify the scattering species. It is this Raman "signature" which makes the Raman scattering a subject of much interest in current pollution detec-

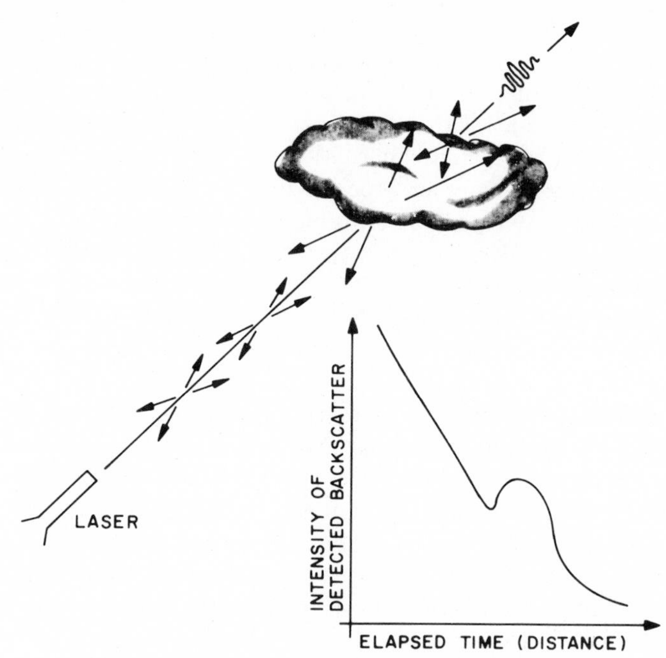

Fig. 78. Pulsed laser scattering experiment.

tion research. Recent reviews have covered some of the theoretical and practical aspects of Raman and resonance scattering (74,75).

A typical atmospheric scattering experiment using a pulsed laser is diagramed in Figure 78. In most experiments the wave train will be short compared to the distance traveled. Furthermore, any lag between the incidence of a photon on the scatterer and the release of the scattered photon can be neglected. The distance to a scatterer is therefore obtained by multiplying half the time between pulse release and pulse reception by the speed of light. The uncertainty in position of the scattering cloud will be half the uncertainty in time resolution of the detection system multiplied by the speed of light. If the pulse duration is 0.1 μsec or less, and the receiver time constant is about 0.1 μsec, the uncertainty in the position of the scattering cloud will be about 15 m.

References

1. L. Goldberg, "The Absorption Spectrum of the Atmosphere," in *The Earth as a Planet,* G. P. Kuiper, Ed., University of Chicago Press, Chicago, 1954, Chapter 9.

2. J. N. Howard, J. S. Garing, and R. G. Walker, "Transmission and Detection of Infrared Radiation," in *Handbook of Geophysics and Space Environments,* S. L. Valley, Ed., McGraw-Hill, New York, 1965, Chapter 10.
3. M. Migeotte, L. Neven, and J. Swensson, "The Solar Spectrum from 2.8 to 23.7 Microns, Part I, Photometric Atlas," in *Memoirs de la Societé Royale des Sciences de Liège,* Special Vol. 1, 1956.
4. Ralph Stair, *Proc. Nat. Air Pollution Symp. 3rd,* Pasadena, Calif., Western Gas and Oil Association, Los Angeles, 1955.
5. David M. Gates, *Proc. Nat. Air Pollution Symp.,* Pasadena, Calif., Western Gas and Oil Association, Los Angeles, 1955.
6. W. E. Scott, E. R. Stephens, P. L. Hanst, and R. C. Doerr, *Proc. Amer. Petrol. Inst. Sect. III,* **37,** 171 (1957).
7. E. R. Stephens, *J. Air Pollution Control Assoc.,* **19,** 181 (1969).
8. J. M. Heuss and W. A. Glasson, *Environ. Sci. Technol.,* **2,** 1109 (1968).
9. J. U. White, *J. Opt. Soc. Amer.,* **32,** 285 (1942).
10. E. Robinson and R. C. Robbins, "Sources, Abundance, and Fate of Gaseous Atmospheric Pollutants," Stanford Research Institute, February 1968.
11. C. B. Ludwig, R. Bartle, and M. Griggs, "Study of Air Pollutant, Detection by Remote Sensors," NASA Contractors Report, CR 1380, Convair Division, General Dynamics Corporation, December 1968.
12. A. P. Altshuller, "Relationships among Pollutants Undergoing Atmospheric Reactions in CAMP Cities in the U.S.," presented at the 63rd Annual Meeting of the Air Pollution Control Association, St. Louis, Mo., June 1970.
13. E. R. Stephens and F. R. Burleson, *J. Air Pollution Control Assoc.,* **19,** 929 (1969).
14. V. J. Schaefer, *Bull. Amer. Meteorol. Soc.,* **50,** 199 (1969).
15. E. Robinson, "Effect on the Physical Properties of the Atmosphere," in *Air Pollution,* A. C. Stern, Ed., Academic Press, New York, 1968, Chapter 11.
16. R. C. Robbins, K. M. Borg, and E. Robinson, *J. Air Pollution Control Assoc.,* **18,** 106 (1969).
17. R. A. McCormick, "Air Pollution Climatology," in *Air Pollution,* A. C. Stern, Ed., Academic Press, New York, 1968, Chapter 9.
18. A. J. Haagen-Smit and L. G. Wayne, "Atmospheric Reactions and Scavenging Processes," in *Air Pollution,* A. C. Stern, Ed., Academic Press, New York, 1968, Chapter 6.
19. L. S. Jaffe, *J. Air Pollution Control Assoc.,* **18,** 534 (1968).
20. *Tables of Wave Numbers for the Calibration of Infrared Spectrometers,* International Union of Pure and Applied Chemistry, Commission on Molecular Structure and Spectroscopy, Butterworths, London, 1961.
21. R. H. Pierson, A. N. Fletcher, E. St. Clair Gantze, *Anal. Chem.,* **28,** 1218 (1956).
22. C. B. Ludwig, presented at the Optical Society Meeting, Philadelphia, Pa., April 1970.
23. P. L. Hanst, E. R. Stephens, W. E. Scott, and R. C. Doerr, *Anal. Chem.,* **33,** 1113 (1961).
24. E. R. Stephens, "The Formation, Reactions, and Properties of Peryoxyacyl Nitrates (PANS) in Photochemical Air Pollution," in *Advances in Environmental Sciences,* Vol. 1, J. N. Pitts, Ed., Wiley, New York, 1969.
25. E. R. Stephens, E. F. Darley, O. C. Taylor, and W. E. Scott, *Proc. Am. Petrol. Inst. Sec. III,* **325** (1960).

26. G. B. Jacobs and L. R. Snowman, *IEEE J. Quantum Electron.,* **QE-3,** No. 11 603 (1967).
27. R. K. Long, *Atmospheric Absorption and Laser Radiation,* Engineering Experiment Station Bulletin 199, Ohio State University, Columbus, Ohio, 129 pp.
28. P. L. Hanst and J. A. Morreal, *J. Air Pollution Control Assoc.,* **18,** 754 (1968).
29. C. K. N. Patel, "Gas Lasers," in *Lasers,* Vol. 2, A. K. Levine, Ed., Marcel Dekker, New York, 1968.
30. D. N. Jaynes and B. H. Beam, *Appl. Opt.,* **8,** 1741 (1969).
31. E. D. Hinkley, T. C. Harman, and C. Freed, *Appl. Phys. Lett.,* **13,** 49 (1968).
32. E. D. Hinkley, *Appl. Phys. Lett.,* **16,** 351 (1970).
33. D. G. Murcray, T. G. Kyle, F. H. Murcray, and W. J. Williams, *J. Opt. Soc. Amer.,* **59,** 1131 (1969).
34. *Life* Magazine, February 9, 1968.
35. A. H. Pfund, *Science,* **90,** 326 (1939).
36. E. R. Stephens, *Appl. Spectry.,* **12,** 80 (1958).
37. D. R. Herriott and H. J. Schulte, *Appl. Opt.,* **4,** 883 (1965).
38. G. A. W. Rutgers, "Temperature Radiation of Solids," *Handbuch der Physik, Band XXVI* Springer, Berlin, 1958, pp. 129–170.
39. *Laser Technology and Applications,* S. L. Marshall, Ed., McGraw-Hill, New York, 1968, 294 pages.
40. *The Laser,* W. V. Smith and P. P. Sorokin, Eds., McGraw-Hill, New York, 1966.
41. *Lasers,* Vols. I and II, A. K. Levine, Ed., Marcel Dekker, New York, 1968.
42. G. B. Jacobs and L. R. Snowman, *IEEE J. Quantum Electron.* **QE-3,** No. 11, 603 (1967).
43. G. Jacobs, P. Morgan, L. Snowman, and D. Ware, General Electric Co. Report, TIS R68ELS-10, January 1968.
44. H. H. Kim, R. A. Paananen, P. L. Hanst, and T. F. Deutsch, *IEEE J. Quantum Electron.,* **QE-4,** No. 11, 908 (1968).
45. P. L. Hanst and J. A. Morreal, *Appl. Opt.,* **8,** 109 (1968).
46. P. L. Hanst, J. A. Morreal, and W. J. Henson, *Appl. Phys. Lett.,* **12,** 58 (1968).
47. J. F. Butler and T. C. Harman, *Appl. Phys. Lett.,* **12,** 347 (1968).
48. H. W. Furumoto and H. L. Ceccon, *Appl. Opt.,* **8,** 1613 (1969).
49. T. M. Hard, *Appl. Opt.,* (1970).
50. M. R. Bowman, A. J. Gibson, and M. C. W. Sandford, *Nature,* **221,** 456 (1969).
51. S. H. Melfi, J. D. Lawrence Jr., and M. P. McCormick, *Appl. Phys. Lett,* **15,** 295 (1969).
52. P. W. Kruse, L. D. McGlaughlin, and R. B. McQuistan, *Elements of Infrared Technology,* Wiley, New York, 1962, 448 pp.
53. R. A. Smith, F. E. Jones, and R. P. Chasmar, *The Detection and Measurement of Infrared Radiation,* 2nd ed., Clarendon Press, Oxford, 1968, 502 pp.
54. M. Dreyfus and D. Hilleary, "Satellite Infrared Spectrometer," *Aerospace Eng.,* February 1962.
55. D. G. Murcray, T. G. Kyle, F. H. Murcray, and W. J. Williams, *J. Opt. Soc. Amer.,* **59,** 1131 (1969).
56. G. A. Vanasse, *Opt. Spectr.,* **34,** April 1970.
57. John Strong, *Appl. Opt.,* **6,** 179 (1967).
58. A. R. Barringer and J. P. Schock, *Proc. 4th Symp. Rem. Sens. Environ.,* **779,** 1966.
59. D. T. Williams and B. L. Kolitz, *Appl. Opt.,* **7,** 607 (1968).
60. Barringer Research Limited, 304 Carlingview Dr., Rexdale, Ontario, Canada.

61. A. H. Pfund, *Science,* **90,** 326 (1939).
62. R. Goody, *J. Opt. Soc. Amer.,* **58,** 900 (1968).
63. R. G. Greenler, in *Concepts of Classical Optics,* John Strong, Freeman and Co., San Francisco, 1958, Appendix O.
64. H. A. Macleod, *Thin Film Optical Filters,* American Elsevier, New York, 1969, 332 pp.
65. R. A. Smith, *Appl. Opt.,* **4,** 631 (1965).
66. P. W. Kruse, *Appl. Opt.,* **4,** 687 (1965).
67. R. Paananen, C. L. Tang, and H. Statz, *Proc. IEEE,* **51,** 63 (1963).
68. K. C. Herr and G. C. Pimentel, *Science,* **166,** 496 (1969).
69. K. C. Herr and G. C. Pimentel, *Science,* **167,** 47 (1970).
70. D. W. Hill and T. Powell, *Nondispersive Infrared Gas Analysis,* Plenum Press, New York, 1968.
71. K. F. Luft, *Z. Techn. Physik,* **24,** 97 (1943).
72. K. F. Luft, G. Kesseler, and K. H. Zörner, *Chemie-Ingenieur-Technik,* **29,** 937 (1967).
73. W. M. Doyle, *Laser Focus,* July 1970.
74. W. Holtzer, W. F. Murphy, and H. J. Bernstein, *J. Chem. Phys.,* **52,** 399 (1970).
75. T. Hirshfeld and S. Klainer, *Opt. Spectr.,* July 1970.

Agricultural Wastes and Environmental Pollution

Jesse Lunin
Chief Soil Chemist,
Soil and Water Conservation Research Division,
U.S. Department of Agriculture,
Beltsville, Maryland

I. INTRODUCTION

Agriculture has made phenomenal strides since the beginning of this century. In spite of the rapid increase in population, agriculture has been able to maintain an abundance of quality food and fiber crops, and has a potential to maintain this abundance for future years. All this has been accomplished with an increasing reduction in the farm labor force. In 1910 in the United States, one farmer produced food and fiber for about seven people; but in 1966, one farmer produced food and fiber for approximately 45 people. The labor force released from agricultural endeavors is utilized now in the production of industrial products and services that contribute to our high standard of living. Yet the relative cost of food to the consumer has not increased. In 1900 the average American housewife spent 40% of the family income on food as compared to only 17% spent in 1967.

Many factors have contributed to this remarkable achievement. Increased use of fertilizers has resulted in greater and more efficient production of crops. Use of pesticides at a rapidly expanding rate has greatly reduced losses from insects and diseases and improved crop quality. The growing use of herbicides has resulted in more effective control of weeds. Introduction of multirow equipment and mechanization of many farm operations have reduced manpower requirements and made these operations more efficient. Intensive production practices for livestock and poultry have sustained costs of animal protein at relatively low levels. Large feed lots for beef production are evident in many areas. Confined intensive production has made poultry one of the least expensive sources of animal protein, whereas in many countries it is still considered a luxury. Improved processing of agricultural products has provided the housewife with a wide variety of processed and prepared foods.

These advances have benefited the consumer in many ways. However, some potential adverse effects have accompanied these advances. Increased intensity of cultivation practices and use of heavy equipment have made the adoption of soil and water conservation practices more difficult, possibly increasing runoff and sediment production from agricultural lands. Concomitant with the beneficial use of agricultural chemicals, such as fertilizers and pesticides, is the potential contamination of the environment from inefficient use of these materials. Intensive production of livestock and poultry has created a problem of animal waste disposal. Expansion of processing plants for agricultural products also has introduced a problem of safe disposal of both effluents and solid wastes from these operations.

Many of these advances in agricultural technology represent potential sources of environmental pollution. These potential hazards can be greatly minimized, but not entirely eliminated. A compromise must be made between the degree of pollution that we can tolerate and the sacrifices in cost, quantity, and quality of food and fiber production that we are willing to accept.

Every responsible citizen is concerned with the quality of his environment. The degree of concern varies among individuals, depending upon their degree of involvement. There are three kinds of pollution: real pollution, political pollution, and hysterical pollution (66). At times the hysterical approach to the pollution problem can be more detrimental than the actual pollution. However, this approach has caused a greater public awareness of the problem and created a degree of urgency for seeking and implementing solutions.

An attempt was made in 1966 to summarize agriculture's role in environmental quality (6). In 1968 (84) a comprehensive report was prepared on the role of wastes in relation to agriculture and forestry. In March 1968 President Johnson delivered a message to Congress in which he expressed deep concern over agricultural pollution. At his request a report was prepared dealing with the control of agricultural pollution (80). By then it was apparent that reliable research data were not available to assess adequately many of the agricultural pollution problems. As a result, a report on a national program of research on agricultural pollution was prepared by a joint task force of the U.S. Department of Agriculture, state universities, and land-grant colleges (33). President Nixon, in his State of the Union Message to Congress on January 22, 1970, indicated that pollution control and abatement is currently, and will be in the future, a matter of high priority.

The term "waste" in relation to agricultural pollution refers to any by-product or residue from agricultural operations that may adversely affect the quality of the environment either directly or indirectly. Changes in environmental quality began when man first started to till the soil and to domesticate animals. Since then, expanding populations and the expansion and intensification of agricultural production has increased the impact on environmental quality. Undoubtedly this trend will continue in the future.

Agriculture often has been accused of being a major polluter of various segments of the environment. Some accusations are justified, but many are based on very limited factual information and often on mere speculation. It is the intent of this discussion to view the current status of our thinking on environmental pollution from agricultural wastes.

II. SOURCE OF AGRICULTURAL WASTES AND POLLUTION POTENTIAL

Many sources of agricultural wastes are cited frequently as pollution hazards. Wadleigh (84) has discussed in detail both the contribution of these wastes to environmental pollution and the effect of wastes on agriculture. Rather awesome figures have been cited regarding tonnages of agricultural chemicals produced and used in the United States, as well as estimates of the amounts of agricultural wastes produced. However, only a small fraction of the agricultural chemicals used actually fall into a "waste" category. Furthermore, not all of the wastes produced become air or water pollutants. They may be absorbed into the environment, either directly or in a transformed state, without constituting a pollution problem.

Many factors determine whether an agricultural waste will constitute a pollution problem. Certain wastes pollute the air in the form of dusts or are problems because of their volatility. Some wastes tend to accumulate in the soil, eventually creating a problem. Wastes that runoff as solutes or are adsorbed in sediment could pollute surface waters downstream. Some agricultural chemicals, because of their solubility, could percolate down through a profile and eventually contaminate groundwater. Finally, certain pesticides leave residues on plant materials or are absorbed by plants and become a problem when the plant is consumed directly, when animal products are contaminated through consumption of plants having residues, or when plant wastes containing residues are disposed of.

The chemical and physical properties of a waste significantly determine its potential to become a pollution hazard. Rates of decomposition or degradation and conditions affecting these processes are significant factors. We are also concerned with the pollutional potential or toxicity of any degradation products or metabolites. For example, the degradation product of a pesticide may be more-or-less toxic than the original compound. Conversion of organic nitrogen in wastes into the nitrate form still represents a hazard; whereas, denitrification of nitrate to N_2 eliminates the pollution potential of this element. Management practices for animal production and waste disposal can greatly lessen pollution hazards. With agricultural chemicals (e.g., fertilizers and pesticides) judicious selection of time, rate, and method of application can mitigate pollution problems. Finally, these factors must all be considered using existing environmental conditions as a frame of reference.

Ramifications of these problems are extensive. Recently, considerable research has been initiated on the evaluation and amelioration of these problems. Some pertinent areas concerning major sources of pollution are discussed in this section with no attempt to summarize or to evaluate all the available information.

A. Air Pollutants

Many people can still remember the dust storms of the early 1930s. Conservation practices have greatly reduced the amounts of windblown soil; yet it is estimated that there are some 55 million acres of cropland in the United States that are still subject to wind erosion. Of the 30 million tons of natural dusts entering the atmosphere annually, a large portion comes from windblown soil.

Detrimental effects of these airborne materials are apparent. Decreased visibility and road maintenance problems are very evident in some areas. Progress is being made in the development of wind erosion control and other remedial practices for agricultural and waste lands. Unfortunately, there are extensive areas of dune land and deteriorated range land that are not as readily controlled.

Other agricultural industries, such as cotton ginning, alfalfa milling, feed lots, harvesting operations, contribute a significant amount to the problem of airborne particulates. Dusts from agricultural lands also may be a source for the transmission of pesticide residues. Smoke from forest fires and burning of crop residues release large amounts of particulate and chemical pollutants into the atmosphere each year. Airborne allergens, such as pollen from weeds, represent a problem of considerable magnitude. Over 12 million people in the United States suffer from hay fever or asthma; therefore a reduction of airborne pollen would be most beneficial.

B. Sediment

It has been stated that sediment, the product of soil erosion, is the number one pollutant. Sediment consists of both organic and inorganic solids detached and transported from their point of origin by water. In addition to depleting our land resources, sediment impairs water quality and creates problems upon deposition. Estimates of sediment movement in the United States include more than 4 billion tons a year, with about one-fourth of this being transported to the sea. Further estimates indicate that about three-fourths of this sediment originates from agricultural lands and the remainder from other sources such as highway construction, stream-bank erosion, and urban and industrial development. A small amount arises from geologic erosion.

Sediment has many detrimental effects. It reduces the storage capacity of reservoirs and conveyance structures. Suspended in water, it impairs quality of aquatic life and causes abrasive damage to machinery, such as turbines and pumping equipment. Sediment-free water is desirable for recreation. Flood-borne sediments may cause considerable crop damage when deposited. Sediment is also a carrier for phosphates and pesticides.

Once sediment becomes a problem in rivers, lakes, or impoundments, remedial measures such as dredging are costly. Furthermore, when dredging a lake or estuary becomes necessary, the effect on the biological characteristics of that body of water is uncertain.

No one questions the severity of the sediment problem. In most cases, adequate technology is currently available to control erosion and to reduce to a significant degree sediment production. The problem is to implement this technology to the fullest extent.

As a result of the adoption of good management practices, soil loss has been materially reduced in many areas. In Iowa (64) sediment loss from level-terraced corn was only 1 ton/acre, as compared to 44.4 tons from field-contoured corn and 1.2 tons from grass. In the Blacklands of Texas (9), sediment yield from a 132-acre watershed with good conservation practices was only 12% that from an adjacent 176-acre watershed without conservation practices. A much larger study was reported on the Chattahoochee River watershed in Georgia (1). Starting in 1930 there was a trend toward improved management practices in the area, including reforestation, fire protection, planting of cover crops, conversion of cultivated land to improved pastures, and other farm conservation practices. As a result, turbidity measurements dropped from 400 ppm in 1932 to less than 100 ppm in 1952.

To control adequately soil erosion, it is necessary to understand the causative, or contributory, factors and how they relate to each other. As an outgrowth of a research program on this problem, a soil loss prediction equation was developed (90). The equation shown permits an estimate of soil loss for a given set of conditions.

$$A = RKLSCP$$

where A = the computed soil loss per unit area.

R = the rainfall factor; the number of erosion-index units in a normal year's rain. The erosion index (EI) is the product of the total kinetic energy of the storm times its maximum 30-min intensity. This index is a measure of the erosive force of a specific rainfall.

K = the soil-erodibility factor; the erosion rate per unit of erosion index for a specific soil in cultivated continuous fallow, on a 9% slope 72.6 ft long.

L = the slope-length factor; the ratio of soil loss from the field slope length to that from a 72.6-ft length on the same soil type and gradient.

S = the slope-gradient factor; the ratio of soil loss from the field gradient to that from a 9% slope.

C = the cropping-management factors; the ratio of soil loss from a field with specified cropping and management to that from the fallow condition on which the factor K is evaluated.

P = the erosion-control practice factor; the ratio of soil loss with contouring, strip-cropping, or terracing to that with straight-row farming, up-and-down slope.

Examining these parameters, we find that R is a climatic factor, and K, L, and S are specific fixed characteristics of the land. Both C and P are management factors which can be altered to reduce the amount of soil loss from a given area to a desired or tolerable level.

Values for the rainfall factor R were determined by Wischmeier (88), from long-term rainfall records. The soil-erodibility factor K was calculated on 23 major soils from data collected on erosion plot studies (57).

TABLE I

Computed "K" Values for Soils on Erosion Research Stations (90)

Soil	Source of data	Computed K
Dunkirk silt loam	Geneva, N.Y.	0.69[a]
Keene silt loam	Zanesville, Ohio	.48
Shelby loam	Bethany, Mo.	.41
Lodi loam	Blacksburg, Va.	.39
Fayette silt loam	LaCrosse, Wis.	.38[a]
Marshall silt loam	Clarinda, Iowa	.33
Mansic clay loam	Hays, Kans.	.32
Hagerstown silty clay loam	State College, Pa.	.31[a]
Mexico silt loam	McCredie, Mo.	.28
Honeoye silt loam	Marcellus, N.Y.	.28[a]
Cecil sandy loam	Clemson, S.C.	.28[a]
Cecil clay loam	Watkinsville, Ga.	.26
Zaneis fine sandy loam	Guthrie, Okla.	.22
Tifton loamy sand	Tifton, Ga.	.10
Freehold loamy sand	Marlboro, N.J.	.08

[a] Evaluated from continuous fallow. All others were computed from row-crop data.

Some of these are shown in Table I. An equation now has been developed for estimating K based on soil properties (89). The crop factor C is rather complex and measures the combined effects of all the interrelated cover and management factors such as type of crop, growth stage, seedbed preparation, residue management, and others. The practice factor, P, for contouring ranges from 0.60 for a 1% slope to 0.90 for slopes over 18%. Contour strip-crop factors would be roughly one-half that for contouring alone, whereas the terracing factor would be equal to 10 to 20% of the contour factor, depending on the type terrace.

Sediment apparently can be controlled by good management practices. Changing patterns of agriculture to meet more efficient, economic production needs require that these management practices be continually reviewed and modified to ensure their effectiveness.

C. Salts and Minerals

All natural waters contain some dissolved salts. The increase in salt content of irrigation return flows over that of the irrigation water, however, has been a source of concern for many years. Approximately 50 to 80% of the water applied to the soil during irrigation is lost by evapotranspiration. Since only pure water evaporates, the salts applied in the water remain in the soil. To maintain a favorable salt balance in the soil, excess water must be added to leach out the accumulated salts, generally through subsurface drainage. Hence there is an increased salt burden in return flows. Leaching of naturally saline soils and saline geologic strata also contribute to the salt burden of groundwater and streams.

Approximately 15 million acres of irrigated land in the Western United States are subject to salinity hazards. It is pointless to cite the great salt burden in the water in the Western United States. The magnitude of the problem is clearly demonstrated by the example (84) of a farmer in the Imperial Valley of California who, by applying 5-acre-ft of Colorado River water, would add at least 6 tons of salt per acre. This six tons of salt must be removed in drainage water to maintain a favorable salt balance in the soil. The resultant effect is to increase the salinity of irrigation water for downstream users. An example of this was demonstrated by Wilcox (87) for the Rio Grande River between Caballo Dam in New Mexico and Fort Quitman, Texas (Table II). The data represent mean values over a 19-year period (1934–1953) at four progressively downstream stations. The weighted mean salt concentration increased downstream, although the total salt burden did not change appreciably because of a decrease in mean flow.

Salinity problems cannot readily be eliminated. Past research has enabled us to deal more effectively with the problem, but a greater endeavor is required to reduce pollution from this source. A recent report (80) outlined four areas in which progress could be made in reducing the degree of water pollution by salts. The first suggests decreasing the salt content of the irrigation supply source. The obvious solution to this is desalination, but the cost is now and will continue to be prohibitive. Practical recourse in this area is the development

TABLE II

Mean Annual Flow, Salt Burden, and Salt Concentration of the Rio Grande at Four Stations for the Period 1934–1953 (87)

Station	Mean flow, kilo acre-ft/yr	Weighted mean salt concentration, tons/acre-ft	Salt burden, kilo tons/yr
Caballo Dam	781	0.70	547
Leasburg Dam	743	0.75	557
El Paso	525	1.07	562
Fort Quitman	203	2.30	467

of systems for managing and distributing water from impoundments and other resources to minimize salt delivery. Reduction of seepage and evaporation losses from reservoirs and ponds is helpful.

A second area deals with improving irrigation and drainage practices. Better control of water application would give greater efficiency of use, and leaching requirements could be met more effectively. Improved subsurface drainage practices prevent salt accumulations in the soil. More efficient soil and water management practices to reduce evaporation losses from water and bare soil surfaces and to minimize the amount of water required for leaching would be most helpful in reducing salt problems. Much progress can be made in the development of new and improved practices, and a great improvement could be achieved through extensive application of existing technology.

Salinity control through use of more salt tolerant plants and the development of plants having low water-use requirements has been suggested. Since much of the problem stems from evapotranspiration losses, suppressing or minimizing transpiration losses has some merit. To date, no economically effective transpiration suppressants have been developed. However, transpirative losses can be minimized by selecting existing crops that have low water-use requirements, breeding new crops and varieties that have low transpiration losses, and elimination of nonbeneficial plants such as phreatophytes.

Finally, the problem of salt accumulation in the return flows must be considered. The diversion of saline return flows to the ocean has been suggested. Feasibility of this approach decreases with increasing distance from the ocean. Exceptions are found, as in the case of the Imperial Valley and Coachella Valley, that can channel their drainage into the Salton Sea. Even this has some disadvantages. The area around the Salton Sea now is being developed as a recreational area. During recent years, as a result of saline return flows and evaporation losses, the salt content of the Salton Sea has increased markedly to a level that could impair its recreational potential. Desalination of brackish waters or saline return flows has been suggested, but existing technology in this area is still too costly. Salt sinks, where saline water can be impounded and evaporated by solar energy, have also been mentioned and are in use in some areas (Australia). These would require large areas of impermeable soils that would preclude infiltration losses under highly saline conditions. Injection systems for disposal of highly concentrated saltwater into underground cavities is another possibility. This is being used successfully by the petroleum industry, as well as for disposal of hazardous industrial plant effluents. Any possibility for groundwater contamination must be avoided.

Salinity represents a problem of considerable magnitude. Solutions will have to be developed by individual farmers, through coordinated action on the part of irrigation districts, and by state and federal government agencies.

D. Animal Wastes

Organic wastes create water pollution problems because of their high biochemical oxygen demand. Moreover, animal wastes are a potential source of nitrogen and phosphorus in water, as well as being a possible source of pathogen contamination.

When animals were raised in small operating units, wastes were not a problem. Furthermore, manure was considered an asset because of its value as a soil amendment. With the advent of intensive production units for beef, hogs, and poultry, distribution of the manure became a problem. Production of chemical fertilizers provided plant nutrients at lower cost than that of equivalent amounts of manure. Feed lots containing 10,000 to 100,000 head of cattle now are being developed. Poultry houses containing several hundred thousand birds are not uncommon. Similar trends are evident with hogs and dairy cattle. Manure elimination from these operations now becomes a disposal problem. Wadleigh (84) estimates that over 1 billion tons of fecal wastes and 400 million tons of liquid wastes are produced annually by domestic animals with about half coming from concentrated production units (Table III).

Two aspects of the problem must be considered. First, pollution hazards from wastes in the feed lots may result from runoff to surface

TABLE III

Production of Wastes by Livestock in the United States (84)

Livestock	U.S. population, 1965, millions	Solid wastes, g/cap/day	Total production of solid waste, million ton/yr	Liquid wastes, g/cap/day	Total production of liquid wastes, million ton/yr
Cattle	107	23,600	1,004.0	9,000	390.0
Horses	3	16,100	17.5	3,600	4.4
Hogs	53	2,700	57.3	1,600	33.9
Sheep	26	1,130	11.8	680	7.1
Chickens	375	182	27.4	—	—
Turkeys	104	448	19.0	—	—
Ducks	11	336	1.6	—	—
Total			1,138.6		435.4

TABLE IV

Variation in Nutrient of Animal Wastes (8)

Element	Cattle, % dry matter	Hogs, % dry matter	Hens, % dry matter
N	0.30–1.30	0.20–0.90	1.8–5.9
P_2O_5	0.15–0.50	0.14–0.83	1.0–6.6
K_2O	0.13–0.92	0.18–0.52	0.8–3.3

waters and deep percolation to groundwater supplies. The second aspect relates to potential pollution hazards resulting from wastes removed from the feed lot for disposal elsewhere, such as application on the land. The pollution implications of animal wastes were summarized recently by Loehr (41).

Many studies have been conducted on treatment and disposal of animal wastes. Occasionally, similar studies have produced different results, giving rise to confusion in interpretation and evaluation of a given practice. One factor responsible for this is that the physical and chemical composition of the wastes often are not adequately characterized, amounts involved are not evaluated adequately, and environmental factors are not taken into consideration. The composition of manure varies among animals, and that produced by a given animal will vary with its breed, age, diet, and environmental factors. Once produced, the characteristics of these wastes will vary with collecting, handling, and storage practices. An example of ranges observed in chemical composition of some wastes are shown in Table IV. In fact, the value of 1.3% for nitrogen may be exceeded on many of today's cattle feed lots.

Runoff from feed lots represents a pollution hazard because of its high oxygen demand. Drainage from feed lots have often killed fish. A recent publication (27) reported 37 cases of fish kills in 1967 attributed to manure-silage drainage. Of these, three major kills were attributed directly to feed lots, and resulted in the loss of 900,000 fish. The nitrogen and phosphorus content of the runoff as a source of nutrient enrichment for receiving streams is of concern.

Many factors affect the amount and composition of runoff from a feed lot. Grub *et al.* (28) list such factors as land slope, surfacing material, feed-lot layout, depth of waste accumulation, and composition of the ration fed. Data showed significant effects of rations on the BOD (biochemical oxygen demand), nitrogen, and phosphorus content of the runoff when cottonseed hulls, all concentrate, and silage were fed. Norton and Hansen (56) attempted to correlate the hydrologic and quality characteristics of the feed lot runoff in order to develop a predictive relationship. Actual measurements of runoff were made from concrete-surfaced and nonsurfaced feed lots (47). Aside from differences in amounts of runoff resulting from the hydrologic characteristics, the type of surface also affected the composition of the runoff. An example of the nitrogen data is presented in Table V; note that concentrations were significantly affected by seasons as well as by surface conditions. Total Kjeldahl nitrogen averaged 500 mg/liter from the concrete lot and 300 mg/liter from the nonsurfaced lot.

Various management and design factors to reduce the pollution potential in feedlots are being investigated. Some are designed to treat directly the solid and liquid runoff from feed lots. Investigations show that animal wastes are amenable to most processes used for domestic and industrial waste disposal. However, the treatment used is a matter of economics. It has been estimated that if conventional domestic waste treatment processes were used, a dairyman would have to pay about $200 per cow per year for waste treatment.

Studies have been conducted on various treatment processes such as anaerobic digestion and the use of anaerobic and aerobic lagoons. Lagoons have been found to be reasonably effective under field conditions. Anaerobic lagoons have a good potential for handling and treating concentrated animal wastes because of their high solids and low water content. Essentially, these units are designed to remove, destroy, and stabilize organic matter. They can serve as a biological control unit, a holding unit prior to land disposal, a control of runoff from confinement areas, or any combination of these. The aerobic oxidation lagoons generally are shallow with a large surface area to maintain aerobic conditions. The high oxygen demand associated with animal waste requires large surface areas and volumes. Loehr (41) estimates that a confinement unit of 1000 head of beef cattle would require an oxidation pond of at least 20 acres. Furthermore, he states that while anaerobic systems

TABLE V

Seasonal Variations in Nitrogen Content of Feedlot Runoff (47)

	Concrete lot, mg/liter			Nonsurfaced lot, mg/liter		
Season	NO_2	NO_3	NH_4	NO_2	NO_3	NH_4
Summer	1.0–6.0	0.1–11	50–139	1.0–7.0	0.1–6.0	26–62
Fall	1.0–5.0	0.4–2.3	20–77	1.0–23	0.5–26	13–45
Winter	—	—	1.3–7.0	—	—	1.0–3.8

can be effective in treating animal wastes, the effluent from these systems requires further treatment. A combination of the anaerobic and aerobic systems could produce an effluent of much superior quality than either process alone. Loehr (42), for example, studied a combined anaerobic-aerobic treatment system for a beef cattle feed lot and found that while treatment greatly reduced the pollution potential of the runoff, the effluent would need further treatment prior to direct disposal into a receiving stream. He states that the effluent would be suitable for land disposal or crop production. As with many animal and human waste treatment systems, the process is frequently successful in reducing the BOD content of the material, but the nitrogen and phosphorus compounds remain as potential pollutants. In fact, reduction of the oxygen demand in the substrate would tend to reduce losses of nitrogen by denitrification and favor the production of nitrates.

Other disposal methods have been investigated such as composting, incineration, wet oxidation, and sanitary land fills. Loehr (41) indicates, however, that no one treatment process or system is universally applicable to all types of animal production units. Systems must be designed to meet specific needs.

Another source of concern is the potential pollution of groundwater from deep percolation under feed lots. An extensive study was made in Colorado to evaluate nitrate distribution in 20-ft cores under feed lots, as compared to other types of land use (73). In this study virgin grassland and dry land fields were assumed to have received no nitrogen; irrigated cropland not in alfalfa to have received 100 lb/acre of nitrogen per year; feed lots to have received as much as 10 tons of organic and urea nitrogen per year. In addition to analyses of 129 soil cores, water samples from the surface of the water table were analyzed. Table VI shows the greatest amount of nitrate accumulated under the feed lot with significant amounts under dry land and irrigated land not in alfalfa. Nitrate gradients in the profile showed accumulations at the soil surface with concentrations decreasing with depth. Nitrate content in the water at the water table surface, and at the 20-ft depth in the soil showed little difference for land-use treatment. The question arises as to what happens to the nitrate. There was evidence that water was moving down in the profile and that some nitrate was probably moving with it. The decrease in nitrate with depth might, therefore, be attributable to denitrification. The greatest decrease in concentration was observed under the feed lots where the investigators also found an abundant energy source for denitrification in the soil water. Thus it appears that denitrification could be a significant factor in determining the fate of nitrogen under barnyards and feed lots.

TABLE VI

Average Nitrate in 20-Ft Profiles and Water at Surface of Water Table[a]

	Number	Soil profiles, NO_3-N, lb/acre	Number	Water table, NO_3-N Mean ppm	Water table, NO_3-N Range ppm
Virgin grassland	17	90	8	11.5	0.1–19
Dry land	21	261	4	7.4	5–9.5
Irrigated land (except alfalfa)	28	506	19	11.1	0.36
Irrigated land (alfalfa)	13	79	11	9.5	1–44
Feed lots	47	1,436	33	13.4	0–41

[a] Adapted from Stewart *et al.*, U.S. Department of Agriculture, ARS 41–134, 1967.

Eventually, the manure from feed lots and the solid and liquid wastes from confined production units must be eliminated. These wastes have been used in the past as soil amendments for crop production. Beneficial attributes of these applications have been more clear cut in some studies than in others. Factors involved in crop responses to organic manures have been described by Bunting (20) in a discussion of results from more than 100 field experiments with these materials at the Rothamsted Experimental Station. Benefits attributed to manures now are suspect on an economic basis. Eby (24) indicated that in 1966 the value of manure averaged about $4.00/ton, while the cost in time, labor, and equipment needed to transport manure from point of origin to point of use might be as much as $6 to $8/ton. However, this practice may well be the least expensive method of disposal and is being used successfully in some areas.

As a disposal method, little information is available in the literature regarding criteria for maximum soil loadings of various animal wastes that will not create a pollution problem or impair the integrity of the soil. It has only been during the last 2 or 3 years that studies really have been designed to investigate this problem. For example, at the Southwestern Great Plains Research Center of the U.S. Department of Agriculture, Agriculture Research Service, solid wastes from feed lots are being applied to the soil at rates ranging from 0 to 400 tons/acre. Crops are being grown to determine the effects on plant growth as well as changes taking place in the physical and chemical characteristics of the soil.

Liquid poultry wastes are being studied at Rutgers University in New Jersey. Applications of slurry in amounts as high as 45 tons/acre, dry weight, are being applied to grass to determine the effect on both plant and soil and to evaluate potential pollution hazards. To facilitate application and to minimize problems associated with surface applications of large quantities of these liquid wastes, special equipment has been developed for injection using a plow-furrow-cover method (62). In Georgia (11) poultry litter was broadcast on a fallow soil up to 40 tons/acre to evaluate nitrogen losses in runoff. In this instance, higher rates of application actually reduced ammonia losses in runoff because the poultry litter served as a mulch, thereby reducing the quantity of runoff and increasing infiltration.

A major factor in the safe application of animal wastes to the land is the disposition of the nitrogenous compounds. It would be desirable to establish a nitrogen balance that would permit an estimate of its excess produced in the system as a result of waste loadings. Such a balance was considered by Webber and Lane (85) for the application of poultry manure to corn, both for utilization by the crop and for disposal purposes. These efforts merely indicate inadequacies in the quantification of various processes in the nitrogen cycle, especially denitrification.

To control effectively pollution from animal wastes, new and improved methods must be developed for animal management and for waste treatment and disposal. Careful consideration must be given to the design of facilities that will be compatible with these new practices. Moreover, the cost of waste disposal will have to be absorbed as part of the production costs, and consideration should be given to converting these wastes to useful products where possible.

E. Agricultural Chemicals—Fertilizers

Fertilizer use has recently become suspect as a potential pollutant. Current alarm over eutrophication has focused attention on fertilizers as a source of nitrogen and phosphorus enrichment of our water resources. Fertilizer consumption has indeed increased rapidly in the United States, especially during the last decade. More acres are being fertilized and increased rates of application are required for more efficient production. The low cost of nitrogen has contributed to its increased use. Concern has been expressed over the fact that over 6 million tons of nitrogen fertilizer are being applied annually in the United States. However, we are consuming approximately 8,672,400 tons of nitrogen per year as protein in our food.* Thus instead of applying an excess, there is actually a deficit, which must be made up from nitrogen in

* C. H. Wadleigh and W. H. Allaway, "Nitrate in Soil, Water, and Food," presented at the Southern Fertilizer Conference, Atlanta, Ga., 1968.

the soil. Waste nitrogen in sewage rarely returns to the field and cannot be counted in the soil balance. However, the possibility of some fertilizer nitrogen finding its way into water supplies is not remote.

Evaluation of the contribution of fertilizers to water quality problems is extremely difficult because of our inability to separate that fraction originating from applied nutrients from that occurring naturally in the environment. Attempts have been made to correlate trends in nitrogen and phosphorus content of streams, lakes, and groundwater with trends in fertilizer consumption. Such correlations necessarily are not valid because of the many other sources of nitrogen and phosphorus which also may be contributory. The potential for fertilizer to pollute water depends upon the time and rate of application; other related factors—soil characteristics, rainfall amounts and distribution, and method of application—must be considered. With an increase in rates of application and total consumption, the effect from this on the nutrient content of water resources is uncertain. There is an abundance of information in the literature dealing with fertilizer-use efficiency. Some investigations have studied losses of applied nutrients, but few studies have been designed to study the significance of these losses in terms of water pollution.

1. *Phosphorus*

It is often said that phosphorus is a major factor in accelerating eutrophication because of the very low threshold value required for stimulation of algal growth. Fish play an important part in a balanced aquatic biological system which would tend to suppress the rate of eutrophication. Yet it has been shown (52) that fish production in many bodies of water was increased greatly by addition of phosphatic fertilizers. This would indicate a low availability of phosphorus in these waters. Since much of the water entering streams, lakes, and impoundments passes over or through soils, consideration must be given to the contribution of soil and fertilizer phosphorus to that water.

Phosphorus is present in all soils to some extent, but frequently is not available to plants, requiring fertilizer supplements to ensure optimum crop growth. Almost 2 million tons of phosphorus were applied to agricultural soils in the United States during 1968. To evaluate accurately the pollution potential of this added fertilizer, one must consider the forms of phosphorus in the soil, and factors affecting movement, chemical reactions, and uptake by plants. Much has been published regarding these factors (60), although little directly oriented toward pollution.

Phosphorus in soils is relatively insoluble for the most part; hence does not move readily in the soil. Applied phosphorus is readily adsorbed, or fixed, in the soil in an insoluble form. The addition of water-soluble phosphatic fertilizers may increase the amount of phosphorus in the soil solution to some extent, but actual concentrations still remain very low (40). Thus phosphorus added as fertilizers seldom moves below the surface foot of soil. Studies (5,36,71) have been made using lysimeters to determine the amount of phosphate occurring in the drainage water, and results indicate that losses from this source are negligible. Addition of phosphate fertilizers to these lysimeters resulted in no appreciable increase in phosphorus loss. Added phosphates are adsorbed rapidly by soil colloidal particles or converted to insoluble compounds such as iron and aluminum phosphates and basic calcium phosphates. Although the amount remaining in the soil solution varies with pH of the soil, soil texture, and other soil characteristics, it is reasonable to assume that there is little or no pollution hazard from phosphatic fertilizers dissolved in drainage waters under normal conditions.

The major pollution hazard is often attributed to phosphates adsorbed on soil particles that find their way into water sources through soil erosion. The actual amount of phosphorus lost per unit of eroded soil depends on the chemical characteristics of the soil, and the amount, type, and method of application of phosphorus fertilizer, and the influence of other added fertilizer materials. Studies on a corn, wheat, and clover rotation (63) showed that phosphorus losses were proportional to the amount of soil eroded and average roughly 1 lb/ton of soil loss from a Dunmore silt loam. In Missouri (46) losses of phosphorus from a 4% slope were 18 lb/acre for continuous corn, 6 lb/acre for rotation corn, and only 0.1 lb/acre from a bluegrass sod.

Long-term studies in Alabama (65) on severely eroded soils indicated that over a period of 26 years, about 82% of phosphorus compound added as superphosphate was lost; whereas, only 60% was lost when rock phosphate was used. Effects of other fertilizer treatments over a 16-year period were demonstrated in a later study in Alabama (26), in which losses due to erosion were those assumed not to be accounted for by the surface 16 in. of the soil or by crop removal. From unlimed soil receiving sodium nitrate, 70% of added phosphorus was lost; only 32% was lost from a limed soil. From a limed soil receiving ammonium sulfate, an average 75% of the added phosphorus was lost. Increased losses where ammonium sulfate was added resulted from the acidic nature of this material. Although these two studies represent phosphorus losses determined indirectly, they are cited to illustrate how various factors can affect the quantities of phosphorus lost from the soil.

Much emphasis has been placed on the role of sediment as a source of phosphorus in water resources. Although this is a significant factor in some instances, it should be mentioned that some sediments, originating from acid subsoils in Southeastern United States, for example, are quite deficient in, and strong fixers of, phosphate. Indeed, these sediments may actually serve as a sink for phosphate ions thereby removing them from solution.

Potential water pollution by phosphatic fertilizers can be minimized by good soil management practices. Incorporation of the fertilizer into the soil will prevent losses from surface washoff. Conservation practices that minimize soil loss will also reduce phosphorus losses.

2. *Nitrogen*

The problem of evaluating nitrogen pollution of water resources from applied fertilizers is more complex than the phosphorus situation. Nitrogen undergoes many transformations in soil that may lead to losses through the formation of gaseous products or soluble forms of nitrogen that are subject to leaching. In addition, not all of the nitrogen entering water supplies comes from fertilizer. Mineralization of soil organic matter, animal wastes, and plant residues also contribute, and in certain parts of the United States there are natural geologic deposits of nitrates.

In a given soil there is a natural pool of nitrogen, largely organic. Nitrogen is lost from the soil through crop removal, leaching, runoff, and formation of gaseous products. It is added through chemical and biological fixation mechanisms, rainfall, crop residues, and fertilizers. There is a vast amount of literature relating to various aspects of the nitrogen cycle, much of which was recently summarized (14). In 1966 Allison (3) attempted to review factors involved in determining the fate of nitrogen applied to soils.

In dealing with pollutional aspects of nitrogen fertilization, the major concern is leaching losses since most nitrogenous fertilizers are quite soluble. Attempts to quantify these losses have involved laboratory, greenhouse, lysimeter, and field-plot studies. In reviewing the literature on leaching losses (3,4), it became apparent that data reported were highly variable, ranging from low to high. Allison pointed out that movement of nitrates is closely associated with movement of water and is dependent upon many variables: (*1*) form and amount of nitrogen present in or added to the soil; (*2*) amount and time of rainfall; (*3*) water retention and transmission characteristics of soil as affected by soil physical properties; (*4*) antecedent soil water content in the soil profile; (*5*) presence of a crop and its growth characteristics; (*6*) evapotranspiration; (*7*) rate of removal of nitrogen by the crop; (*8*) extent of upward movement of nitrogen in the soil during a drought; and (*9*) extent to which nitrogen is leached below the root zone, especially to the groundwater.

It is extremely difficult to evaluate quantitatively leaching losses under field-plot conditions. Use of lysimeters has some limitations, but they do permit accurate measure of the amount of leaching. Lysimeter studies in California (16) indicated that leaching losses from 200 lb/acre of nitrogen applied as calcium nitrate were about 50 lb/acre of nitrogen per year. Where citrus is grown, nitrogen is applied at rates up to 300 lb/acre. Since citrus uses only 50–70 lb/yr of nitrogen, the balance of the fertilizer applied is subject to leaching losses, but not all is lost in this manner.

A lysimeter study using Lakeland sand in South Carolina (5) illustrates the complex effect of fertilizer and cropping practices on nitrogen leaching losses. Over a 5-year period one lysimeter was kept fallow; four were cropped to millet and mixed crops; and three to crotelaria and millet. The data in Table VII represent a balance sheet which shows the interactive effects of crop and fertilizer management practices. It is apparent that the millet and mixed crops had the greatest amount of nitrogen applied, yet showed the smallest leaching losses. The amount of fertilizer used alone is no indication of pollution potential. Smith (68) has pointed out that high rates of nitrogen fertilizer applied over a period of years can result in an accumulation of nitrates in the soil. The degree of accumulation varies with soil texture, being greatest in fine-textured soils and least in coarse-textured ones. Obviously, nitrates are readily leached from the coarse-textured soil. The possibility exists that nitrate accumulations in deep soil profiles may pose a groundwater pollution problem sometime in the future.

Losses due to erosion and runoff are usually considered to be small. It has been estimated that 24.2 lb/acre of nitrogen are removed annually from cropped areas in the United States by erosion (4). Undoubtedly

TABLE VII

Nitrogen Balance Sheet Expressed as lb/acre, for 5–Year Period (1933–1938) (5)

Crop	N added	N in leachate	N in crops	Soil N (gain or loss)
Fallow	25	154	0	−28
Millet and mixed crops[a]	679	18	318	+223
Crotelaria and millet[b]	263	100	67	+83

[a] Average of 4 lysimeters.
[b] Average of 3 lysimeters.

much of this is lost as particulate organic matter. Barrows and Kilmer (13) indicate that losses of soluble nitrogen salts are generally very low. A study in Georgia was reported (86) where 200 lb/acre of N as ammonium nitrate were applied to the surface of plots on a 5% slope of Cecil sandy loam. One hour after application, 5 in. of rain were applied artificially at the rate of 2.5 in./hr. Runoff from fallow plots contained 2.3% of the fertilizer applied; whereas only 0.15% was lost from sodded plots. In a similar study in Indiana (50) 2.5 and 3.2% of the nitrogen applied was lost from sod and fallow plots, respectively. Studies on the composition of Playa lakes in Texas (43) indicate little evidence of nutrients in surface runoff attributable to fertilizers. There was little difference in nitrate content of the water where the watershed consisted of 80% irrigated cropland, compared to a native grass watershed.

Nitrate losses in tile drain effluents from irrigated cropland may be a significant factor. A study in the San Joaquin Valley of California (32) indicated that nitrate losses in tile drain effluents and tail water ranged from 9 to 70% of fertilizer applied. Concentrations of nitrate nitrogen in the tile effluents ranged from 1.8 to 62.4 ppm. In contrast, in a field study in the Imperial Valley of California (45) nitrate-nitrogen concentrations in tile effluents from a cotton field fertilized with 280 kg/hectare of anhydrous ammonia ranged from 0.48 to 2.06 ppm. Differences in results obtained for these two studies could be due to different fertilizer practices, presence of natural nitrate sources, or different drainage conditions. In the latter study there was evidence of some denitrification as indicated by a reduction in nitrate concentration as the soil solution approached the water table. In Wisconsin (48) loss of nutrients by deep drainage from agricultural watersheds was determined by evaluating base flow in streams. Data from 36 drainage areas covering 643 miles^2 showed average concentrations of 1.1 lb/acre of nitrogen and 0.10 lb/acre of phosphorus. Contribution from fertilizers in this area would be very small.

Loss of applied fertilizer through leaching, erosion, volatilization, and immobilization will be greatly affected by climatic conditions, soil factors, and crop management practices. Time, rate, and method of fertilizer application are also significant factors. The effect of soil and climatic variables on potential loss of fall-applied fertilizers was discussed by Nelson and Uhland (54).

Occurrence of high nitrogen or phosphorus concentrations in surface or groundwaters associated with land intensively cropped is not proof of pollution from fertilizers. Other sources of these nutrients also must be considered. It is frequently difficult, however, to specifically evaluate the contribution of various nutrient sources within a watershed to the overall content in surface and groundwaters. Nevertheless, increasing rates of application and acreages fertilized do present a potential threat to water quality. It will be necessary constantly to improve and to develop soil, water, and crop management practices that will ensure maximum economic use and efficiency of the fertilizer applied and, at the same time, reduce runoff and percolation losses.

F. Agricultural Chemicals—Pesticides

The term "pesticides" is generally used to include insecticides, herbicides, fungicides, nematocides, and rodenticides. This group of compounds has received considerable publicity in the last few years as a

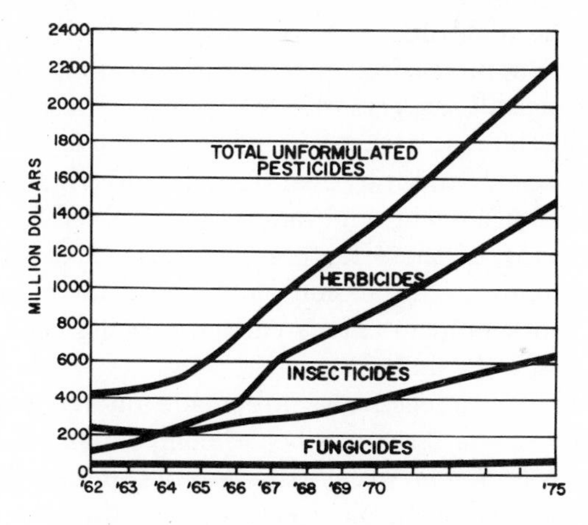

Fig. 1. Trend of pesticide sales (55).

potential health hazard. These pesticides become a problem when they, their metabolites, or their degradation products, remain in the environment after their desired purpose has been accomplished or when they reach some part of the environment other than the intended target.

The quantity of organic pesticides used in the United States has skyrocketed since 1940 (49). Recent and projected trends for pesticide sales are shown in Figure 1. It is estimated now that pesticide use will increase at an annual rate of approximately 15%; insecticide use will probably double by 1975; and use of herbicides will increase at a more accelerated pace.

When one considers that these pesticides represent some 45,000 different formulations of over 900 compounds, problems of monitoring, identifying, and generally studying their fate in the environment assume tremendous proportions. When used with reasonable caution, serious problems generally do not occur. More often, it is misuse that creates problems. Accidental spillage on the farm, industrial wastes from pesticide manufacturers, residues on agricultural by-products and wastes, drift in air (especially when applied by aircraft), and losses from fields in runoff or adsorbed on sediments following heavy rains are all contributory.

Determining the effect of pesticides on environmental quality is probably the most complex problem of all the potential pollutants listed in the "agricultural waste" category. This results from both the large number of compounds involved and the complexity of these compounds. Furthermore, the situation is complicated by the possibility that these compounds may form complex metabolites or degradation products. These also must be identified and their level of toxicity, if any, determined. There is an increasing mass of information appearing in the literature regarding all aspects of pesticides. These include residue levels in soils and plants, content in air and water, toxicity to plants and animals, chemical and physical factors affecting persistence and movement in the environment, and many other related factors. Reviews of literature on many of these factors were published in 1960 (78) and 1969 (59). The following discussion points out some of the complexities of the problem and significant areas currently being considered, without attempting to cite specific data.

Frequently, studies on pesticides are made on the basis of class of compound because of the large number of compounds being used. Some of these are shown in Table VIII. In studying the activity and decomposition of pesticides, the chemical structure of the compound, or class of compounds, has been shown to be a significant factor (34). Physical

TABLE VIII

Types of Pesticide Compounds

Insecticides	Herbicides	Fungicides
Chlorinated hydrocarbons	Phenoxyalkyl acids	Dithiocarbamates
Organophosphates	Aromatic acids	Copper compounds
Carbamates	Nitriles and phenols	
Arsenicals	Carbamates	
	Ureas	
	Amides	
	Heterocyclics	
	Miscellaneous inorganic	

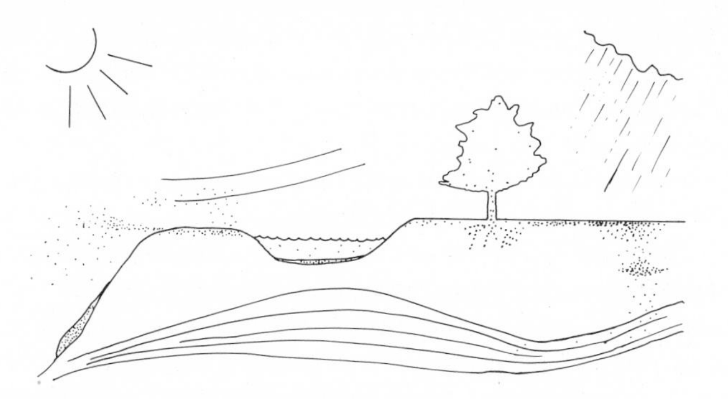

Fig. 2. Cycling of pesticides in the environment.

and chemical properties of each compound also determine the degree to which it will persist or degrade in the environment.

Pesticides move in the environment in many ways. Cycling of these compounds in the environment is illustrated in Figure 2. Transport of pesticides in air have been reported (22). Introduction into the aerial environment may occur as droplets or vapor arising from spraying operations, volatilization of compounds from soils or plants, or burning of materials containing residues. Movement in the air may occur in the vapor phase or adsorbed on dust particles and is greatly affected by meteorological conditions. Cohen and Pinkerton (22) found average concentrations of 0.19, 0.018, and 0.025 ppb of DDT, DDE, and BHC, respectively, in Ohio rainfall. Dust samples contained 0.6, 0.5, 0.04 ppm of DDT, chlordane, and 2,4,5-T, respectively. Pesticides also are carried in water and may contaminate both surface and groundwaters. In surface waters, contamination occurs by pesticides both in solution and adsorbed on entrained sediments; whereas, groundwaters are affected primarily by that in solution. Serious pollution problems result largely through misuse and disposal of wastes. Some contamination of water occurs as a result of aerial spraying such as that done in forests. Pesticide losses in surface runoff from cultivated fallow land have been reported (10). The quantity of pesticide lost in runoff depends upon the time after field application, solubility, flotation characteristics, volatility, penetration into the soil, and degree of adsorption by soil colloids. Although water pollution from cultivated areas is a major area of concern, there is some evidence of pollution of waters in forests (23,76).

Several factors have been listed as affecting movement in soils (59): the nature of the pesticide, soil type, soil moisture, soil temperature, presence of cover crops, degree of cultivation, mode and rate of application and formulation, and the degree of degradation that takes place in the soil. Aside from volatilization and degradation losses, a possibility exists that some pesticides percolate down through the soil and appear in drainage water (15).

Several chemical and physical processes determine the persistence and movement of pesticides. Biogradation of pesticides is a most significant factor in determining persistence in the environment. Alexander (2) lists the following conditions required for biodegradation: (*1*) an organism capable of metabolizing the chemical must exist in the soil; (*2*) the compound must be in a form suitable for degradation; (*3*) the chemical and the organism must achieve contact; (*4*) the compound must be capable of inducing formation of the appropriate enzyme concerned in the detoxication; (*5*) environmental conditions must be suitable for the microorganisms to proliferate and the enzymes to operate.

Adsorption and desorption are significant processes in determining the movement of pesticides. There is evidence that adsorption also affects the bioactivity of a pesticide in soil. Two types of adsorption have been described (7): one is physical and the other is chemical. Physical adsorption results from van der Waals forces, and chemical adsorption may be due to coulombic forces or the formation of hydrogen bonds. Many factors affect adsorption. The type of colloids is a most significant factor. The degree of adsorption varies with type of clay minerals and the presence of organic matter. The specific surface area and ion exchange capacity are significant characteristics of soil colloids. Soil moisture and temperature also have been shown to affect the degree of adsorption and desorption of pesticides. In addition, the physical properties of the soil and the nature of the formulation must be taken into consideration.

Other significant chemical and physical processes affect movement and degradation of pesticides. Occurrence in air has been attributed to spray drifts, but losses from soil and plant surfaces through volatilization were evident. It has been demonstrated that these losses may be significant for some chlorinated hydrocarbons and certain herbicides (35). Such factors as pesticide concentration, soil-water content, and temperature are significant factors in determining volatilization losses (70). Diffusion of pesticides in soils in either the vapor or nonvapor phase is a significant factor in determining movement and persistence. The water content of the soil, temperature, and soil structure influence the diffusion process (25). Photochemical decomposition of pesticide residues on soil and plant surfaces must be taken into consideration (53). In addition, there are processes such as precipitation, oxidation, hydrolysis, and catalytic decomposition that may alter the state of a pesticide in the soil.

Many environmental factors affect both the degradation and movement of pesticide residues in the environment (82). Reference already has been made to the effect of temperature both in the soil and air. Water is not only a significant vehicle for transport of pesticides, but the content in the soil greatly affects many of the physical and chemical processes. Sunlight, wind velocity, humidity, and precipitation characteristics are quite influential factors. The nature of the substrate also determines movement and persistence of pesticides. In the soil, factors such as structure, texture, organic matter content, type of colloids, and electrolyte concentration in the soil solution must all be considered.

Pesticide residues are of major concern because they represent a potential health hazard. Pesticides in plant products consumed by man may occur as metabolic products in the plant or as residues on plant surfaces. The amount of residue occurring on a plant product depends upon many factors (82), such as the nature of the compound and time, rate, and method of application. Climatic conditions such as sun, wind, and rain exert a considerable influence. Plant characteristics, such as foliage density, geometry of the canopy, degree of protection of edible portions, and the biochemical and physiological characteristics of different portions of the plant, must be taken into consideration. In dealing with plant products it is possible that harvesting and processing techniques can reduce hazards from pesticide residues (82). In addition pesticides may be absorbed by plants from the soil, but amounts vary for different crops and even among varieties. Evidence of this has been cited for vegetable crops and alfalfa (59).

Questions have been raised regarding the toxicity of pesticides for livestock. Much of the work done prior to 1960 was summarized by Radeleff and Bushland (61), who state that the majority of pesticides are not hazardous to livestock under conditions of ordinary usage or even misuse to a moderate extent. The toxicity hazard of pesticides to livestock is affected by such factors as age, species, formulation, lactation, emaciation, and stress. Although pesticide residues on forage may not be toxic to livestock, these materials are accumulated in the tissues and excreted in the milk, thereby creating a potential hazard to people consuming these products. Fish often have been cited as being notorious bioaccumulators of pesticides in water. This can impair their value as food, impair biological functions, or result in fish kills. In 1967 there were 32 reports of fish killed in various waters of the United States (27) attributable to pesticides.

Little has been said thus far about pesticides containing trace elements. Yet these elements can effect plant growth and be absorbed in the food chain when they accumulate in soils (38). Of these, copper and arsenic are of major concern. Copper compounds have been used for many years as a fungicide. Copper, itself, may accumulate in soils either adsorbed on colloidal surfaces or as organometal complexes. In the cupric form this element is not readily available to plants. In acid soils or where reducing conditions prevail, the cuprous form is favored, and this is more readily available to plants. Arsenic also tends to accumulate in soils where arsenical pesticides are used. Arsenates are immobilized rapidly in soils in a manner similar to phosphates. However, under certain conditions arsenic is available to plants in limited amounts. Generally, high concentrations of available arsenic in soils will appreciably decrease yields of plants before it becomes a hazard through plant uptake.

This discussion portrays the complexity of problems arising from the use of pesticides. Many studies have been made on all aspects of the problem, but it should be kept in mind that not all pesticides are used by agriculture. Hence when evaluating pesticide survey data from various bodies of water, urban and industrial contributions must be taken into consideration. To evaluate agriculture's contribution, a monitoring survey was established by the U.S. Department of Agriculture (51). In 1964 six 1-mile^2 study areas were established in different parts of the United States to determine existing pesticide residue levels and changes in those levels that might occur during the study. The following summary of results was reported in 1969.

> Soil analyses indicate that residue levels of the more persistent pesticides such as DDT (including the TDE and DDE isomers), dieldrin, and endrin did not change appreciably during the study period. Residues in soils were generally higher in the fall than in the spring, particularly when applications of the respective pesticides were made between the spring and fall sampling. As might be expected, the largest residues in soils were usually found in the fields where the greatest amounts had been used.
>
> Analyses of paired crop and soil samples show a wide variety of pesticides and amounts recovered. Residues of combined DDT, endrin, and dieldrin were the most frequently found, however. Residues of at least one of these three chemicals were detected in 88 percent of the 17 crops sampled at the seven study areas. DDT was found in 76 percent of the crop samples; endrin and dieldrin were each found in 30 percent of the crop samples. The only other pesticides found in the crop samples were methyl and ethyl parathion and dicofol, but these chemicals were detected in only a few scattered samples at Yuma, Ariz.
>
> Residue levels found in crop samples were generally below 0.1 ppm. Somewhat higher levels were found in crop plant samples, but were princi-

pally the result of foliar applications before sampling. It does appear, however, that translocation of pesticides from the soil into crops does occur in the areas where samples were collected. Some residues were found in crops and crop plants taken from fields that were not treated with the pesticides found during the period of study.

Water sample analyses indicate that only very small amounts of pesticides were present in any of the sources sampled. By and large, the levels detected were below 1 part per billion. DDT, its metabolites TDE and DDE, and dieldrin were the most prevalent pesticides in water. Residues of any pesticide were most frequently found in contained surface sources and in exit water at Yuma Ariz., indicating that pesticides are carried into water sources from cropland by normal drainage or irrigation. Quick-runoff water data also indicate that pesticide residues are picked up from the soil by runoff water and are carried into water sources with sediment.

Analysis of sediment shows residues in the magnitude of decimal fractions of a part per million to a high of 4.90 ppm. DDT and its isomers, TDE and DDE, were the principal pesticides found in sediment from any of the areas where it was collected. Dieldrin and endrin, however, were also found in sediment from Greenville, Miss.; Stuttgart, Ark.; and Theodore, Ala. These pesticides evidently entered the ponds in sediment carried from fields by drainage water.

Pesticides undoubtedly constitute a hazard in the environment. Relatively large amounts have been found in the fatty tissues and other organs of fish. Small residues on plant materials may result in an accumulation of pesticide in the fatty tissues of birds and animals consuming those plants. It is beyond the scope of this discussion to evaluate the significance of these accumulations. However, a great need exists for determining realistic levels of pesticides that can affect human health. There are some potential hazards in using pesticides, but one must also take into consideration benefits to human health and the need for efficient production of quality food and fiber crops.

Continuing research efforts (67) can help minimize these hazards through: (*1*) improving conventional chemicals and methods of application to minimize residue hazards; (*2*) the development and improvement of nonconventional chemical, biological, physical, cultural, and related methods of control; and (*3*) the development of insect and disease resistance in plants and animals.

G. Processing Wastes

Although not directly a part of farming operations, wastes from the processing of agricultural products are considered within the scope of agricultural operations, including waste products from processing forest products, slaughtering and processing meat animals, processing dairy products, canning fruits and vegetables, and processing grain products. In many instances usable products have been developed from some of the solid wastes. The nonusable solid wastes often present a disposal problem and cause annoying fly and odor problems. Burning these wastes creates air pollution problems. New processing techniques have resulted in a decrease in waste production in some industries. Effluents from processing plants present a difficult problem in relation to disposal.

Processing plant effluents are usually objectionable because of their oxidative demand. The contribution of various processing industries to pollution in terms of BOD is shown in Table IX. Data on population equivalent represent the pollutional equivalent of the waste compared to the BOD of normal sewage (30). Aside from large concentrations of organic matter, these effluents may also contain some quantities of dissolved salts and suspended colloidal material. Many of these wastes now cannot be eliminated directly into receiving streams unless prior treatment renders them safe.

Solid wastes resulting from processing agricultural products are difficult to eliminate. Application to land could cause foul odors and fly problems. Not all solid wastes are amenable to incorporation into the soil because of limited degradability and the possibility of creating unfavorable physical or chemical conditions in the soil. Burning of these materials creates air pollution problems. Some wastes have been used for land fills and some for animal feed. Attempts have been made to compost or digest solid wastes, but again, the disposal of a product results. The development of marketable by-products is desirable if feasible. Since these wastes vary greatly in their characteristics, disposal methods must be designed specifically for a given product.

TABLE IX

Estimated Loadings of Selected Agricultural Processing Industries (30)

Processing industry	Annual production, million lb	Lb BOD/ 1000 lb processed	Potential daily load, 1000 lb	Potential daily pop. equivalent, millions
Canneries	—	—	1,300	8.0
Dairy products	—	—	1,982	11.93
Meat processing	59,400	14.0	2,000	13.0
Poultry	8,200	10.0	225	1.3
Potato processing	18.4	89.7	347	2.11
Sugar refining	95,000	3.0	800	4.8

Disposal of processing plant effluents can be accomplished in two ways: by land disposal by spray irrigation, or by suitable pretreatment methods which render the effluent safe for disposal into receiving streams. Many treatments have been used singly and in combination with others, including sedimentation, filtration, use of both aerobic and anaerobic lagoons, and chemical methods (oxidation and precipitation). Attempts have been made to use percolation and evaporation lagoons to treat completely and discharge effluents. Treatments must be evaluated carefully to avoid pollution of surface or groundwaters.

Land disposal of effluents has considerable promise, but the design of such systems is still more of an art than a science (31). Many factors must be considered which will permit this type of disposal either on a seasonal basis, as with canneries, or as a continuous operation. The composition of the effluent must be such that no residues will accumulate in the soil that could deteriorate the soil or become toxic to plant growth. Where effluents are used to irrigate food crops, there must be no toxic component capable of entering the food chain. Effluents must be pretreated to remove any suspended solids that may clog sprinkler nozzles or reduce the infiltration capacity of the soil.

When applied to the soil, effluents are dissipated through both percolation and evapotranspiration. Careful site selection for soil type and topography is essential to ensure maximum infiltration and minimum runoff and erosion. Plant cover is essential for water dissipation through transpiration and to prevent a decrease in percolation from the surface sealing action of heavy irrigation application rates on the soil. Consideration must be given to the plant tolerance to continuous wet conditions. Climatic factors are significant since they affect evapotranspiration losses. The amount and seasonal distribution of rainfall can affect the loading rate. In northern climates, snow cover and frozen soil present a problem where year-round disposal is required. Land disposal can be a very satisfactory means of renovating effluents, but all related factors must be taken into consideration when designing a system.

III. EVALUATING POLLUTION FROM AN AGRICULTURAL WATERSHED

Water pollution from agricultural wastes indicates that the water quality at a given point has been impaired for a specific use. The water quality at that point reflects the integrated effects of the entire contributing watershed. Unfortunately, many researchers have looked in depth only at small segments of the problem, whereas resource planners have attempted to extrapolate fragmented bits of information to large watersheds to explain their pollution problems. There is an urgent need to bridge this gap and provide the information necessary to permit a reasonably accurate evaluation of pollution from agricultural sources on a watershed basis.

We have acknowledged that agricultural wastes do contribute to water pollution. In addition, we have stated that some of the accusations are made on the basis of estimates made on rather limited factual information. An example of this was found in the Lake Erie Report (39), which states:

> If an estimated rate of 250 pounds of total phosphorus per square mile per year is used to calculate the agricultural contribution, almost six million pounds are contributed to Lake Erie per year from this source. The nitrogen input from runoff is at least ten times this amount.

If true, this would constitute a significant source of pollution to Lake Erie. These values represent only an estimate of the integrated pollution contribution from the various land use systems which may occur in an agricultural watershed. No distinction can be made as to what part results from natural processes and what part results from farming operations or nonagricultural activities.

In some river basin studies estimates of agriculture's contribution to nutrient pollution of rivers are based on fertilizer use data and the population pollutional equivalent of all the farm animals in the basin. Apparently, it is assumed that a large portion of the fertilizer and animal wastes find their way into streams or other water resources. Such assumptions are unwarranted. Estimates are made of pollutional contributions from rainfall and municipal and industrial wastes; the balance by difference is attributed to agricultural sources.

In the spring of 1964 a major fish kill reported in the Mississippi River was attributed to endrin in runoff from treated agricultural lands. With no justification for such an assumption apparent, the Agricultural Research Service, with the assistance of the Corps of Engineers, made a study of the sediments in the Mississippi River. Results indicated that the major source of the endrin originated at the location of a manufacturing plant producing this material (12). This is but one example of how erroneous conclusions can be drawn in the absence of factual data.

High nitrate content has been reported in the groundwater of the San Joaquin Valley (21). Values ranging 5–53 ppm of nitrate-nitrogen and 0.05–0.70 ppm of phosphate-phosphorus have been reported in tile drain flows from different parts of the valley.* This is an area of irrigated agriculture that uses considerable amounts of fertilizer, and this is often referred to as a major contributing factor. It is extremely difficult to evaluate the contribution of fertilizer to this high nitrate content since other sources also contribute. This would include nitrate in rainfall, nitrogen fixation in soils, sewage effluents, decaying plant materials, and naturally occurring nitrate deposits in the substrate. Development of remedial measures would be more effective if a reasonably accurate nitrogen balance could be established in this valley that would pinpoint sources.

Another example is that of the Sangamon River in Illinois (29). Monthly nitrate values range from 760 lb/day in October to 188,000 lb/day in April. A point is made that there are only 7000 to 8000 people served by this part of the Sangamon watershed. If one assumes an average daily per capita nitrate contribution from waste treatment plant effluents of approximately 0.11 lb, this would amount to between 750 to 850 lb/day of nitrate on a regular basis throughout the year. This corresponds to the total nitrate load in October, but is less than 0.5% of that for April. Some would use data such as these as evidence of pollution from fertilizers and animal wastes. These data merely show that there are nitrate sources other than municipal sewage. This nitrate may originate from other urban or industrial sources, from agricultural sources, or from other nonurban areas; it may result from either man's activities or natural sources. One fact is evident: more factual information is necessary.

Research on agricultural wastes, such as pesticides, animal wastes, plant nutrients, has been conducted in the laboratory and greenhouse to evaluate chemical and biological transformations. Field studies have been conducted using lysimeters and small plots in which the amount and chemical composition of runoff and leachates were determined. These investigations have attempted to study in depth various segments of the pollution problem. However, few have attempted to study in detail an entire watershed to integrate the various specific effects which contribute to the quality of water leaving that watershed. This is essential; yet the gap is evident from the examples cited above.

To evaluate sources contributing to the pollution of a given water resource, the entire watershed must be taken into consideration as a dynamic system. As pointed out by Bormann and Likens (17), the

* L. R. Glandon and L. A. Beck, "Monitoring Nutrients and Pesticides in Subsurface Agricultural Drainage," presented at the Meeting of the American Geophysical Union, San Francisco, Calif., December 16, 1969.

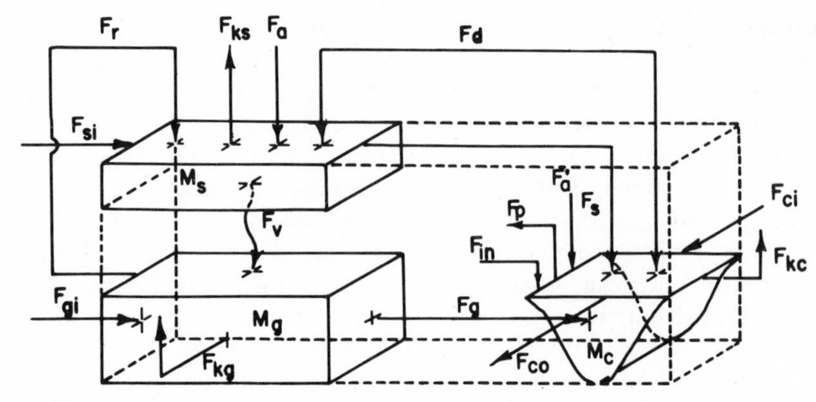

Fig. 3. A proposed water quality model. Courtesy P. C. Woods (91).

watershed ecosystem provides an ideal setting for studies of ecosystem dynamics. If one is to quantify the input and output of a given pollutant, it is essential also to quantify the hydrologic input and output for a specific watershed.

There are considerable water quality data available in the literature for surface waters, groundwaters, and lakes. Survey data of this type are very useful in monitoring water quality. Should high values for a water quality deterrent be obtained, however, the data may not always be adequate to pinpoint the source of pollution. This is especially true for agricultural watersheds. On the other hand, studies on pollution from agricultural sources have included only small segments of the problem. Very often concentrations of pollutants in water or soils have been measured without giving due consideration to flow volumes or fluxes. More specific data are necessary to evaluate fully the effects of various agricultural management practices. We must evaluate a system.

Mathematical models are good tools for evaluating systems. Several models have been proposed for studying hydrologic systems. Recently, Woods (91) proposed the water quality model shown in Figure 3, with appropriate unit notations described in Table X. This model is designed to meet the following basic requirements:

1. Compatibility with a "dynamic" hydrologic model of the same system.
2. Facility for accommodating time-dependent decay functions.
3. Facility for time-delay of quality constituents brought about by interaction with the physical media through which the constituents must pass.

TABLE X

Water Quality Model Notation

Parameter	Description
F_a	Pollutant applied directly to land element.
F_a^1	Pollutant applied directly to surface water element.
F_{ci}	Pollutant inflow to surface water element.
F_{co}	Pollutant outflow from surface water element.
F_d	Pollutant diverted and applied with irrigation water to an adjacent land element.
F_g	Pollutant flow between groundwater element and surface water element.
F_{gi}	Extraneous pollutant flow between groundwater elements.
F_{in}	Pollutant inflow to surface water element from extraneous surface sources such as small tributary streams.
F_{kc}	Pollutant extraction (decay, volatilization) from surface water element.
F_{kg}	Pollutant extraction (decay, volatilization) from groundwater element.
F_{ks}	Pollutant extraction (decay, volatilization) from soil zone element.
F_p	Pollutant contained in extractions from surface water element for export to other hydrologic units.
F_r	Pollutant recirculation from drains for irrigation use in the same unit.
F_s	Pollutant contained in surface return water from irrigated element to surface water element.
F_{si}	Pollutant in surface flow to irrigated element such as might be carried by intermittent channels, small streams, or a diversion from another unit.
F_v	Pollutant in verticle flow between soil zone and groundwater.
M_s	Pollutant stored in soil zone element.
M_g	Pollutant stored in groundwater element.
M_c	Pollutant stored in surface water element.

This model is composed of three elements: a surface storage element such as a stream, lake, or other body or surface water; a groundwater storage element; and a land element which may extend from the soil surface and vegetation downward to include some perched water and some transient water.

Movement of a given pollutant in this model could be described by the generalized equation:

$$F_x = K_x Q_x C_x$$

where F_x is the mass rate of flow of the constituent; K_x is a distribution coefficient; Q_x is any flow in the hydrologic model; and C_x is the material concentration. Each of these quantities could be space and time dependent in the proposed model.

The model described above takes into consideration the movement of a pollutant into and out of a given system. It also characterizes the movement of a pollutant within the three elements of the system. Attempting to use this approach to solve a watershed situation becomes somewhat complex, largely because a watershed is composed of many "land elements." Both the hydrologic and pollution characteristics of the land element will vary with land use differences, soil chemical characteristics, soil water retention and transmission properties, vegetative cover, and treatments applied.

Evaluation of water quality emanating from a watershed must integrate the effects of the various land elements contained therein. Let us use the hypothetical watershed shown in Figure 4 as an example. Pollutional contribution from the wooded areas (*1*) will probably be negligible. The contribution of the cultivated areas (*2*) might give rise to sediment production and perhaps loss of some pesticides or nutrients in runoff. The presence of a barnyard around the homestead (*3*) adds the contribution from animal wastes. Assuming some of the runoff from

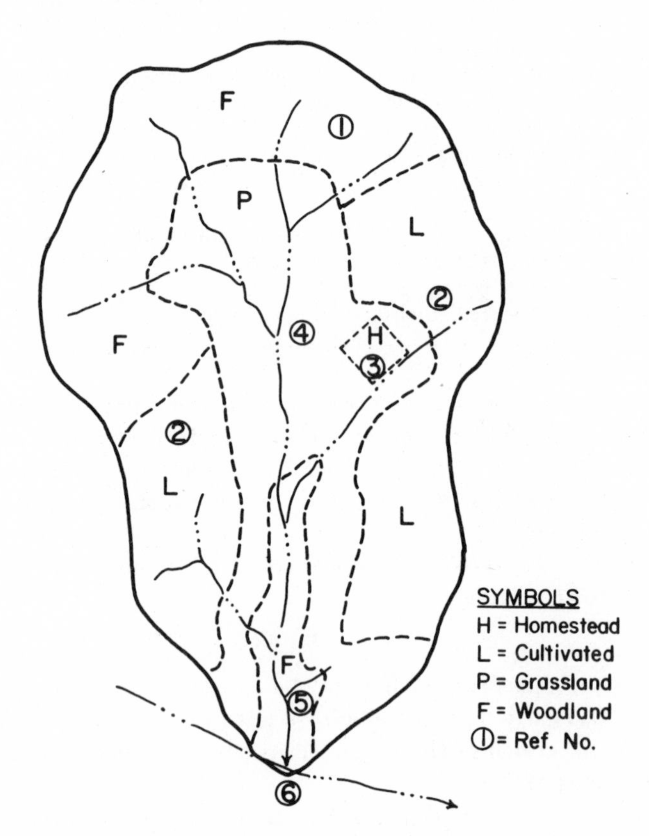

Fig. 4. Hypothetical watershed.

cultivated areas (*2*) passes over the grassland (*4*), this grassland can act as a sink rather than a source, by trapping some of the sediment and preventing some of the runoff from reaching the waterway. The forested area (*5*) might be a broad bottom area that could also serve as a sink for some pollutants. During heavy storms, there could also be some runoff from some of the grassed area which, if heavily grazed, could also contribute animal wastes to the pollution load. The effects of each of these land use areas on the amount of water and dissolved solutes reaching groundwater and appearing as base flow must be taken into consideration. Finally, the quality of water leaving the watershed (*6*) is the integrated effect of surface runoff and base flow contributions from the various land elements in the area. There may be one or many land use elements within the smallest definable natural watershed.

Evaluation of water quality from a watershed must include flow data as well as concentration of specific pollutants. This is true whether one is sampling a small element within a watershed or the entire watershed. Sampling becomes a highly significant factor, especially during storms that produce runoff. Variations in pollutant concentration should be determined on the basis of a storm hydrograph to enable calculation of the total pollutional contribution from a given land area during that storm. Measurements of base flow, when added to storm runoff, will give the contribution for a longer period of time. Variation in flow rate and pollutant concentration must be assessed accurately throughout the year. A study of this type was done at Coshocton, Ohio, on a 303-acre watershed by Taylor.* An example of the type of data obtained is shown in Figure 5, illustrating variations in the concentration of phosphorus in the water, volume of flow, and pounds of phosphorus in the water leaving the watershed. Integrated over a period of a year, it was determined that 0.031 lb/acre of soluble phosphorus left the area. This study is extremely interesting, since it illustrates the importance of relating pollutant concentration to flow volumes.

The significance of water quality leaving a watershed is dependent upon what happens downstream and its final quality at point of use. If water continues to flow into a river, pollutants may be diluted by better quality water from other watersheds or the pollution load may be increased. If water flows into a lake or an impoundment, certain pollutants may accumulate and seriously affect aquatic biota. The same quality water, continuing in a stream or river, could have a lesser pollutional effect.

* A. W. Taylor, U.S. Soils Laboratory and North Appalachian Experimental Watershed, unpublished data, U.S. Department of Agriculture, Agricultural Research Service.

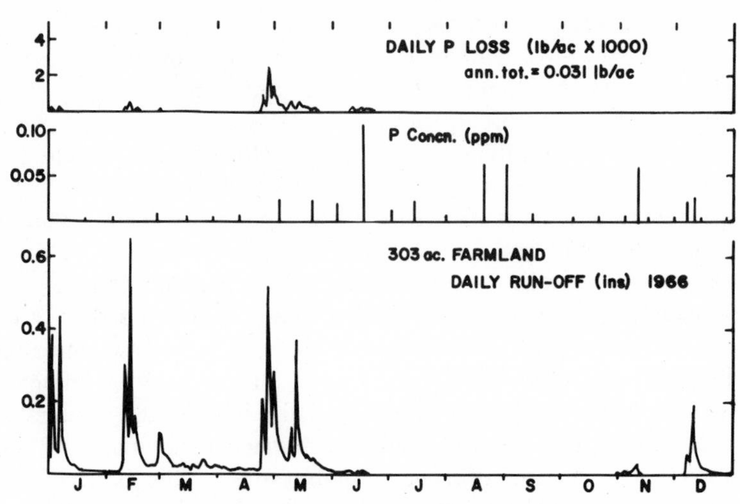

Fig. 5. Runoff and phosphorus loss from a formed watershed in Coshocton, Ohio. From A. W. Taylor, unpublished data.

Pollutional loadings from municipal and industrial sources usually do not vary greatly throughout the year. The pollutional contribution from an agricultural watershed may be highly variable. Cultivation practices, time and method of application of agricultural chemicals, character of the vegetative cover, and harvesting practices all affect the temporal aspects of the problem. Even for a given set of land use and management practices, the pollution potential will vary with climatic conditions. For these reasons, sampling and analytical techniques are critical in correctly evaluating temporal changes, as well as the total, of the pollutional contribution from an agricultural watershed.

IV. FACTORS AFFECTING POLLUTION FROM AN AGRICULTURAL WATERSHED

Many factors determine the pollutional contribution to water leaving an agricultural watershed. Some are inherent in the land itself; others result from man's activities that alter the watershed system. Since pollutants move in water, watershed characteristics that affect water movement will also influence the movement of pollutants.

Land use is a major factor in determining movement of water, hence the movement of any pollutants present. Timbered land generally produces little runoff from rainfall except where steep slopes and very shallow soils occur. However, runoff from snow melt in these areas can be appreciable. Unless fertilizer or pesticides are applied to these areas, the pollution potential is small. Sediment production is negligible. Some nitrogen and phosphorus may enter water courses as a result of decomposition of organic residues, but the amount is small compared to other types of land use (75). There is some evidence, however, that clear cutting of a forest greatly increases nitrate losses (18). Good grassland also has a low pollutional contribution unless large amounts of animal wastes occur as a result of intensive grazing. The amount of runoff and sediment production from grassland will depend upon the vegetative cover, management practices, and soil and topographic characteristics. As indicated previously, in a hypothetical watershed, timber and grassland actually may serve as a sink for pollutants rather than a source. In general, pollution problems arise when land is intensively cultivated or when used for intensive livestock production.

Crop rotations and management systems can greatly affect movement of pollutants within a watershed. Conservation practices that minimize runoff and sediment production also minimize movement of potential pollutants. For example, level terraces for corn in Iowa greatly reduced both sediment losses and runoff when compared to field-contoured corn, or even grass, as shown in Table XI (64). Level terraces restricted surface movement of water and greatly increased deep percolation. The pollution potential for soluble materials is much less in water moving down through the soil than in surface runoff. In some semiarid areas use of terraces for soil and water conservation may create salinity problems. Proper fertility treatments can provide good soil cover and reduce runoff and nutrient losses. Smith (69) has shown that during two rainstorms the nitrate loss in runoff and erosion was actually less where 177 lb/acre of nitrogen had been added in a continuous corn system than in rotation systems with little or no fertilizer addition.

TABLE XI

Data Summary for Treynor, Iowa, Watersheds, 1965 (5)

	Field-contoured corn	Grass	Level-terraced corn
Drainage area, acres	82.6	107.0	150.0
Rainfall, in.	44.4	44.3	44.9
Water yield, in.	13.65	9.22	13.07
Surface runoff, in.	10.69	4.61	2.52
Base flow, in.	2.96	4.61	10.55
Sediment yield from sheet erosion, tons/acre	44.4	1.2	1.0

The presence of growing plants in general may affect movement of wastes. During summer months water losses through evapotranspiration tend to prevent movement of water and solutes below the root zone. In contrast, during dry seasons water from below will move into the root zone where salts or other soluble pollutants tend to accumulate. In some instances plants can be used to absorb excessive nutrients from waste waters that are applied to the land. Good plant cover tends to increase infiltration of water into the soil, thereby minimizing runoff and sediment production.

Associated with good management practices is the judicious use of fertilizers and pesticides. Both groups of chemicals could be potential pollutants within a watershed if improperly used. Fertilizers that are properly applied in amounts designed to give maximum use efficiency and economic return normally do not present a problem. However, the low cost of nitrogen fertilizers is often an incentive to farmers to overfertilize, thereby creating an excess of nitrogen in that portion of the environment. This may create a pollution hazard depending upon management practices employed and other environmental factors. Similarly, pollution from pesticides is usually a problem when the manufacturer's recommendations are not followed, and containers and wastes are disposed of carelessly. Hence any potential pollution from these chemicals will be determined primarily by the manner in which they are used.

Animal management practices influence the quality of water leaving a watershed. There are three categories to be considered. The first is that which occurs in pastures. If pastures are not stocked too heavily, no great hazard exists. Little runoff occurs and leaching losses are minimal. The second relates to heavily stocked feed lots where cattle are confined in areas ranging from 150 to 450 ft^2/head. In the past these were constructed on slopes to permit water to carry away the wastes. This is no longer permissible, and feed lots must be designed to minimize surface water pollution through runoff and groundwater pollution through deep percolation. Inevitably, the accumulated solid wastes in these feed lots must be eliminated. Land is a natural repository for these wastes. Manure can be used as a soil amendment and applied at appropriate rates, or maximum loadings can be applied solely as a means of disposal. In the latter case application rates must not be high enough to destroy the integrity of the soil or create a pollution hazard. The third category involves confined animal enterprises such as dairies, poultry production, and egg factories. Since the barns and poultry houses in these large operations must be cleaned frequently, unlike the feed lot situation, wastes must be disposed of on a regular basis. Liquid wastes are treated by means of oxidation ditches and lagoons. Solid wastes and sometimes liquid wastes are eliminated directly on the land. Where the ground is frozen or covered with snow during the winter, disposal becomes a greater problem. The type of animal enterprises and management practices used will determine to a large extent the pollution potential from animal wastes within the watershed.

Physical and chemical properties of soil are important attributes of a watershed. They not only have a great influence on movement of water, but also upon the retention or movement of chemicals. Water retention and transmission characteristics of the soil determine water movement, hence movement of solutes. Soil properties are important in determining particle detachment and sediment production. The fate of chemicals in the soil is significantly affected by the biological characteristics of the soil. Kunze (37) and Bailey and White (7) have stressed the importance of the interactions of clay minerals and other soil colloids with pesticides in soil and find rather complex relationships. Organic matter also has been demonstrated to be a significant factor in adsorption and retention of pesticides in soils (81). Adsorption of pesticides and phosphates on soil colloidal material is a significant mode of transport for these chemicals.

Climate and seasonal factors exert a great influence on the movement of wastes in a watershed. Rainfall frequency, intensity, and distribution determine to some extent the amount of water infiltration into the soil profile, occurrence of surface runoff, particle detachment, and amount of sediment transport. Antecedent soil water content prior to precipitation greatly influences the effect of a given precipitation event. Snow cover and frozen ground tend to decrease infiltration and increase surface runoff. This is a significant factor where solid animal wastes must be applied to the land in the winter, or where disposal of waste effluents by sprinkler irrigation must be carried on throughout the year (58). Climatic conditions affect evapotranspiration rates and, as previously mentioned, movement of water and solutes in soils. Temperature is a significant factor in determining volatilization losses of certain wastes, and it greatly affects the rate of chemical and biological transformations of nitrogenous compounds and other organic and inorganic pollutants in the soil, on the soil surface, and on plants. Wind is responsible for movement of soil particles, dusts, and pollen into the air from an agricultural watershed. It also determines movement of volatile waste compounds and serves as a mechanism for transport of such pollutants as pesticides in rain and in dust particles. Undoubtedly, climate greatly affects the movement of wastes within and from a watershed.

In irrigated areas, salinity is usually the major water quality problem as discussed earlier in this report. Generally, the salt content of the drainage water exceeds that of the irrigation water applied in arid areas. Some concern has been expressed over movement of nitrates in irrigation and drainage waters from one area to another. It is true that high nitrate concentrations have been found in drainage waters from some irrigated areas, but not all of this is the result of fertilization. Bower and Wilcox (19) analyzed nitrate data from an Upper Rio Grande study cited earlier as a salt balance study. They found that over a 30-year period nitrogen fertilizer use in three irrigated areas adjacent to the river increased from a very low level to a very high one. Yet the overall nitrate concentration of the river did not increase during this period, indicating no significant stream pollution from the applied fertilizer. Expanding use of herbicides for control of aquatic and bank weeds in irrigation and drainage systems has caused some concern over hazards from residues in the water. In a recent review of the subject Timmons *et al.* (79) concluded that careful use of registered herbicides should result in no serious problems. Most herbicides dissipate as they move downstream. There is little evidence to indicate injury to, or contamination of crops produced when herbicides are properly used.

Concern has been expressed regarding the presence of nitrates and phosphates in waters and their relationship to the eutrophication problem. Animal wastes and fertilizers are considered to be the major sources of nitrogen and phosphorus in waters leaving an agricultural watershed, but it is essential that natural sources of nitrogen and phosphorus also be considered. For example, a certain amount of nitrogen occurs in precipitation. In reviewing literature, Stevenson (72) found values reported ranging from 0.7 to 18.7 lb/acre/year with a majority of values in the 4–7 lb range. Examples of much higher values were found for some areas. Another source of nitrogen is that from natural nitrate deposits. Mansfield and Broadman (44) state that conditions under which nitrate deposits may occur are so widely distributed that minor amounts might be expected to occur in any part of the United States. Indeed, Smith (69) refers to nitrate deposits in the caves of Missouri in discussing nitrate content of groundwater in that state. In further discussing ways in which nitrogen is added to the soil, Stevenson (72) lists biological fixation by both symbiotic and nonsymbiotic organisms as being significant but also mentions nonbiological fixation and adsorption of nitrogenous gases from the air. The quantity of nitrogen fixed by biological activity may vary from a few pounds per acre to over 100 lb/acre, depending upon many factors.

Stout and Burau (74) point out that soil organic matter is the largest single land reservoir of fixed nitrogen. In a comparative study of nitrate movement in deep profiles of well-aerated, permeable California soils, they concluded that soil fertility is a major factor in determining the concentrations of nitrate that exist below the root zone. Their studies showed that a fertile soil, cultivated or noncultivated, has a potential nitrate-supplying power which could contribute as much as 100 ppm of nitrate. Stanford* estimates that during the last 70 years, loss of organic nitrogen from our soils through denitrification and mineralization has approached 20 million tons a year. Other organic sources such as peat and mucks and decayed plant material also are potential natural sources of nitrate in water when mineralization takes place.

Phosphorus, one of the principal nutrients controlling the growth of algae, is also a natural constituent of all soils. The amount varies widely in soils, but the largest amounts, ranging from 800 to 2700 lb/acre, are found in Western states. Phosphorus is generally immobile in soils and that applied as fertilizer is rapidly rendered insoluble. It becomes a pollutional problem through soil erosion, and it has shown that phosphorus losses are proportional to soil loss (63). Phosphorus is tightly adsorbed on soil colloids, and the amount in solution in a given turbid stream is a function of the equilibrium between the water and the suspended sediments. It was previously stated that these sediments could serve as either a source or sink for dissolved phosphates. Taylor (77) states, however, that the amount of biologically active phosphorus in a sediment-laden stream is a small fraction of the total present. Mineralization of organic matter may also contribute to the phosphorus content of water. Furthermore, although we are not considering farm animal wastes here, wastes from wild animals and birds must be considered as natural sources of nitrogen and phosphorus.

It is difficult to find an agricultural watershed that does not have people living on it. The number of people may range from a few farm families to a small community. Hence domestic wastes must be considered in determining the pollutional output of that watershed. Numerous estimates have been made of the nitrogen and phosphorus contents of these wastes. Values of 1.7 and 7.0 lb of phosphorus and nitrogen, respectively, per capita per year have been suggested for private domestic waste disposal systems.

Attempting to evaluate the pollutional output of an agricultural watershed is, therefore, a complex problem. It is a dynamic system, having both fixed and variable inputs, and it is controlled to a large extent by the unpredictable qualities of the climate. The physical attributes of the watershed do not change appreciably, but farm management practices can and do change. The effects of agricultural wastes on pollution

* Stanford, personal communication.

will vary seasonally, according to existing management systems. Alteration of these systems can change the pattern of these pollution problems. A thorough understanding of processes involved in the movement and disposition of agricultural wastes can lead the way to the development of management practices that can minimize the pollution hazard.

V. SUMMARY

In a message to Congress on July 21, 1969, President Nixon expressed concern over problems associated with population growth, both on a national and worldwide basis. He assigned high priority to "research into new techniques and to other proposals that can help safeguard the environment and increase the world's food supply." Increasing the efficiency of food production involves the development and adoption of new technology that may pose a threat to environmental quality. Unfortunately, concern for environmental quality in the past has lagged behind the development of this advanced technology in the United States. We have adopted more intensified systems of livestock production, increased our use of agricultural chemicals, and expanded mechanization at a rapid rate, but only in recent years have we given consideration to the impact of this technology on environmental quality. As a result, our research effort to date has neither enabled us to assess adequately the effect of agricultural wastes on overall environmental quality nor to efficiently develop or modify technology to minimize pollution hazards.

Definition of a problem is the first step toward its solution. It is necessary to identify, quantify, and evaluate the significance of agricultural wastes in environmental pollution before efficient management practices can be developed for control and abatement.

When speaking of a water resource, such as a river, lake, or aquifer, one must evaluate the inputs of all the individual sources within the watershed that contribute to that resource. Distinction should be made as to which contributions are derived from sources unaffected by man, those which result directly from agricultural operations and those arising from man's activities not directly related to agriculture. The water resource and its watershed must be treated as a system. When evaluating the pollutional contribution of a given practice, its impact on water quality at the point of use is a significant factor. Effects on soils, plants, and animals should also be taken into consideration.

It is not possible to eliminate completely all agricultural wastes as a source of pollution. It is possible however, to minimize these to a level where they no longer constitute a hazard. To do this we need specific definitions of what constitutes a hazard in soils, water, plants, animals, or air from the various agricultural sources. Once criteria are established, management practices can be developed to achieve desired levels of pollution control and abatement. It may not be possible to achieve these controls without some cost to the farmer, and these costs may have to be absorbed as part of overall production.

There is a need for improving existing management practices and developing new ones to make advances in production efficiency compatible with safeguarding the quality of our environment. Many problems can be solved through curbing runoff and soil loss. Existing erosion control practices, if universally adopted, would be quite effective, but these must be constantly modified to meet intensification of agricultural operations. Technology for increasing fertilizer use efficiency would greatly minimize pollution hazards from plant nutrients. Much can be done to improve the design of facilities for intensive animal production and to develop technology for more safe and economic disposal of animal wastes. Alternatives for the use of persistent insecticides are currently being investigated. The use of herbicides will continue to increase and new compounds developed must be evaluated for safety. New and improved methods for pesticides applications would help decrease pollution hazards.

References

1. F. A. Albert and A. H. Spector, "A New Song on the Muddy Chattahoochee," in *Water, U.S. Dep. of Agric. Yearbook* 205–210, 1955.
2. M. Alexander, "Biodegradation of Pesticides," in *Pesticides and Their Effect on Soils and Water,* Amer. Soc. Agron. Spec. Publ., **8,** 78–84 (1966).
3. F. E. Allison, "The Fate of Nitrogen Applied to the Soil," in *Adv. Agron.,* **18,** 219–258.
4. F. E. Allison. "Evaluation of Incoming and Outgoing Processes that Affect Soil Nitrogen," in *Soil Nitrogen,* Amer. Soc. of Agron. Monograph **10,** 573–606 (1965).
5. F. E. Allison, E. M. Roller, and J. E. Adams, "Soil Fertility Studies in Lysimeters Containing Lakeland Sand," U.S. Department of Agriculture Bulletin 1199, 1959.
6. *Agriculture and the Quality of Our Environment,* N. C. Brady, Ed., AAAS Publ. 85, 1966.
7. G. W. Bailey and J. L. White, "Review of Adsorption and Desorption of Organic Pesticides by Soil Colloids with Implications Concerning Pesticide Bioactivity," *J. Agr. Food Chem.,* **12,** 324–332 (1965).
8. S. Baines, "Some Aspects of the Disposal and Utilization of Farm Wastes," *J. Proc. Inst. Sewage Purif.,* 578–588 (1964).
9. R. W. Baird, "Sediment Yields from Blacklands Watersheds," *Amer. Soc. Agr. Eng. Trans.,* **7,** 454–456 (1964).
10. A. P. Barnett, E. W. Hauser, A. W. White, and J. H. Holladay. "Loss of 2,4-D in Washoff from Cultivated Fallow Land," *Weeds,* **15,** 133–137 (1967).
11. A. P. Barnett, W. A. Jackson, and W. E. Adams, "Apply More, Not Less, Poultry Litter to Reduce Pollution," *Crops Soils,* **21,** 7 (1969).

12. W. F. Barthel, D. A. Parsons, L. L. McDowell, and E. H. Grissinger, "Surface Hydrology and Pesticides—a Preliminary Report on the Analysis of Sediments of the Lower Mississippi River," in *Pesticides and Their Effect on Soils and Water,* Amer. Soc. Agron. Spec. Publ. **8,** 128–144 (1966).
13. H. L. Barrows and V. J. Kilmer, "Plant Nutrient Losses from Soils by Water Erosion," *Adv. Agron.,* **15,** 303–316 (1963).
14. W. V. Bartholemew and F. E. Clark, Eds., *Soil Nitrogen,* Amer. Soc. Agron. Monograph 10, 1965.
15. J. W. Bigger, "Factors Affecting the Appearance of Pesticides in Drainage Water," in *Agricultural Waste Waters,* Water Resources Center, University of California, Report, **10,** 194–211 (1966).
16. F. T. Bingham, "Effects of Fertilizers and Amendments on Water Quality," in *Agricultural Waste Waters,* Water Resources Center, University of California, Report, **10,** 79–81 (1966).
17. F. H. Bormann and G. E. Likens, Nutrient Cycling," *Science,* **55,** 424–429 (1967).
18. F. H. Bormann, G. E. Likens, D. W. Fisher, and R. S. Pierce, "Nutrient Loss Accelerated by Clear Cutting of a Forest Ecosystem," in *Symp. Primary Productivity and Mineral Cycling in Natural Ecosystems,* University of Maine Press, pp. 187–196, 1967.
19. C. A. Bower and L. V. Wilcox, "Nitrate Content of the Upper Rio Grande as Influenced by Nitrogen Fertilization of Adjacent Irrigated Lands," *Soil Sci. Soc. Amer. Proc.,* **33,** 971–973 (1969).
20. A. H. Bunting, "Experiments on Organic Manures, 1942–49," *J. Agr. Sci.,* **60,** 121–140 (1963).
21. *Delano Nitrate Investigation,* California Department of Water Resources, Bull. No. 143–46, 1968.
22. J. M. Cohen and Pinkerton, "Widespread Translocation of Pesticides by Air Transport and Rain-Out," in *Organic Pesticides in the Environment, Adv. Chem. Ser.,* **60,** 163–176 (1966).
23. G. L. Crouch and R. F. Perkins, Surveillance Report—1965 Burns Report, Pacific Northwest Region—USDA Forest Service, 1968.
24. H. T. Eby, "Two Billion Tons of—What?" *Compost Sci.,* **7,** 7–10 (1966).
25. W. Ehlers, W. J. Farmer, W. F. Spencer, and J. Letey "Lindane Diffusion in Soils: II. Water Content, Bulk Density and Temperature Effects," *Soil Sci. Soc. Amer. Proc.,* **33,** 505–508 (1969).
26. L. E. Ensminger and J. T. Cope, Jr., "Effect of Soil Reaction on the Efficiency of Various Phosphates for Cotton and Loss of Phosphorus by Erosion," *Agron. J.,* **39,** 1–12 (1947).
27. Federal Water Pollution Control Administration, "Pollution-Caused Fish Kills," 8th Annual Report, 1967.
28. W. Grub, R. C. Albin, D. M. Wells, and R. Z. Wheaton," The Effect of Feed Design and Management on the Control of Pollution from Beef Cattle Feed Lots," in *Animal Waste Management,* Cornell University Conference, pp. 217–224 1969.
29. R. H. Hermeson and T. E. Larson, "Interim Report on the Presence of Nitrates in Illinois Surface Waters," *Proc. 1969 Illinois Fertilizer Conf.,* pp. 33–36 January 1969.
30. S. R. Hoover and L. B. Jasewicz, "Agricultural Processing Wastes: Magnitude of the Problem," in *Agriculture and the Quality of Our Environment,* AAAS Publ. **5,** 187–203 (1966).

31. "In-Plant Treatment of Cannery Wastes—A Guide for Cannery Waste Treatment Utilization and Disposal," *Calif. State Water Resources Control Board Publ.,* **38,** 78 (1968).
32. W. R. Johnston, F. Ittihadieth, R. M. Daum, and A. F. Pillsbury, "Nitrogen and Phosphorus in Tile Drain Effluent," *Soil Sci. Soc. Amer. Proc.,* **29,** 287–289 (1965).
33. Joint Task Force—U.S. Department of Agriculture, state universities, and land-grant colleges, "A National Program of Research for Environmental Quality—Pollution in Relation to Agricultural and Forestry," 1968.
34. D. D. Kaufman, "Structure of Pesticides and Decomposition by Soil Micro-organisms," in *Pesticides and Their Effects on Soil and Water,* Amer. Soc. Agron. Spec. Publ., **8,** 85–94 (1966).
35. P. C. Kearney and D. D. Kaufman, *Degradation of Herbicides,* Marcel Dekker, New York, 1969.
36. V. J. Kilmer, O. E. Hayes, and R. J. Muckenhirn, "Plant Nutrient and Water Losses from Fayette Silt Loam as Measured by Monolith Lysimeters," *Agron. J.,* **36,** 249–263 (1944).
37. G. W. Kunze, "Pesticides and Clay Minerals," in *Pesticides, and Their Effect on Soil and Water,* Amer. Soc. Agron. Spec. Publ., **8,** 49–70 (1966).
38. J. V. Lagerwerff, "Heavy-Metal Contamination of Soils," *Agriculture and the Quality of Our Environment,* AAAS Publ., **5,** 343–364 (1966).
39. *Lake Erie Report,* U.S. Department of the Interior, Federal Water Pollution Control Administration, Great Lakes Region, August 1968.
40. W. L. Lindsay, M. Peech, and J. C. Clark, "Solubility Criteria for the Existence of Variscite in Soils," *Soil Sci. Soc. Amer. Proc.,* **23,** 357–360 (1959).
41. R. C. Loehr, "Pollution Implications of Animal Wastes—A Forward Oriented Review," prepared for the Robert S. Kerr Water Research Center, Ada, Okla., July 1968.
42. R. C. Loehr, "Treatment of Wastes from Beef Cattle Feed Lots," in *Animal Waste Management,* Cornell University Conference, 225–241 1969.
43. F. B. Lotspeich, V. L. Hauser, and O. R. Lehman, "Quality of Waters from Playas on the Southern High Plains," *J. Water Resources Res.,* **5,** 45–58 (1969).
44. G. R. Mansfield and L. Boardman, "Nitrate Deposits in the United States," *U.S. Geol. Surv. Bull.* **838** (1932).
45. B. D. Meek, L. B. Grass, and A. J. MacKenzie, "Applied Nitrogen Losses in Relation to Oxygen Status of Soils," *Soil Sci. Soc. Amer. Proc.,* **33,** 575–578 (1969).
46. M. F. Miller and H. H. Krusekopf, "The Influence of Systems of Cropping and Methods of Culture or Surface Runoff and Soil Erosion," *Missouri Univ. Agric. Expt. Sta. Res. Bull.* **177** (1932).
47. J. R. Miner, R. I. Lipper, L. R. Fina, and J. W. Funk, "Cattle Feed lot Runoff—Its Nature and Variation," *J. Water Pollution Control Federation* **38,** 1582–1591 (1966).
48. N. Minshall, M. S. Nichols, and S. A. Witzel, "Plant Nutrients in Base Flow of Streams in Southwestern Wisconsin," *J. Water Resources Res.* **5,** 706–713 (1969).
49. L. E. Mitchell, "Pesticides: Properties and Prognosis," in *Organic Pesticides in the Environment, Adv. Chem.,* **60,** 1–22 (1966).
50. P. G. Moe, J. V. Mannering, and C. B. Johnson, "Loss of Fertilizer Nitrogen in Surface Runoff Water," *Soil Sci. Soc. Amer. Proc.,* **104,** 389–394 (1967).

51. "Monitoring Agricultural Pesticide Residues 1965–1967," *U.S. Dept. Agr. ARS,* 81–32 (1969).
52. C. H. Mortimer and C. F. Hickling, "Fertilizers in Fishponds—A Review and Bibliography," *G. Brit. Min. Fisheries Publ.,* **5** (1954).
53. A. R. Mosier and W. D. Guenzi, "Photochemical Decomposition of DDT by a Free Radical Mechanism," *Science* **164,** 1083–1085 (1969).
54. L. B. Nelson and R. E. Uhland, "Facts That Influence Loss of Fall-Applied Fertilizers and Their Probable Importance in Different Sections of the United States," *Soil Sci. Soc. Amer. Proc.* **19,** 492–496 (1955).
55. J. Neumeyer, D. Gibbons, and H. Frask, "Pesticides," *Chem. Week,* 38–68, April 12, 1969.
56. T. E. Norton and R. W. Hansen, "Cattle Feed Lot Water Quality Hydrology," in *Agricultural Waste Management Proc.,* Cornell University Conference, 203–216 (1969).
57. T. C. Olsen and W. H. Wischmeier, "Soil Erodibility Evaluations for Soils on the Runoff and Erosion Stations," *Soil Sci. Soc. Amer. Proc.,* **27,** 590–592 (1963).
58. L. T. Parizek, *et al.* "Waste Water Renovation and Conservation," *Penna. State Univ. Studies,* **23** (1967).
59. "Pesticides in the Environment," in *The Chemical Basis for Action,* a report of the American Chemical Society, 193–244 (1969).
60. W. H. Pierre and A. G. Norman, Eds., "Soil and Fertilizer Phosphorus," Amer. Soc. Agron. Monograph **4,** (1953).
61. R. D. Radeleff and R. C. Bushland, "The Toxicity of Pesticides for Livestock," in *The Nature and Fate of Chemicals Applied to Soils, Plants and Animals, U.S. Dept. Agr. ARS,* 20-9, 134–159 (1960).
62. C. H. Reed, "Specifications for Equipment for Liquid Manure Disposal by the Plow-Furrow-Cover Method," in *Animal Waste Management, Proc.,* Cornell University Conf., 114–119 (1969).
63. H. T. Rogers, "Plant Nutrient Losses from a Corn, Wheat and Clover Rotation in Dunmore Silt Loam," *Soil Sci. Soc. Amer. Proc.,* **6,** 263–271 (1941).
64. K. E. Saxton and R. G. Spomer, "Conservation Effects on the Hydrology of Loessal Watersheds," *Trans. Amer. Soc. Agr. Eng.,* **11,** 848 (1968).
65. G. D. Scarseth and W. W. Chandler, "Losses of Phosphate from a Light-Textured Soil in Alabama and Its Relation to Some Aspects of Soil Conservation," *Agron. J.,* **30,** 361–374 (1938).
66. "Severity of Lake Erie's Pollution Debated," *Chem. Eng. News,* **47,** 43 (May 19, 1969).
67. W. C. Shaw, "Pests and Their Control," *J. Amer. Soc. Sugar Beets Technologists,* **14,** 10–18 (1966).
68. G. E. Smith, "Nitrates and Water-Facts and Fancy," *Proc. 16th Annual Calif. Fertilizer Conf.,* 34–45 (1968).
69. G. E. Smith, "Fertilizer Nutrients in Water Supplies," in *Agriculture and the Quality of Our Environment,* AAAS Publ. **85,** 173–186 (1967).
70. W. F. Spencer, M. M. Claith, and W. J. Farmer, "Vapor Density of Soil-Applied Dieldrin as Related to Soil-Water Content, Temperature and Dieldrin Concentration," *Soil Sci. Soc. Amer. Proc.,* **33,** 509–511 (1969).
71. R. S. Stauffer, "Runoff, Percolate and Leaching Losses from Some Illinois Soils," *Agron. J.,* **34,** 830–835 (1942).
72. F. J. Stevenson, "Origin and Distribution of Nitrogen in Soil," in *Soil Nitrogen,* Amer. Soc. Agron. Monograph **10,** 1–42 (1965).

73. B. A. Stewart, F. G. Viets, G. L. Hutchinson, W. D. Kemper, F. E. Clark, M. C. Fairbourn, and F. Strauch, "Distribution of Nitrates and Other Water Pollutants under Fields and Corrals in the Middle South Platte Valley of Colorado," *U.S. Dept. Agr., ARS,* 41–134 (1967).
74. P. R. Stout and R. G. Burau, "The Extent and Significance of Fertilizer Buildup in Soils as Revealed by Verticle Distribution of Nitrogenous Matter between Soils and Underlying Water Reservoirs," in *Agriculture and the Quality of Our Environment,* AAAS Publ., **85,** 283–310 (1967).
75. R. O. Sylvester, "Nutrient Content of Drainage Water from Forested, Urban and Agricultural Areas," *Public Health Rept.* (*U.S.*), SEC-TR-W61-3, 80–87 (1960).
76. R. F. Tarrant, "Pesticides in Forest Waters—Symptom of a Growing Problem," *Forest Sci.,* 159–163 (1966).
77. A. W. Taylor, "Phosphorus and Water Pollution," *J. Soil Water Conserv.,* **22,** 228–231 (1967).
78. "The Nature and Fate of Chemicals Applied to Soils, Plants and Animals," *U.S. Dept. Agr., ARS,* 20-9 (1960).
79. F. L. Timmons, P. A. Frank, and R. J. Demint, "Herbicides Residues in Agricultural Water from Control of Aquatic and Bank Weeds," *Proc. Conf. Concerning the Role of Agriculture in Clean Water,* Ames, Iowa, November 1969.
80. U.S. Department of Agriculture and Office of Science and Technology, a report to the President, "Control of Agriculture—Related Pollution," Washington, D.C., 1969.
81. R. P. Upchurch and W. C. Pierce, "The Leaching of Monuron from Lakeland Sand Soil, Part I. The Effect of Amount, Intensity and Frequency of Simulated Rainfall," *Weeds,* **5,** 321–330 (1957).
82. C. H. Van Middelem, "Fate and Persistence of Organic Pesticides in the Environment," in *Organic Pesticides in the Environment, Adv. Chem.,* **60,** 228–249 (1966).
83. F. G. Viets, "Soil Use and Water Quality—A Look into the Future," *J. Agr. Food Chem.* (1969).
84. C. H. Wadleigh, "Wastes in Relation to Agriculture and Forestry," *U.S. Dept. Agr. Misc. Publ.,* 1065 (1968).
85. L. R. Webber and T. H. Lane, "The Nitrogen Problem in the Land Disposal of Liquid Manure," in *Agricultural Waste Management,* Cornell University Conference, 124–130 (1969).
86. A. W. White, A. P. Barnett, and V. J. Kilmer, "Nitrogen Fertilizer Loss in Runoff from Cropland Tested," *Crops Soils* **19,** 28 (1967).
87. L. F. Wilcox, "Analysis of Salt Balance and Salt Burden on the Rio Grande," *Symp. Probl. Upper Rio Grande,* Albuquerque, N.M., New Mexico State Engineers Office, 39–44, (1967).
88. W. H. Wischmeier, "A Rainfall Erosion Index for a Universal Soil-Loss Equation," *Soil Sci. Soc. Amer. Proc.,* **23,** 246–249 (1959).
89. W. H. Wischmeier and J. V. Mannering, "Relation of Soil Properties to Its Erodibility," *Soil Sci. Amer. Proc.,* **33,** 131–137 (1969).
90. W. H. Wischmeier and D. D. Smith, "Predicting Rainfall-Erosion Losses from Cropland East of the Rocky Mountains," *U.S. Dept. Agr., Agr. Handbook,* **282,** Agr. Res. Serv., (1965).
91. P. C. Woods, "Management of Hydrologic Systems for Water Quality Control," contribution 121, Water Resources Center, University of California, Berkeley, Calif. (1967).

Remote Sensing for Air Pollution Measurements

HERMAN SIEVERING

Department of Electrical Engineering,
University, of Illinois at C-U,
Urbana, Illinois

I. INTRODUCTION

Air pollution detection and monitoring generally is carried out on a local basis. Gas sampling is most often used followed by chemical or spectroscopic analysis. However, these techniques cannot measure low level temperature, humidity, and pollutant gas vertical height profiles—necessary adjuncts to the prediction of air pollution loading. Continuous monitoring (nighttime as well as daytime) in combination with vertical soundings is rarely used in air pollution research or forecasting.

Meteorological conditions conducive to high pollution load must be better understood. For example, a significant parameter is the temperature height profile and concomitant determination of the temperature inversion height. The development of an SO_2 model for the city of Chicago has been hindered by the lack of precise hourly data on temperature profiles.

It is apparent that day and night data regarding the concentration, spatial distribution, and diffusion and fall-out rates of pollutants under meteorological influences would be most beneficial. Conversely, the effects of pollution on meteorological and climatological parameters must be determined. For example, the consequences of stratospheric pollution caused by supersonic transport flights can only be surmised.

Progress on these and many associated problems has been impeded by the lack of direct measurements of atmospheric parameters in three dimensions, over large volumes, and with sufficient resolution in time and space. Remote sensing methods appear to offer solutions to this heterogeneous group of problems. However, of the many techniques available which does one choose? The transmitter alone—basic to any remote sensing method—raises a number of questions. Should it be active or passive; continuous or pulsed; ground-based, airborne, or on board a satellite? The first portion of this study gives a brief description of the various remote sensing methods. The second portion considers their applicability to air pollution measurements.

II. ATMOSPHERIC REMOTE SENSING TECHNIQUES

Until recently, the study of wave propagation in the atmosphere has emphasized the "deleterious" consequences of that propagation. Generally studies have considered the analysis of the degradation of electromagnetic waves with passage through an air mass and the minimization of this degradation. Conversely, remote sensing is concerned with maximizing the air mass information that one can glean from analysis of the "degraded" electromagnetic wave. For this reason, the theory of remote sensing is still in an early stage of development.

Remote sensing—encompassing any observation in which the sensor is not in direct contact with the object of study—can be divided into two classes: active and passive. An active system involves transmission of a signal, interaction with a target, and quantitative as well as qualitative detection. A passive system involves the detection of natural emanations of the target and in no way interacts with it.

Because passive systems require only moderate power and size, they have been used extensively in spacecraft. Active systems, requiring large amounts of power and massive components, are practical from stations on the earth's surface.

A. Active Systems

Active systems have potentially greater information content due to one's control over the probing signal. Present active sensing techniques can be grouped loosely into five categories:

1. Radar
2. Laser Radar
3. Acoustic

4. Microwave line-of-sight propagation
5. Optical and near infrared line-of-sight propagation.

1. *Radar*

The radar technique is characterized by energy scattered off dielectric inhomogeneities present in a limited atmospheric volume. Measurements of the echo power, Doppler-shift, polarization, and attenuation along the path allow interpretation of the state of the probed medium.

The primary radar measurements are reflectivity and Doppler shift. Reflectivity measurements can be related to the size and number of particles giving rise to the echoes. Doppler shift measurements allow for the measurements of atmospheric motions. Two secondary measurements which can give an indication of particle shape should be mentioned: bistatic radar (i.e., angular separation of transmitter and receiver), and concomitant measurement of polarization effects. Recent theoretical work indicates that particle size distributions may also be obtained from bistatic radar. Measurement of the path attenuation due to absorption by atmospheric constituents may allow determination of total or height-discriminated concentrations. At normal radar wavelengths only water vapor concentration measurements appear feasible.

Recent attention has been given to a study of scattering from sharp gradients of refractive index. Radar determinations of the cross section per unit volume of a refractively perturbed medium can be used to determine the intensity of refractive index fluctuations. While some powerful radars have had sufficient sensitivity to detect the relatively weak refractive index perturbations which commonly mark interfaces across which the mean refractivity changes sharply, none has had both sufficiently high resolution and sensitivity to detect the fine-scale structure of the scattering medium. The combination of great sensitivity and ultrahigh resolution recently has been achieved by Atlas *et al.* (1). This has been made possible by the use of a unique frequency-modulated, continuous wave radar developed by Richter (1).

A second area of recent interest is large-scale radar mapping. A major reason for its importance is the need for the collection of data on a scale and with a coverage either prohibitively expensive or nearly impossible by conventional means. Meteorological measurement requirements can no longer be met by any system of *in situ* sensors. Weather prediction as well as maximum allowable pollution concentrations requires large-scale quantitative data. Both satellite radar and correlation of many surface-based stations suggest themselves as possible approaches. However, the first suffers from the inherent problem of discriminating ground returns from atmospheric returns.

2. *Laser Radar*

Laser radar is very similar in operation to radar, for the backscattered signal as a function of range is observed upon transmission of a short pulse of radiation. The major difference is in the operating wavelength. A Q-switched ruby or neodymium laser of 6943 and 10,600 Å respectively, is the source. Being in the optical range of the spectrum, backscattered energy is collected at a telescope, passed through collimating optics and a narrow-band optical filter centered at the laser wavelength, and detected by a photomultiplier tube.

Backscatter from the atmosphere can occur in one of three ways: off atmospheric molecular constituents, off natural aerosols and pollutant particulates, and off refractive index variations. The signal level return from refractive index variations is far below that from either atmospheric constituents or pollutant particulates. In the lower atmosphere (below 25 km) both aerosol-pollutant and molecular backscatter are observed. Above 25 km the scatter is mainly molecular and thus allows for air density determination.

As in radar, bistatic laser radar permits measurement of the polarization dependence of the scattered signal. Theoretically, these measurements can give information about the concentration and size distribution of the scatterers.

Schotland *et al.* (2) suggest that Doppler techniques could be applied to determine temperature from the width of the molecular scattering spectra as well as wind velocity from the mean Doppler shift.

Recent laser radar experiments have been concerned with shifted wavelength signal returns. Two approaches appear to be particularly promising: (*1*) absorption, in which the difference in signal return from a laser radar "tuned" to and away from atmospheric absorption lines is observed; and (*2*) Raman scattering, in which the scattered energy is frequency shifted, and the shifted wavelength signal return unique to a specific atmospheric constituent is observed.

3. *Acoustic*

In remote probing of the atmosphere by electromagnetic waves, the interaction with the gases of the lower atmosphere (excluding water vapor) is generally very weak. Thus sensitive, sophisticated equipment is required to measure the interactions. However, the interaction of acoustic (longitudinal pressure) waves with the atmosphere is very strong. Thus the propagation of these waves is affected strongly by atmospheric wind, temperature, and humidity. Until recently (3,4), this fact has been little exploited. In part, because of concern for identification of acoustic waves of natural origin for which the power radiated is quite low, studies have focused on passive remote sensing for turbulence measurements (5).

The potential of active remote sensing with acoustic scattered energy for atmospheric information has been considered by Little (3). Under the conditions of complete measurement of scattered power as a function of wavelength and scattered angle, atmospheric information on the following could be obtained:

1. The intensity of temperature fluctuations directly proportional to the refractive index,
2. The intensity of velocity fluctuations as a function of direction and height, (these three-dimensional data are useful in determining atmospheric turbulence and diffusion),
3. The mean wind speed and direction from Doppler shift or bistatic acoustic measurements,
4. The time history of temperature inversion layers including intensity.

The major limitation in the use of acoustic probing is its short-range capability; 1500-m heights is a probable limit on range for meaningful atmospheric information content. Of course, it complements radar in that very short ranges (down to 30 m) can be realized.

4. *Microwave Line-of-Sight Propagation*

Wave propagation from one location to another is affected by the refractive index. For electromagnetic waves greater than 1 mm, the refractive index is a function of temperature, pressure, and water vapor content.

The perturbation due to refractive index fluctuations is well enough understood to allow: (*1*) total oxygen and water vapor concentrations integrated across the propagation path, and (*2*) the amplitude of refractive-index fluctuations and the spectral distribution and internal rate of change of refractivity.

Although most measurements are carried out on a horizontal basis, satellite-to-ground measurements have been made in a more limited context. When absorption becomes comparable to refractive effects, the above is no longer true. However, absorption effects have application in the determination of water vapor content.

5. *Optical and Near Infrared Line-of-Sight Propagation*

Two major atmospheric measurements obtainable by this technique are turbulence and wind structure, and path-averaged temperature and gas densities. Thermal inhomogeneities give rise to the ability to measure turbulence structure as do refractivity irregularities in giving rise to wind structure measurements. Measurement of optical path length gives a straightforward manner of obtaining average temperature. To measure gas densities, tunable transmitters are needed. Two transmission wavelengths, one on and one off an absorption line, are compared to give the path-averaged density of an atmospheric gas. Because of the very high sensitivity of this technique, it appears feasible that very rare gases and very low pollutant gas concentrations may be observed using short paths.

B. Passive Systems

Several sources for passive system operation exist. The transmitted and scattered radiation from the sun, the moon, or the earth can be observed by a passive system. This system can utilize the natural emanations from gas or aerosol constituents of the atmosphere. Passive systems can be grouped into two categories:

1. Infrared optical
2. Microwave radiometric.

1. *Infrared Optical*

The absolute temperature of a body and its coefficient of emissivity determine the radiating properties of any object. At earth surface temperatures of approximately 300°K, the intensity maximum of thermal radiation falls between 10 and 20 μ. Thus infrared probing appears to be the obvious choice for radiation measurements of the earth's surface or atmosphere. Conversely, at wavelengths below about 3μ, solar energy reflected from the surface of the earth is greater than the thermal emission from the earth and atmosphere. Generally, thermal radiation can be neglected below about 2 μ, and optical probing assumed to be measuring only reflected solar energy.

Most applications of infrared optical probing have emphasized global meteorological measurements. Thus satellite systems have received the most attention. The theory of infrared probing has dealt mainly with atmospheric thermal structure inference by inversion of radiation measurements (8) and earth surface or cloud-type properties by multiwavelength observation (9). Since interest here is focused on atmospheric measurements, there will be no further discussion of the second category.

The most useful passive infrared technique is the inference of vertical profiles of temperature, water vapor content, and ozone content (10). The accuracy of atmospheric inferences by inversion techniques is, however, not as great as one might at first suspect. The accuracy of inversion techniques is normally limited by the number of viewing channels of observing equipment (n pieces of information allowing inference of n characteristics of atmospheric structure) and the accuracy (i.e., signal-to-noise ratio of the observational data). However, as Twomey (11) points out, the number of pieces of information obtained is usually far less than the number of viewing channels available.

A second important method of passive infrared probing is that of interferometry. In theory, infrared interferometry (also called Fourier spectroscopy) is also an inversion technique. Instead of sensing at preselected wavelengths or bands of wavelengths, the sensor records a pattern of interference maxima and minima that can be converted to the more conventional spectrum of wavelength versus intensity. Numerous spurious effects are present, some of which cannot easily be corrected.

2. *Microwave Radiometric*

The microwave radiometric method of probing the atmosphere measures the properties of thermal emission and absorption by the atmosphere itself. The intensity, spectrum, and polarization of the radiation as measured by the microwave radiometer (now among the most sensitive electromagnetic detectors) can provide atmospheric constituent profiles, temperature profiles, and, possibly, pollutant profiles as well.

Microwave probing is generally subdivided into ground- and satellite-based measurements. Ground-based measurements may utilize either emission by the atmosphere or absorption of an external source of microwave emission. The theoretical potential of satellite sensors for water vapor and oxygen profiles appears good, particularly when their all-weather capabilities are considered. However, no experiments have been flown. Thus satellite microwave detection must be considered as conjecture.

Ground-based emission measurements provide most microwave atmospheric data. Data near 5-mm wavelength measure thermal emission by molecular oxygen. By shifting the wavelength through an oxygen absorption band, the probing range can be varied. This allows for temperature profile determination. Emission observations at wavelengths near 1.35 cm detect water vapor thermal emission and thus a capability for water vapor profile data.

Absorption measurements can give detectable signals with smaller amounts of a constituent present than can emission measurements. Thus in addition to water vapor, oxygen, and ozone measurements, trace constituents may be observed through the use of absorption techniques. Examples might be SO_2, CO, NO, and NO_2. The major limitation in the use of this technique is the need for an external emission source. Thus atmospheric observations are line-of-sight path to the sun.

A technique which utilizes both passive and active systems is crossed-beam correlation. It employs two detecting systems with view fields that intersect. Turbulence-induced fluctuations result in fluctuations in the detector output signals. By cross-correlating detector signals, information relating to the intensity of fluctuations, wind speeds, and spectrum of turbulence can be obtained. Frequently, the magnitude of the correlated signal is far down from that of the uncorrelated signal. Under this condition active probing systems must be used.

III. APPLICATIONS TO AIR POLLUTION MEASUREMENTS

The present capabilities of remote sensing as described in Section II indicate that emphasis has been placed on measurement of meteorological parameters without much consideration for its use in air pollution analysis. Moreover, research is being conducted in a fragmented fashion. That remote meteorological measurement schemes are in a sophisticated stage of development is certain, but techniques which measure the same or interdependent parameters must be better coordinated. A facility at which the various measurement schemes can be correlated would be highly desirable. Although greater consideration for the impact of meteorological measurements on air pollution parameters is needed, the measurement schemes themselves are progressing well.

What of the use of remote sensing for direct air pollution measurements? Until recently, little attempt had been made to sophisticate this purpose. A survey of pollutants by type and by their scattering, absorption, and emission characteristics is in order.

Air pollutants can occur in the form of gases, solid particles, or liquid aerosols. These forms can exist either separately or in combinations. Gaseous pollutants constitute about 90% of the total mass emitted to the atmosphere, while particulates comprise the rest. Given suitable conditions, many of the primary pollutants will participate in reactions in the atmosphere that produce secondary pollutants, e.g., photochemical smog. The most important pollutant types are the following:

1. Particulates including carbon fly ash, lead, zinc oxide, and arsenic.
2. Nitrogen compounds, including NO, NO_2, NO_3, HNO_2, and HNO_3.
3. Sulfur compounds, including SO_2, SO_3, and H_2SO_4.
4. Carbon compounds, including CO, CO_2, and CH_4.
5. Oxygen compounds: in particular O_3.

In addition, hydrogen chloride and other halogen compounds, ammonia, and peroxylacetyl nitrate (PAN) are important pollutants to be measured and monitored.

A. Pollutant Particulates

The types of atmospheric particles and their size ranges are shown in Figure 1. Aerosol particles, including pollutant particulates, are not equally distributed in size. The atmosphere's carrying capacity is far greater for smaller particles than for larger ones. A typical size distribution plot is shown in Figure 2, with an approximation to the curve also indicated. The approximation to the curve is the Junge distribution $dN/d(\log r) = \gamma r^{-\nu}$. For most lower atmospheric distributions $\nu = 3$. Scattering intensity is a function of the particulate number concentration, size distribution, shape, and absorption–emission characteristics. Pollutant particulates are generally passive scatterers and because of their high relative number densities and random shape can be assumed to exhibit no shape characteristics in the scattering process. An ensemble of random-shaped particles is equivalent to the same number of spherical-shaped particles of some mean diameter. The question then remains as to which wavelength will optimize pollutant particulate signal detection.

In meteorological radar measurements of cloud water droplets, raindrops, snow crystals, and hailstones, optimum wavelengths are on the order of 1 cm. The drops detected have a geometric-mean diameter of 0.1 cm. The ratio of wavelength to diameter is thus of order 10. Since the geometric-mean diameter of atmospheric particulates is

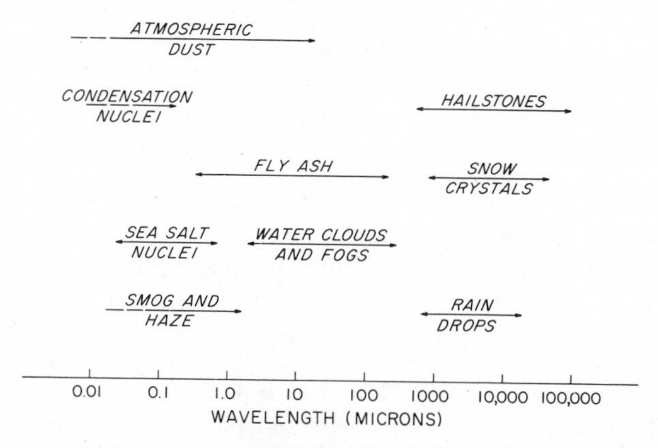

Fig. 1. Relation of transmitter wavelengths to diameters of atmospheric particles.

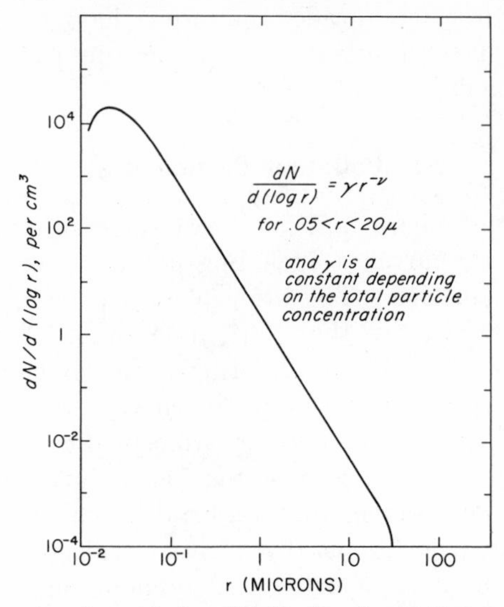

Fig. 2. Aerosol size distribution in continental air.

of the order of 0.1 μ, it would follow that an optimum transmitter wavelength for pollutant particulate remote detection is 1 μ. For those familiar with electromagnetic wave scattering processes, this result would be expected. When the wavelength of electromagnetic radiation is on the order of the scattering particle diameter, the scattering cross section per particle may be as much as 10^{18} larger than at a wavelength far removed from the particle diameter.

Assuming an optimal wavelength near 1 μ, several measurement techniques appear possible:

1. Passive satellite optical probing
2. Passive or active correlation techniques
3. Optical infrared line-of-sight measurements
4. Laser radar.

Passive satellite probing and line-of-sight measurements have limited use because of the small value of volume extinction coefficient. The extinction of a monochromatic beam propagating in a scattering medium is given by

$$dI(\lambda) = -B_{\text{ext}}(\lambda, n)I(\lambda)\,dr \qquad (1)$$

where the intensity $I(\lambda)$ of the beam is defined as the energy per unit bandwidth, at wavelength λ, transmitted per second through a unit area normal to the propagating direction; dr is the amount of scatterer in a volume of unit cross-sectional area and length dl; and $B_{ext}(\lambda, n)$ is the volume extinction coefficient (per unit length) for scatterers with index of refraction $n = m_1 + im_2$. Long and Rensch (13) have calculated B_{ext} for a Junge distribution with a particulate concentration of $10^9/m^3$ and relative humidity of less than 90%. Results are shown in Figure 3. At a wavelength of 1 μ (B_{ext} being 0.02/km), an approximate 0.002/100 m signal change in the first several hundred meters can be expected. With the relatively short paths used in line-of-sight measurements this technique will not allow for particulate measurements. Passive satellite probing can give total-path particulate attenuation but cannot determine distribution along the path. Moreover, this assumes

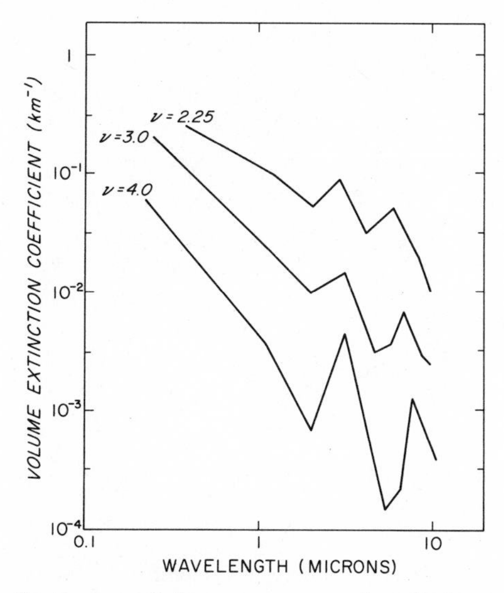

Fig. 3. Volume extinction coefficient versus wavelength for aerosol model with relative humidity <90% and aerosol concentration of $10^9/m^3$ (13).

a wavelength at which attenuation is caused by particulate scattering—a difficult criterion. Thus satellite probing may be useful in particulate measurements to the extent of mapping total atmospheric loading with no determination of height or time variations.

A consideration of the magnitude of backscatter coefficient indicates that more complete particulate measurements may be obtained by utilization of backscattered energy. The energy transfer equation relating source power to backscattered power received from a scattering volume (13) is

$$\Delta P_R(\lambda, 180°) = WAT_1T_2\omega b_{sca}(\lambda, m)\, \Delta l \tag{2}$$

where $\Delta P_R(\lambda, 180°)$ is the received backscattered power in watts from the scattering volume of length $\Delta l(m)$; W is the irradiance incident on the scattering volume with area $A\,(\mathrm{W}/_{m^2})$; T_1 is the transmittance between transmitter and scatter volume; T_2 is the transmittance between scatter volume and detector; ω is the solid angle subtended by detector (stere); and $b_{sca}(\lambda, m)$ is the backscatter coefficient (km stere)$^{-1}$ for scatterers with index of refraction $n = m_1 + im_2$. The backscatter coefficient can be written as

$$b_{sca}(\lambda, m) = \frac{1}{K^2} \int_{r_1}^{r_2} |S(n, \lambda, 180°)|^2 N(r)\, dr \tag{3}$$

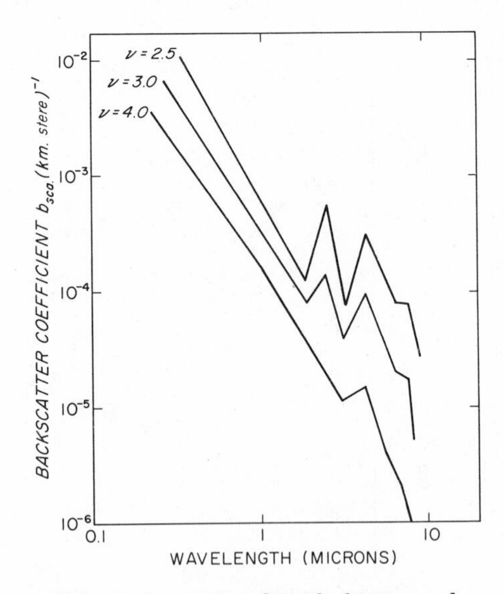

Fig. 4. Backscatter coefficient versus wavelength for aerosol model with relative humidity <90% and aerosol concentration of $10^9/\mathrm{m}^3$ (13).

where K is the free-space propagation constant (m^{-1}); $S(n, \lambda, 180°)$ is the scattering amplitude function for a single particle (14); and $N(r)$ is the scatterer size density distribution with lower and upper radius limits of r_1 and r_2, respectively. An exact solution for $S(n, \lambda, 180°)$ can be found using the classical boundary value method of Mie in which an infinite set of eigenfunctions is used to represent the scattered field (15). Calculations of b_{sca} were made by Long and Rensch (13) using the exact Mie theory and a Junge-size distribution with $r_1 = 0.1\ \mu$ and $r_2 = 10.0\ \mu$. The results are shown in Figure 4. Note again that a concentration of $10^9/\mathrm{m}^3$ and relative humidity of less than 90% were used.

At 1 μ wavelength, b_{sca} is approximately 2.5×10^{-4} (km stere)$^{-1}$ for $\nu = 3.0$. In a typical laser radar, $W \approx 1$ MW; $A \approx 1$ m^2, and $T_1 = T_2 = 0.98$ at 1 km range, $\omega \approx 10^{-6}$ and $\Delta l \approx 1$ m. Thus $\Delta PR(\lambda, 180°)$ equals approximately 0.1 μW. As this is far above the noise level of red-sensitive multistage photomultiplier tubes, detection of pollutant particulates appears very conducive to the laser radar technique. Note that vertical profiles can be obtained because of the relatively large signal returns from highly resolved (10 m or less because of the short pulses of laser light used) scattering volume heights.

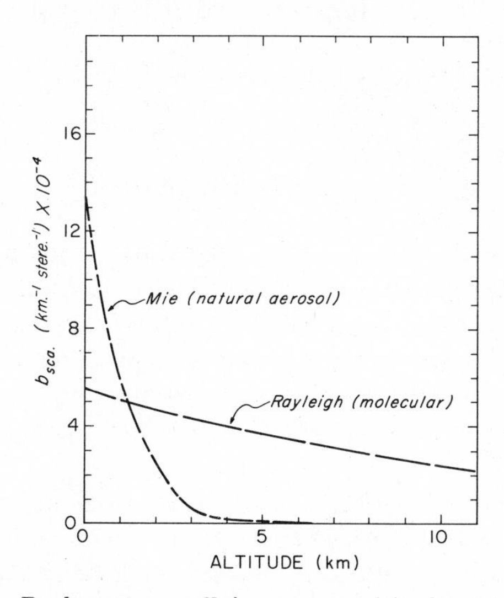

Fig. 5. Backscatter coefficient versus altitude at $\lambda = 0.7$ μ.

As alluded to in Section II.A.2., a problem does however arise. The backscatter power has been discussed only with regard to natural aerosols and pollutant particulates (i.e., particles for which Mie scattering applies). Since r_1 was taken as 0.1 μ, no consideration was given to atmospheric molecular constituent scattering. Lawrence *et al.* (16) have calculated the variation in backscatter coefficient versus height for a clear atmosphere of natural aerosols and for a Rayleigh atmosphere (i.e., scattering particles much smaller than the wavelength as is true for the molecular constituents). Results are shown in Figure 5.

In a highly polluted atmosphere the Mie scattering curve may be from 20 to 40 times the magnitude that is shown in Figure 5. Thus two questions arise. How does one discriminate backscatter contributed by pollutant particulates from that by atmospheric molecular constituents and that by natural aerosols? Some researchers (17,18) have assumed the theoretical curves of Figure 5 as fitting the real atmosphere so that "background scattering" can be subtracted from the total observed scatter to give pollutant particulate concentrations. Under this assumption the particulate load density can be related (17) to the corrected backscatter coefficient as:

$$Q = 4.96(B - 5.8) \tag{4}$$

where Q is the load density in $\mu g/m^3$ and B is the corrected backscatter coefficient in m-stere^{-1}.

In laser radar the total backscatter coefficient can be directly related to system parameters:

$$B = (\text{system constant})R^2X \tag{5}$$

in which R is the range and X is the vertical deflection at an oscilloscope whose sweep is initiated at laser firing. The deflection at any one point on the oscilloscope display then can be related to an instantaneous scattering volume at known range. Since a typical laser radar operates with a 40-sec pulse length, instantaneous scattering return results from a 6-m vertical length volume. Thus pollutant particulate vertical profiles can be realized.

Several experiments (16,19,20) indicate the real atmosphere does not always fit the theoretical curves of Figure 5. One method of overcoming this problem is experimental. A background "clear-air" signal level can be approximated simply by averaging data taken at times when the atmosphere is suspected to be "most clean." Although this may appear crude, it is the most widely used technique. An example of such data analysis appears in Figure 6 and 7. Figure 6 is an oscilloscope trace when a shelf of pollutants was present at the 400-ft region. Figure 7 is the corrected plot of particulate loading based on data from Figure 6 (17).

It has been suggested that the temperature inversion height can be monitored by observation of pollutant particulates. Atmospheric processes usually dilute and disperse atmospheric contaminants. However, the presence of an atmospheric inversion is conducive to pollutant accumulation. Temperature inversions allow much less vertical mixing through the inversion than situations for which there is a temperature decrease with height close to the adiabatic lapse rate. Thus particulate loading should fall sharply at the inversion. Figure 8 taken from Johnson (18) shows such a sharp drop in signal return near 400 ft. Although

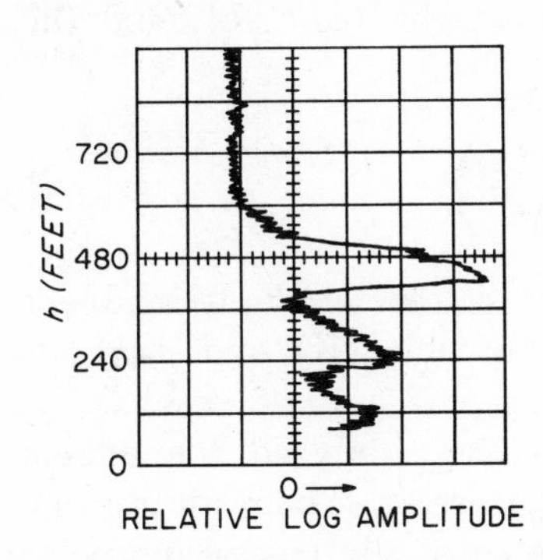

Fig. 6. Sample laser radar return (17).

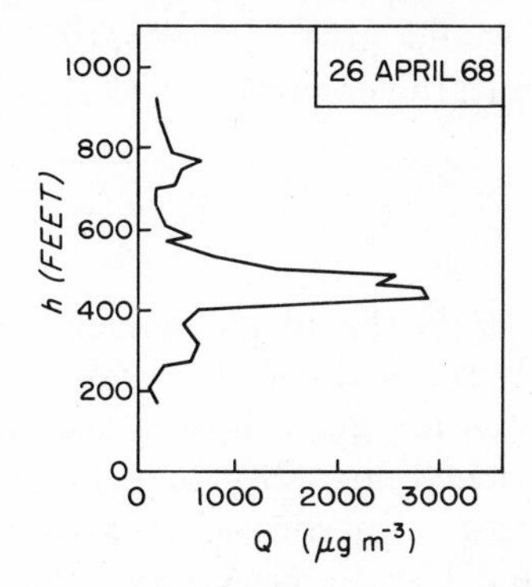

Fig. 7. Plot of particulate loading versus height (17).

this is probably the temperature inversion height, laser radar returns rarely display such a clear demarcation as that in Figure 8.

Multiwavelength laser radar may prove very useful in obtaining true pollutant particulate profiles. Discussion of this method and its applicability will be deferred to Section III.B. Theoretical work to examine methods for pollutant particulate-aerosol-molecular constituent discrimination is being carried out by several investigators (21–23). One benefit of any discrimination technique would be a simple method for temperature profile determination. Molecular constituent (Rayleigh) scattering being directly proportional to the density of atmospheric molecular scatters (ρ_r), the laser radar signal return from Rayleigh scattering

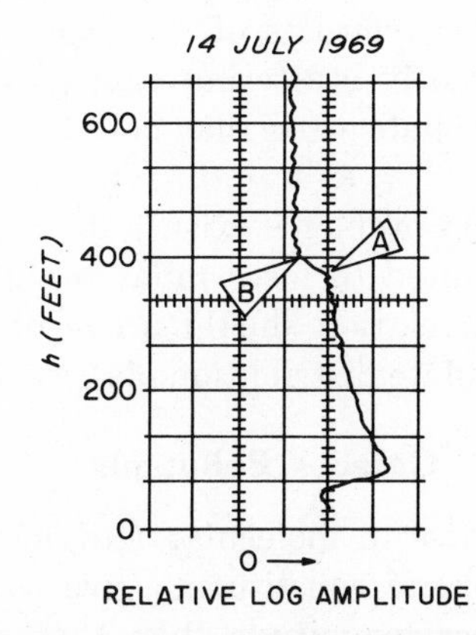

Fig. 8. Sample laser radar return showing transition from relatively polluted air below the inversion base to cleaner air above (13).

alone would give the variation of ρ_r with height $(\Delta\rho_r/\Delta h)$, and thus temperature versus height (T_h):

$$T_h = -\frac{Mg\rho_r(h)}{R}\left(\frac{\Delta\rho_r(h)}{\Delta h}\right)^{-1} \quad (5)$$

where M is the mean molecular mass (kg); g, the acceleration due to gravity (m/sec); R, the gas constant (J/kg mole °K) and $\rho_r(h)$ and $\Delta\rho_r(h)/\Delta h$ are given by the laser radar data.

Deirmendjian (23) and Penndorf (24) have analyzed the angular scattering properties of atmospheric Mie scatterers. Their studies indicate that bistatic laser radar (25–27), by measuring intensity and polarization of light scattered at angles other than 180°, may allow for the required discrimination. Measurements of the ellipticity of angularly scattered light that is initially linearly polarized unambiguously can be accounted for by aerosols and pollutant particulates because molecular (Rayleigh) scattering does not give rise to elliptical polarization (assuming no multiple scattering). Furthermore, the ellipticity of angular scattering is sensitive to the size distribution and shape of aerosols and pollutant particulates thus allowing for possible discrimination of these two scatterer types. However, again referring to Twomey (11), bistatic laser radar cannot be expected to determine unique-size distributions or type of scatterer.

Recent work (22) indicates that another technique (possibly used in conjunction with bistatic measurements) may achieve the desired discrimination. The laser radar equation, eq. 2, assumes the scattering process to be an independent one. Consequently, eq. 2 is correct only under the conditions of incoherent scattering. By recognizing that the laser radar scattering process consists of an incoherent scattering addition of a number of coherently scattering aggregates, a corrected laser radar equation results (22), indicating that pollutant particulate scattering can be discriminated.

Although many interesting problems exist, remote sensing for pollutant particulates appears best suited to laser radar probing. The additional use of laser radar cross-correlation should prove of further significance in accurate pollutant particulate distribution observations.

B. Gaseous Pollutants

Gaseous pollutants give rise to molecular (Rayleigh) scattering. Because nonpollutant atmospheric constituents give rise to Rayleigh scattering that is of far greater magnitude than that caused by pollutant gases, scattering processes cannot be expected to allow discrimination and monitoring of these gaseous pollutants. However, absorption and re-emission processes may allow pollutant gas monitoring.

The absorption of energy by various chemical materials is based on Beer's law. The monochromatic transmission, $\tau(\lambda)$, is given by

$$\tau(\lambda) = \frac{I(\lambda)}{I_0(\lambda)} = e^{-k(\lambda)pl} \tag{6}$$

in which $I(\lambda)$ is the transmitted portion of the monochromatic energy and $I_0(\lambda)$ is the incident on the gas layer. The partial pressure of the absorbing gas is p and the thickness of gas layer is l. The absorption coefficient, $k(\lambda)$, is a function of the gas observed, the operating wavelength, and the total pressure present in the gas layer. The partial concentration of a gas then can be determined, if the following factors are known: (*1*) the signal ratio at a particular wavelength at which the ratio is not one that is measured, and (*2*) which gas is present, its corresponding absorption coefficient, and the measurement of the path length.

If more than one gas is present, quantitative analysis requires a knowledge of which gases are present. A completely independent signal ratio must then be determined for each gas present. Although scattering has been assumed not to occur, compensation can be made (28).

Ludwig, Bartle, and Griggs (29) have tabulated the low resolution percentage radiance changes on passage through total clean and polluted atmospheres. Figure 9, taken from Ludwig *et al.*, shows the results assuming a temperature profile of 1°C change/100 m, with increasing temperature to 400 m, decreasing temperature to 10 km, constant temperature to 25 km, and increasing temperature above this height. Shown in the Figure are the assumed concentrations for various pollutants. The uppermost curve indicates the radiance change when all assumed pollutants are present. The other curves indicate the contribution to radiance change of the individual pollutants. Of present interest is that Ludwig *et al.* have concluded from these calculations that differences between the earth's temperature and that of pollutants are significant in the quantitative determination of pollutant concentrations by radiance measurements. Temperature profiles are also necessary if quantitative measurements are to be achieved. Most important is the fact that considerable spectral interference can be seen to exist between individual pollutants that are present in a typically polluted atmosphere. It would appear that higher spectral resolution experiments may allow for unique determination. An added benefit of high resolution experiments is larger values of absorption coefficients, possibly allowing shorter path measurements.

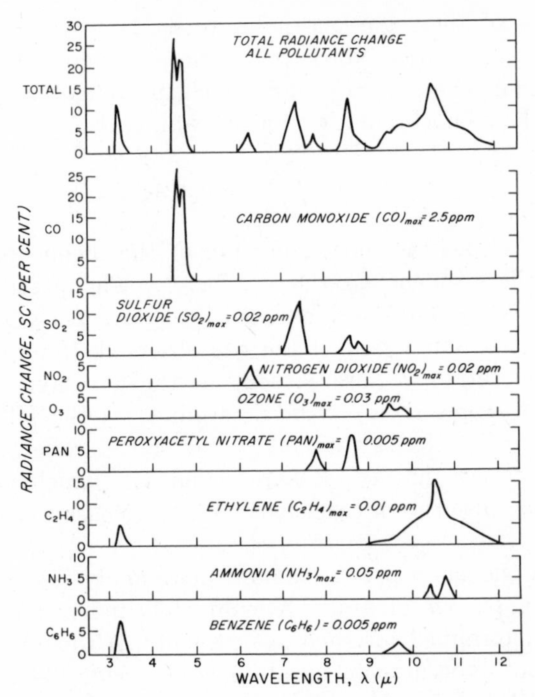

Fig. 9. Percentage radiance changes between clean and polluted atmosphere.

The most complete compilations of high resolution absorption spectra for pollutant gases are to be found in the articles of Long (30) and Hanst and Henson (31). These tabulations indicate that almost all pollutants have strong high resolution absorption spectra in the near infrared between 2 and 25 μ. A number of pollutants also have absorption spectra in the microwave region. Because of pressure-broadening of these absorption lines, considerable overlapping of pollutant spectra occurs. Thus overall pollution content may be determined by microwave measurements but no quantitative data on specific pollutants can be expected. Although overlapping of spectra occurs in the infrared as well, high resolution instrumentation available in the infrared should still permit discrimination of the various pollutants.

As example spectra, Figures 10 and 11 show the sulfur dioxide spectrum from 7.2 to 7.6 μ and ozone spectrum from 9.2 to 10.2 μ, respectively. These are from Hanst and Henson (31). Figure 12, taken from Long (30), shows the CO line at 4.609 μ. The infrared spectra shown in Long and Hanst do not interfere with the strong absorption spectra of water and CO_2.

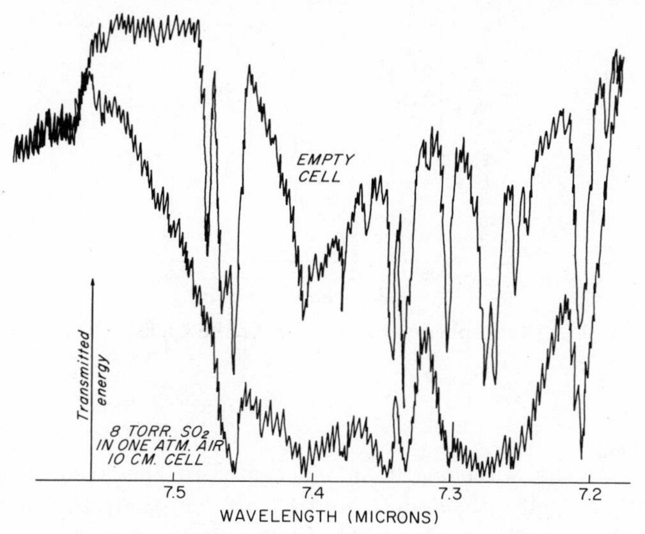

Fig. 10. Absorption of SO_2 from 7.2 to 7.6 μ.

In order to obtain a value for the partial pressure of a pollutant, the absorption coefficient must be known. At constant temperature and at pressures above one-tenth atmosphere, the absorption coefficient, $k(\lambda)$, will depend on pressure P, as

$$k(\lambda) = \frac{(\lambda\lambda_0)^2 C_1 P}{(\lambda_0 - \lambda)^2 + (\lambda\lambda_0 P)^2 C_2} \tag{7}$$

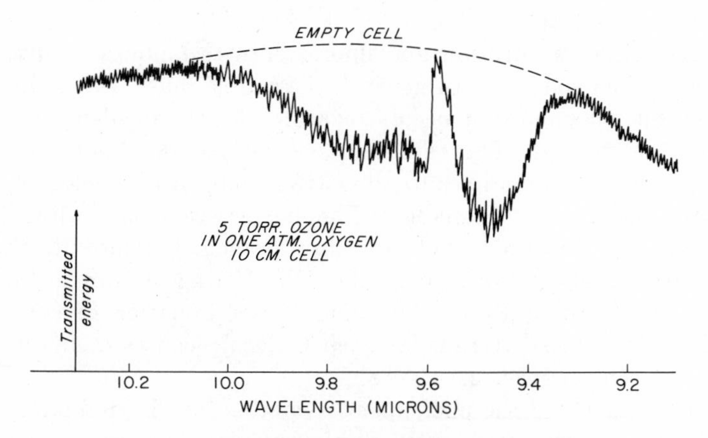

Fig. 11. Absorption of Ozone from 9.2 to 10.2 μ.

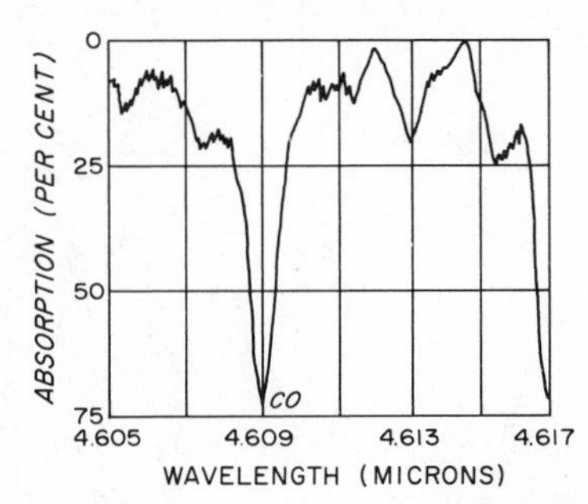

Fig. 12. Atmospheric absorption near the 4.609-μ CO line.

in which λ_0 is the wavelength at the center of the absorption line, λ the operating wavelength, and C_1 and C_2 are constants that are characteristic of the absorbing gas. The $k(\lambda)$ will be directly proportional to pressure far from line center. At the center of the line ($\lambda = \lambda_0$), $k(\lambda)$ will be inversely proportional to pressure. This effect must also be considered in any measurement procedure that might utilize the absorption properties of pollutant gases.

Pollutant gas concentration measurements utilizing absorption may be made through fluorescent scattering. Fluorescent scattering can occur only when the wavelength of incident radiation is in an absorption line or band of the scattering medium. Transition to a higher state or higher energy may then occur, followed by relaxation and emission of light at wavelengths equal to, greater than, or less than the incident wavelength. The longer wavelength energy, called Stokes fluorescence, is greater by a factor of 10^3.

Both the direct absorption and fluorescence techniques require wavelength-controlled sources. In general, the source must be tunable. One scattering phenomenon that occurs regardless of the incident wavelength is Raman scattering. The radiation wavelength does not have to be matched with absorption lines to allow transitions of the molecule. Thus the source need not be tunable. The longer wavelength lines are of higher intensity, and the wavelength displacement is related to the rotation-vibration spectrum of a molecule. The listing of Raman lines may be found in Herzberg (32). The unique determination of constituents allowed by Raman scattering for each molecule causes a unique wavelength displacement spectra.

After one has noted the phenomena available for discriminatory pollutant gas observation, some of the Section II sensing methods can immediately be eliminated. Several methods still appear as candidates:

1. Passive satellite infrared probing
2. Infrared line-of-sight absorption analysis
3. Laser radar.

In essence passive satellite infrared probing has been scrutinized earlier in this section in dealing with radiance measurements. Reference to Figure 9 shows the percentage radiance changes to be expected for low resolution satellite measurements. As Ludwig *et al.* indicate, passive satellite techniques can be used for total pollutant load measurements. Of course this is useful, for data on horizontal distributions on a world basis is necessary. However, the difficulties Ludwig *et al.* have indicated in obtaining even these data are cause enough to consider other techniques. As alluded to earlier, the problems inherent in obtaining unique pollutant data by low resolution measurements are indeed fierce.

In attempting to use high resolution absorption spectroscopy, a major restriction has been the limited amount of energy available from sources of radiation. For infrared line-of-sight absorption analysis of trace constituents of gas mixtures, a long path is usually required. Because radiation from light sources is naturally divergent, the receiver signal after long paths causes a weak signal. If data on localized samples are desired, a White cell may be used, which folds the beam over a long path without losing energy through beam divergence. However, even if such cells are used, the detection sensitivity is energy limited. With the development of the laser not only are energy limitations no longer a problem but also long-path real atmosphere line-of-sight absorption analysis has become feasible. The laser energy is confined to a very narrow range of wavelengths. Thus the high resolution needed for absorption analysis of spectra such as those shown in Figures 10, 11, and 12 is easily available with the laser as source; in fact, the laser energy is generally so narrow in spectra that a lack of knowledge of the exact location of the laser spectra with respect to the pollutant spectra can cause significant errors in pollutant concentration values. Furthermore, it should be obvious that one may not be able to find lasers that emit at wavelengths coincident with large pollutant absorption coefficients. Hanst (33) has considered this problem for a number of pollutants resulting in the data appearing in Table I.

Perusal of Table I indicates that the carbon dioxide and iodine lasers will be most useful for pollutant detection. With both lasers having numerous possible wavelength outputs, the appropriate one is selected by utilization of an absorbing-gas cell which enhances the relative gain

TABLE I

Pollutant Absorption for Laser Lines

Pollutant	Laser line, μ		Concentration in ppm for 5% signal change over 1-km path
C_2H_2	Ne	13.76	0.03
C_2H_4	CO_2	10.53	0.10
C_4H_{10}	I_2	3.43	0.05
PAN	He Ne	?	0.10
NO	I_2	5.5	1.0
CO	I_2	4.86	2.0
O_3	CO_2	9.5	0.15
SO_2	CO_2	9.1	1.0
NH_3	CO_2	10.6	0.10

of the desired output wavelength with respect to other wavelengths. Long paths are needed to obtain significant signal changes. Restriction to horizontal ground-based measurements is not necessary. The tunable laser source and corresponding detector system could be mounted on a rooftop in a central location and aimed at a corner reflector mounted on a helicopter. As the remote reflector is carried past critical areas, the changes in absorption could be used to identify and to locate pollution sources. Larger regions could be covered by mounting the laser and detector entirely in the helicopter. The path capability is shorter in this case for the laser beam illuminating the ground gives rise to a diffuse remote source. The major criticism of infrared line-of-sight measurements is the inability to discriminate pollutant concentration along the path. The total path absorption is measured with no capability for a profile of concentration along the path. However, by using tunable or multi-wavelength laser radar this problem can be overcome.

The ruby laser normally used in laser radar does not fall on any atmospheric or pollutant absorption lines. Figure 13 taken from Long (30) shows the solar spectrum near the ruby laser wavelength. One method of tuning a laser radar is indicated by the lower scales in Figure 13. These show the relationship of wavelength output versus temperature of the ruby rod. By utilizing a ruby laser in a temperature-controlled water bath as the laser radar source, emission at and around the 6943.8-Å atmospheric water vapor line can be achieved. In addition to vertical profiles on water vapor distribution a second useful measurement may be obtained. If laser radar emission much narrower in wavelength spread than atmospheric absorption lines can be obtained, low-

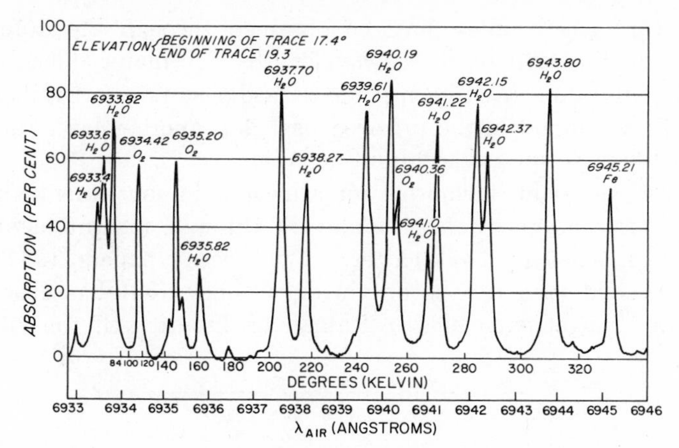

Fig. 13. Solar spectrum near 0.6943-μ ruby laser line.

level temperature profiles can be constructed from laser radar signal returns and one temperature measurement along the path (34). Whether laser radar emission can be made narrow enough to scan absorption lines is still uncertain. For example, the water vapor line at 6943.8 Å has a width of 0.1 Å at half intensity. Thus a laser line of approximately 0.01 Å width is necessary. Tiffany (35), using a solid sapphire etalon for laser mode control, has achieved a half-width of <0.015 Å. However, field experiments do not allow for the ideal conditions that Tiffany experienced in the laboratory.

If laser radar is to be effectively used in gaseous pollutant measurements, lasers with high pulsed power at pollutant gas absorption lines (i.e., 2 to 25-μ range) must be used. Some mention has been made of stimulated Raman emission to shift the source wavelength to the near infrared. In shifting the 6943-Å ruby laser emission, generally one finds too low an energy conversion to the infrared wavelength, too broad a laser emission spectrum, and not enough control over the wavelength output. By making use of the fast-developing dye laser, these difficulties may be overcome.

First reported in 1963 (36) and best described by P. Sorokin (37), efficient conversion ($\geq 5\%$) with little pulse distortion and accurate tuning capabilities presently exist. By mounting a dye laser system on a laser radar, a system easily can be designed to give equal power levels at the normal ruby and tunable dye wavelengths. By locating the dye laser output at a water vapor or pollutant gas absorption line and the normal ruby being in the clear of absorption, the humidity, temperature, and pollutant gas profiles may be obtained through the difference in signal return for each of these wavelengths. A major difficulty is the fact that no dyes have yet been made to lase past 1.5 μ. Until efficiently lasing dyes are found in the infrared past 2 μ, another laser radar technique must be used for gas pollutant detection.

The most successful technique for atmospheric constituent and pollutant gas observation by laser radar is through measurement of the Raman component of backscatter. Two recent papers (38,39) have shown that field data can be obtained. Cooney (38) has observed the first Stokes (longer wavelength) Raman backscatter off the vibrational

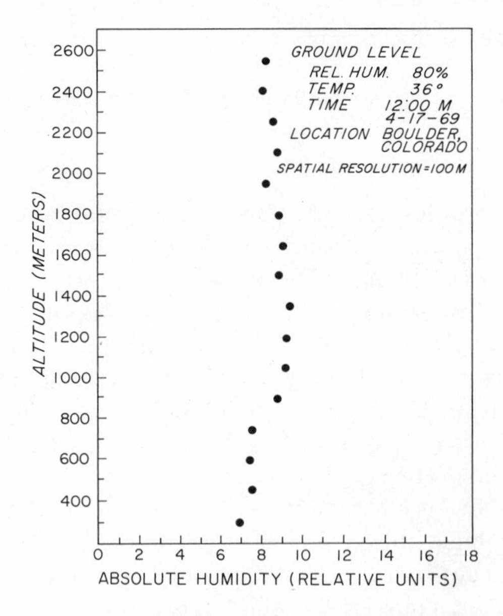

Fig. 14. Vertical profile of absolute humidity at Boulder, Colorado, on April 17, 1969.

levels of atmospheric N_2 and H_2O to > 2 km heights. Noting that the Raman backscatter cross section has a $1/\lambda^4$ dependence, ruby laser emission was first frequency doubled, so that signal emission occurred at 3472 Å, and backscatter return in the case of water vapor occurred at 3976 Å. An added benefit is that the spectral response of the detector peaks near this wavelength. A profile on absolute humidity obtained by Cooney is shown in Figure 14.

Kobayasi and Inaba (39) have detected the Raman component of sulfur dioxide and carbon dioxide backscatter using laser radar. They did so using the 6943-Å ruby emission as source and observing SO_2

signal return at 7545 Å and CO_2 return at 7683 Å. This experiment demonstrates that laser radar can be used not only to uniquely identify gaseous pollutants but also to obtain the measurement of their spatial distribution on a real-time basis. With further improvements in lasers and use in mobile platforms (including satellites), the laser radar technique appears the best candidate for reliable, quantitative, spatial distribution, and real-time gaseous pollutant remote sensing.

References

1. D. Atlas, J. Metcalf, J. Richter, and E. Gossard, *J. Atoms. Sci.*, **27**, 903 (1970).
2. R. M. Schotland, J. Bradley, and A. M. Nathan, Technical Report 67-2, ECOM-02207-F (1967).
3. C. G. Little, "On the Potential of Acoustic Echo Sounding Techniques for the Study of the Atmosphere," Session IV paper at the Sixth Symposium on Remote Sensing of Environment, Ann Arbor, Mich., 1969.
4. L. G. McAllister and J. R. Pollard, "Acoustic Sounding of the Lower Atmosphere," Session IV paper at the Sixth Symposium on Remote Sensing of Environment, Ann Arbor, Mich., 1969.
5. R. K. Cook, "Subsonic Atmospheric Oscillations," Reports of the 6th International Congress on Acoustics, Tokyo, Japan, 1968.
6. D. L. Fried, *J. Opt. Soc. Amer.*, **57**, 175 (1967).
7. R. W. Lee and A. T. Waterman, *Proc. IEEE*, **54**, 454 (1966).
8. W. L. Smith, *Appl. Opt.*, **9**, 1993 (1970).
9. Sessions I, V, VI, VII, IX of the Sixth Symposium on Remote Sensing of Environment, Ann Arbor, Mich., 1969.
10. D. Q. Wark, "Atmospheric Temperature Determinations from the Nimbus III Satellite," Session IV paper at the Sixth Symposium on Remote Sensing of Environment, Ann Arbor, Mich., 1969.
11. S. Twomey, *J. Atmos. Sci.*, **27**, 515 (1970).
12. C. E. Junge, *J. Meteorol.*, **11**, 323 (1954).
13. R. K. Long and D. B. Rensch, *Appl. Opt.*, **9**, 1563 (1970).
14. H. C. van de Hulst, *Light Scattering by Small Particles*, Wiley, New York, 1957.
15. K. Bullrich, "Scattered Radiation in the Atmosphere and the Natural Aerosol," in *Advances in Geophysics*, Vol. 10, Academic Press, New York, 1964.
16. J. D. Lawrence, M. P. McCormick, and S. H. Melfi, "Optical Radar Studies of the Atmosphere," Session II, Session IV paper at the Sixth Symposium on Remote Sensing of Environment, Ann Arbor, Mich., 1969.
17. E. W. Barrett and O. Ben-Dov, *J. Appl. Meteorol.*, **6**, 500 (1967).
18. W. B. Johnson, "Lidar Applications in Air Pollution Studies," in *Stanford Res. Inst. Report Series*, Menlo Park, Calif., 1968.
19. L. Elterman, Environmental Research Paper 285, AFCRL-68-0153, 1968.
20. J. M. Rosen, Atmospheric Physics Report, School of Physics and Astronomy, University of Minnesota, Minneapolis, Minn.
21. F. S. Harris, *Tellus*, **21**, 223 (1969).
22. H. Sievering and R. Mittra, "Investigation of the Radar Equation for the Coherent Scattering of Laser Radar Waves by Atmospheric Constituents, Cloud Droplets, and Pollutant Particulates," Session II paper at the Third Laser Radar Conference, Ocho Rios, Jamaica, 1970.

23. D. Deirmendjian, *Appl. Opt.,* **3,** 187 (1964).
24. R. Penndorf, *J. Opt. Soc. Amer.,* **52,** 402 (1962).
25. B. Herman, "Determination of Aerosol Size Distribution from Lidar Measurements," Session II paper at the Third Laser Radar Conference, Ocho Rios, Jamaica, 1970.
26. E. Palmer, "Aerosol-Particle Parameters from Light Scattering Data," Session II paper at the Third Laser Radar Conference, Ocho Rios, Jamaica, 1970.
27. J. A. Reagen, "A Bistatic Lidar for Measuring Atmospheric Aerosol Distributions," Session II paper at the Third Laser Radar Conference, Ocho Rios, Jamaica, 1970.
28. J. T. Beard, "The Quantitative Determination of Gaseous Air Pollutants by Long-Path Infrared Transmission Spectroscopy," presented at Air Pollution Control Association's 61st Annual Meeting, St. Paul, Minn.
29. C. B. Ludwig, R. Bartle, and M. Griggs, NASA CR-1380, General Dynamics Corporation Report, San Diego, Calif., 1969.
30. R. K. Long, *Ohio State Univ. Eng. Exp. Stat. Bull. 199,* 1965.
31 P. L. Hanst and W. J. Henson, *Air Pollutant Infrared Spectra and Absorption Coefficients at Laser Wavelengths,* NASA Electronics Research Center, Cambridge, Mass., 1969.
32. G. Herzberg, in *Molecular Spectra and Molecular Structure,* Vol. 2, "Infrared and Raman Spectra of Polyatomic Molecules," Van Nostrand, Princeton, N.J., 1960.
33. P. L. Hanst, *Laser Focus,* **6,** 19 (1970).
34. H. Sievering, R. Mittra, and R. Semonin, "Use of Laser Radar for Air Pollution Vertical Profile Measurements," Session K paper at the 8th Environmental Engineering Conference, George Washington University, Washington, D.C., 1970.
35. W. B. Tiffany, *Appl. Opt.,* **7,** 67 (1968).
36. A. Lempicki and H. Samuelson, *Appl. Phys. Lett.,* **4,** 133 (1963).
37. P. Sorokin, *Sci. Amer.,* **220,** 30 (1969).
38. J. Cooney, *J. Appl. Meteorol,* **9,** 182 (1970).
39. T. Kobayasi and H. Inaba, *Appl. Phys. Lett.,* **17,** 139 (1970).

Antibacterial Drugs as Environmental Contaminants*

W. G. Huber
College of Veterinary Medicine,
University of Illinois,
Urbana, Illinois

* Research by the author has been supported in part by a grant from the U.S. Public Health Service, National Center for Urban and Industrial Health (FD 00054).

I. INTRODUCTION

The antibacterial drugs used for man, his animals, and his crops consist to a significant degree of antibiotic drugs and to a lesser degree of chemotherapeutic substances, such as sulfonamides, nitrofuran derivatives, arsenicals. Of special importance is the fact that many of the nonmedical uses involve antibiotic drugs that are used also to treat diseases of man.

Use of antibiotic drugs is widespread. Some uses are of questionable efficacy and may be contributing to environmental contamination and public health hazards. Antibacterial drugs benefit man and his animals when they are used intelligently to treat specific diseases. The benefits are fewer and the hazards are greater when the drugs are used indiscriminately as a panacea for disease problems (animals or man) or for production problems involving nutrition and management of livestock and poultry.

Numerous nonmedical applications of antibiotics are to be found. For this study, the term "nonmedical" refers to the use of antibacterial drugs for a purpose other than the treatment or control of a disease entity (e.g., to act on rate of gain, feed efficiency, or egg production in livestock or poultry, crop application, and food preservation).

The effects of antibacterial drugs on our environment can be measured in terms of direct residual toxicity, such as an acute allergic response to penicillin and the ecologic responses of pathogenic and nonpathogenic bacteria. To minimize the contaminating effect of antibiotic drugs, current uses must be evaluated for efficacy. Many of the antibacterial drugs were approved for use by federal regulatory authorities prior to the enactment of legislation requiring the submission of adequate efficacy data. It was not until 1962 that the Harris-Kefauver Amendment to the Food, Drug, and Cosmetic Act enabled by law the Food and Drug Administration to require adequate data for establishing efficacy. Most of the antibacterial drugs were on the market prior to 1962. An evaluation of the effectiveness of currently used antibiotics is of importance. If a particular antibiotic drug is not effective, its use should be prohibited rather than trying to develop a means of controlling its undesirable environmental effects. An evaluation of antibiotics must be made to determine their economic feasibility. If their uses are of marginal benefit to man's welfare, it would be unwise to continue their usage, especially if they endanger man's health or environment.

II. CLASSES OF ANTIBACTERIAL DRUGS

A. Antibiotic Drugs

The antibiotic drugs are the most common of the antibacterial drugs used for animal purposes. The antibacterial agents with greatest usage can be chemically classified as follows: the tetracyclines (e.g., oxytetracycline, chlortetracycline); the penicillins; the aminoglycosides (e.g., streptomycin, dihydrostreptomycin, and neomycin); the macrolides (e.g., erythromycin, oleandomycin, and tylosin); the polyenes (e.g., nystatin); the polypeptides (e.g., polymyxin B bacitracin); and assorted structural groups (e.g., chloramphenicol, hygromycin B, and novobiocin).

1. *Procaine Penicillin G*

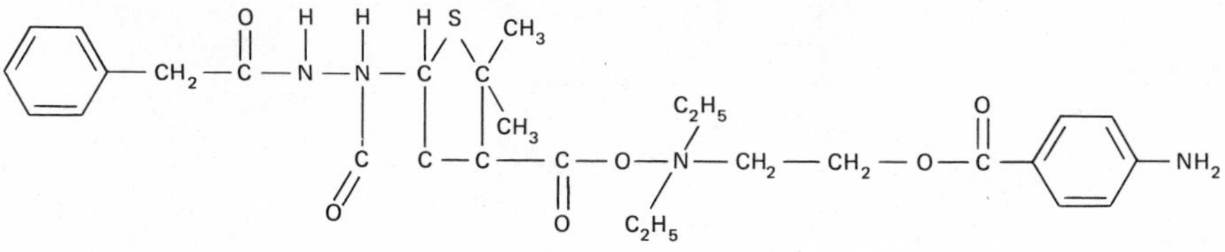

2. *Streptomycin*

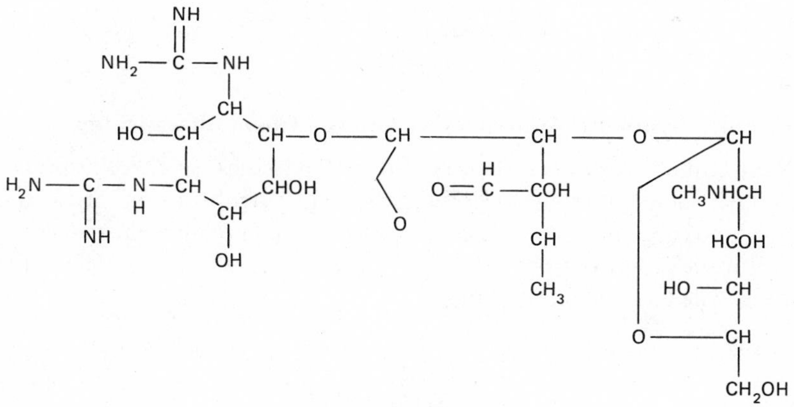

3. *Oxytetracycline*

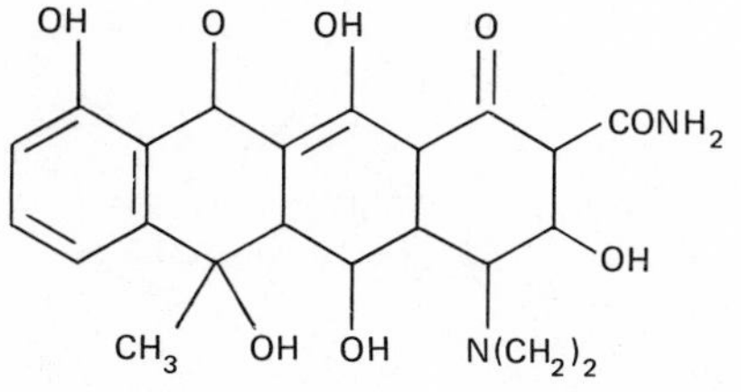

4. *Chlortetracycline*

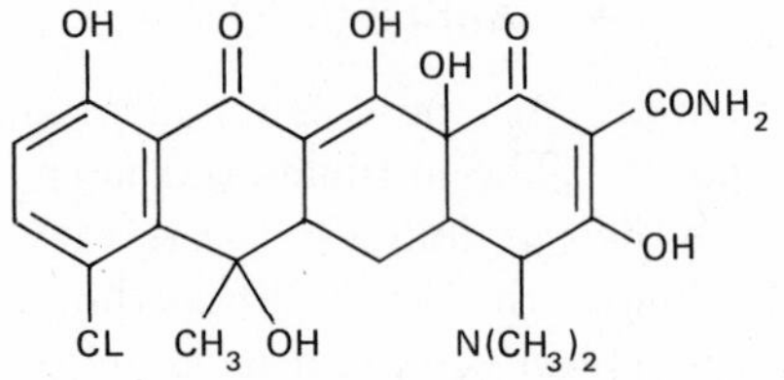

OH O OH OH O
CONH$_2$
OH
CL CH$_3$ OH N(CH$_3$)$_2$

5. *Nitrofurazone*

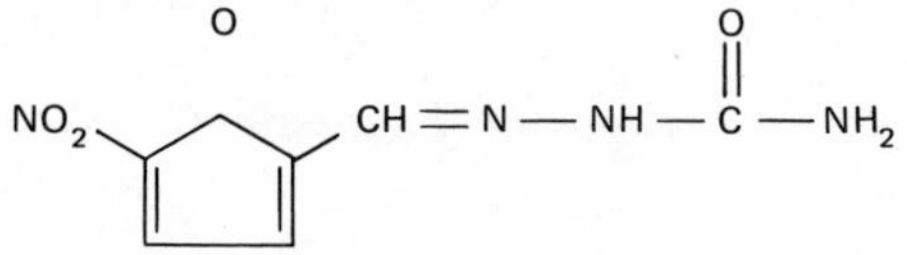

O
O
NO$_2$ CH = N — NH — C — NH$_2$

Nitrofurazone

6. *Nitrofurantoin*

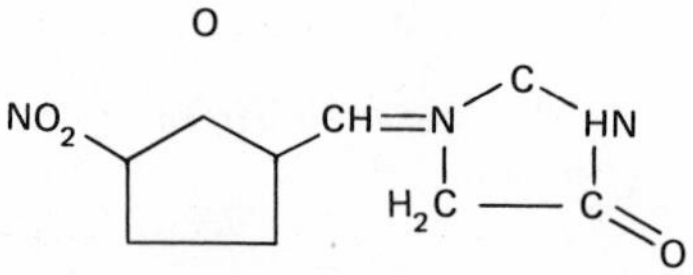

O
C
NO$_2$ CH = N HN
H$_2$C — C = O

Nitrofurantoin

B. Synthetic Antibacterial Drugs (Chemotherapeutics)

Nonantibiotic or chemotherapeutic sulfonamides (e.g., sulfamethazine or sulfathiazole), nitrofuran derivatives (e.g., nitrofurantoin or nitrofurazone), and organic and inorganic arsenical compounds are drugs that constitute a group that have antibacterial activity yet are not traditionally classified as antibiotic drugs.

1. *Sulfathiazole*

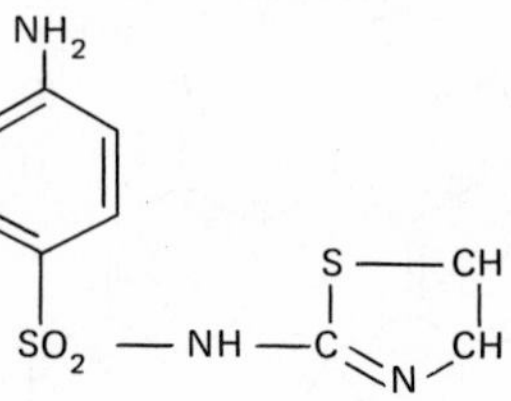

NH$_2$
S — CH
SO$_2$ — NH — C = N — CH

2. *Sulfamethazine*

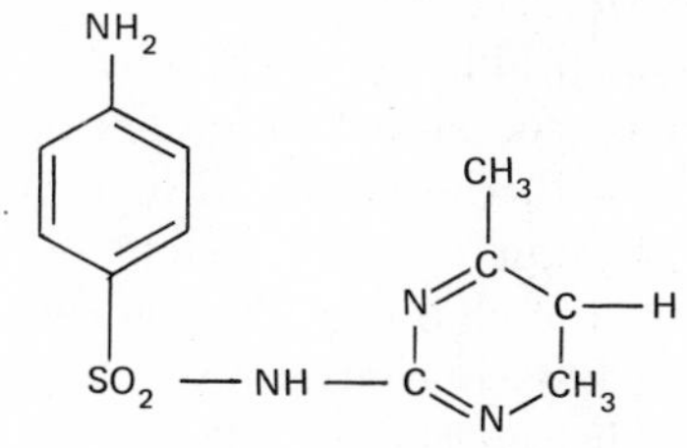

NH_2

CH_3

C

N C—H

SO_2 — NH — C CH_3

N

III. USAGE PATTERNS OF ANTIBACTERIAL DRUGS

A. Animal Feed Additives

The addition of antibacterial drugs to the feed and drinking water of livestock and poultry is a routine practice. The amount administered orally may range from subtherapeutic dosage to concentrations two or three times greater than the recommended therapeutic amount. During 1966, $215 million was spent on animal feed additives and $115 million on animal health pharmaceuticals in the United States. More than half of the antibiotics manufactured in the United States are used for agricultural purposes, primarily as animal feed additives. Many domestic animals are exposed to antibacterial drugs from birth until slaughter.

Antibacterial drugs are used extensively in all types of animal feeds. In the United States in 1968, 87 million tons of feed were produced, of which 60 million tons were prepared commercially. Forty of the 60 million tons of commercially prepared feed contained medication. Of the 40 million tons of medicated feed, 30 million tons required withdrawal times before the animal consuming medicated feed should be sent to the slaughterhouse.

There are many problems with medicated feed withdrawal times. The livestock and poultry producers incur additional expense in money and labor when the feed has to be changed a few days prior to slaughter for large numbers of animals. If the farm operation is sizable and mechanized, the problem is magnified. The standards of postmortem meat inspection currently employed by the U.S.D.A. are not sufficiently sophisticated to detect a representative estimate of residual antibiotics in animal feeds or tissues. A detection program could be established similar to that used for safeguarding the fluid milk supply. Successful detection programs for contaminants in milk have been used since the early 1960s.

In the United States, 77.8% of meat and eggs is produced by animals that have been fed medicated feeds. Because antibacterial drugs are used on the majority of the livestock and poultry raised in the United States, residual and ecologic problems may result.

Antibacterial drugs have been added to animal feeds at low concentrations, less than 100 ppm, with the intent to increase average daily gain and to reduce the cost per unit of body weight or to improve feed efficiency. In some instances the claims of disease prevention or of therapy have been given for these low concentrations. Adequate documentation is not available for many disease claims. Environmental and nutritional factors play a part in determining whether small amounts of antibacterial drugs will produce a beneficial effect (1). A rate in gain in body weight may occur more readily if the antibacterial drugs are used when: (*1*) poor quality rations are fed, (*2*) the rations will allow improved utilizations to occur, (*3*) poor sanitation exists or the premises are not properly cleaned, (*4*) disease exists—clinical or subclinical, (*5*) there are "runt" or undersized animals in the herd, and (*6*) the animals are in a stage of rapid growth.

An increase in the rate of gain has been the most consistent finding following the use of small amounts of antimicrobial feed additives. However, observations indicate that the body weight of animals not receiving antibiotic feed additives may equal the body weight of animals receiving the additives. The antibiotic-treated group has an average daily gain rate that slows as market weight is approached, whereas the nontreated group usually reaches market weight within a few days of the treated group (2). Antibacterial drugs do not stimulate or promote growth. However, they may permit normal growth to occur if the animals are in an adverse environment.

In order to use antibacterial drugs with a minimum effect on the environment, additional research is needed to quantitate the most beneficial periods of rapid gain for the various classes of livestock and poultry. The restriction of drug usage to proven beneficial periods would be a wiser approach than the current practices of feeding the antibacterial drugs from birth to death. The use of these drugs for nonmedical husbandry purposes would be most beneficial if they were used at the proper time and then discontinued when the benefit became marginal or disappeared. This restriction would result in a reduction of the antibiotic pressures on the bacteria in the animal and the environment. Furthermore, it would be desirable to use separate antibiotics for nonmedical husbandry purposes from those used to treat diseases of man and animals. The use of antibiotics and other antibacterial agents for nonmedical purposes greatly diminishes their therapeutic effectiveness against organisms, such as the gram-negative enteric organisms that develop tolerance to the drug (3). Bacterial resistance development can be related to the extent that the antibiotics are used for medical and nonmedical purposes. Continued widespread use of valuable therapeutic drugs for nonmedical purposes seems unwise, especially, if nontherapeutic agents are equally effective for the same animal production purposes.

There is a need for greater scientific sophistication for measuring the effects of medicinal drugs used for nonmedical purposes as animal feed additives. The gross measurements of body weight and feed efficiency are influenced by numerous factors. The variation due to subclinical diseases, parasite loads, genetic makeup, nutrient intake, and biological availability of the nutrients contribute to the determination of body weight. Information to establish the mechanisms of action would help explain the inconsistent results regarding feed additive usage that is supported only by empirical observations. The use of antibacterial drugs as feed additives to enhance gain and feed efficiency is complicated, lacking basic pharmacologic information regarding dose and response measurements. A recent study of the amounts of antibiotic drugs present in manufactured animal feeds was made. These drugs were listed as active ingredients, yet numerous assayed feed samples were found to contain less than the labeled amounts of the antibiotic. One thousand thirty-one samples were found to contain less than the stated amounts of the antibiotics and only 594 satisfied the quantities listed on the label. Thus 50% of the feed samples tested did not contain the labeled amount of antibiotic; observations were made that no significant differences could be detected in the feed lot performance. This observation may not be valid with regard to the benefit of such additives because of lack of suitable controls or because the animals may have performed in an acceptable manner without ingesting the recommended amounts of antibiotic drugs. A comparison of studies on domestic animals shows that an improvement in rate of gain has been reported more frequently and consistently than improvement in feed efficiency for animals fed antibacterial substances (2). The monetary benefits associated with the rate of gain will determine the longevity of these practices when adequately controlled comparisons can be made.

B. Therapeutic Administration

The antibacterial drugs have been used most effectively as therapeutic agents. Their benefits are well established and documented for a majority of bacterial diseases in animals and man. A dynamic relationship exists between the therapeutic antibacterial drug and the ability of the microorganism to resist or remain sensitive to the drug. The antibacterial drug provides maximum benefit when the disease is accurately diagnosed, the sensitivity pattern of the pathogen determined, and the antibacterial agent administered in the correct amount and frequency. When the antibacterial drugs are used in this manner, their benefits are easily established and well supported by scientific principles.

C. Subtherapeutic Administration

Antibacterial drugs frequently are used in subtherapeutic amounts for medical and nonmedical animal purposes. They have been used in attempts to prevent specific diseases and to aid in the control of other diseases. When used in this manner their benefits are very elusive and nebulous. Evidence indicates that much of the subtherapeutic administration contributes to the development of resistant strains of bacteria, resulting in the creation of resistant organisms that are refractive to therapeutic amounts of many antibacterial drugs (4).

A search of the literature regarding subtherapeutic administration of antibacterial drugs for disease purposes reveals that most of the affirmative reports are testimonials or case reports. There is a paucity of well-designed experimental trials utilizing adequate control procedures and statistical analysis. However, the antibacterial drugs have been used extensively in this manner and have been subjected to massive advertisement efforts to the extent that their usage has been designated as an "industrial myth" (5).

A review of the morbidity and mortality records of animal diseases prior to and after the advent of antibacterial drugs used extensively in subtherapeutic amounts as feed additives failed to demonstrate a reduction in bacterial diseases produced by organisms originally sensitive to commonly used antibiotics. In such animal diseases (e.g., porcine bacterial enteritis), the morbidity and mortality rates have increased. However, there are some animal diseases that can be controlled with chemoprophylaxis, but therapeutic dosages are required.

The risk of ecologic damage must be weighed against beneficial effects when considering chemoprophylaxis. Not only must the ratio of toxicity to positive protection be considered, but also the natural frequency of the disease must be considered. If a drug has a tenfold advantage when its dosage produces a beneficial effect compared with a dosage that is toxic or harmful ecologically, it would be satisfactory. However, if the drug is used preventively with only 5% of the population likely to contract the disease, if the drug is not used the ratio shifts to 2:1 against its use.

The subtherapeutic administration of antibacterial drugs increases the probability that resistant pathogens will develop. Some of the pathogens may be responsible for zoonotic diseases (diseases transferred from animals to man or from man to animals). The pathogens may become refractory to therapeutic amounts of the drug because of previous exposure to subtherapeutic amounts for nonmedical purposes.

Antibacterial drug residues may occur if the so-called preventive drugs are not withdrawn from the feed for the necessary time, allowing the animal to excrete these residues from body fluids and tissues.

Antibacterial drugs have been used as preservative agents when added to certain foodstuffs; however, in the United States such uses have been discontinued, primarily for economic reasons. Poultry have been dipped in solutions of chlortetracycline or oxytetracyline, and fish have been placed in water and ice containing antibiotics in attempts to prolong the shelf life and delay postmortem decomposition. Antibacterial drugs also have been applied to vegetables and fruits to control various plant diseases.

IV. ENVIRONMENTAL HAZARDS

A. Drug Intoxication and Sensitization

The environmental hazards resulting from the extensive use of antibacterial drugs for nonmedical animal production purposes and animal health are twofold: (*1*) drug sensitization or intoxication when man consumes contaminated meat, and (*2*) ecologic problems caused by bac-
produced by other residual chemical substances. The antibacterial drugs
and bacterial populations that have undergone selective changes because of antibiotic pressures.

Man may become sensitized to drugs in several ways. Antibacterial drug usage in domestic livestock and poultry may account for one important source if food-containing drug residues are consumed via the food chain. If a person is sensitive to a specific drug, its subsequent use is obviated; if it must be used, an undesirable allergic response is likely to occur.

The direct toxic effects of antibacterial agents are similar to reactions produced by other residual chemical substances. The antibacterial drugs may create additional environmental hazards because they are able to render bacterial populations insensitive to the commonly used therapeutic drugs.

Currently, much controversy exists regarding the use of antibacterial drugs as animal feed additives. Those supporting unlimited usage warn that one must not move too quickly or stringently on hypothetical dangers associated with or related to the use of antibacterial feed additives. This might interfere with needed progress in the development of food technology (6). Support for this attitude would be the lack of adverse effects indicating success of the drugs. However, the essential question is whether adverse effects are lacking or whether they are being ignored. One way to "lack adverse effects" is not to look for them. Little attention has been directed to the assessment of possible adverse effects related to the use of antibacterial additives to animal feed in the United States. However, research workers in England have conducted much research dealing with the health hazards to man and have proposed that the usage of antibacterial drug additives to animal feeds be limited (5).

1. *Prevalence of Drug-Sensitive People*

Estimates indicate that there are 17–20 million people in the United States who may be sensitive to antibiotics or chemotherapeutic drugs; the exact number of people has not been established (7). A report concerning a Connecticut hospital involved 1000 hospital patients (8). The patients were numbered chronologically, based on their admission to the hospital, and tested to determine the prevalence of drug allergies or hypersensitivities. Of 1000 patients, 149 or 15% were found to be hypersensitive to various drugs. Penicillin hypersensitivity was the most common with 72 or 7.2% of the patients showing a positive reaction. Sulfonamide hypersensitivity was observed in 16 people, meperidine in 12, codeine in 12, horse serum in 7, barbiturates in 5, morphine in 5, procaine in 4, adhesive tap and acetylsalicylic acid in 3 each, and other compounds in 10 people. Some of the hypersensitive reactions may have been initiated by the consumption of antibiotic residues on or in foods.

However, hypersensitive reactions may be more prevalent than information in the scientific literature has indicated. Reporting of untoward reactions following the administration of drugs is complicated with obstacles ranging from ineffective methods of reporting adverse drug reaction to the possibility of litigation by the patient or the heirs.

If a chemical epidemiologic investigation determines that a form of sensitization is found to be relatively common in a segment of a population, it would be desirable to avoid contamination of the food or the sensitizing substance. Staple foods such as milk, eggs, and meat that constitute a large and consistent portion of the diet require special attention.

2. *Factors Determining Residual Hazards*

The presence of antibacterial drug residues in or on food may have many effects on the bacteria found on food and in the gastrointestinal tract of man. The following factors can determine the hazard potential and should be considered:

1. The frequency of ingesting the antibacterial drug is important. It is possible that if a person purchases a side or quarter of beef or pork containing a residue of an antibacterial drug that the exposure will be more continuous than if the contaminated carcass was processed and marketed through a supermarket and consumed by a dozen or more families.
2. The amount and kind of antibacterial residue consumed varies with the extent of the hazard. Some antibacterial drugs and their metabolites or degradation products are potent sensitizers even in extremely small amounts (e.g., penicillin). Other antibacterial drugs (e.g., oxytetracycline) have minimal sensitizing properties rarely recorded.
3. The effects that antibacterial drugs may have on the normal food flora and food storage must be considered. Some drugs when added to foodstuffs will interfere with the sensitivity of subsequent organoleptic tests used in the inspection of food. Residual antibacterial drugs may result in the development of antibiotic resistance in the normal nonpathogenic bacterial flora and then transfer the resistance to pathogenic bacteria. Antibacterial drugs may prolong shelf life for some foodstuffs if refrigeration storage systems are not available; however, good processing and storage facilities greatly minimize the need for antibacterial drugs and reduce the opportunity for untoward drug reactions.
4. Antibacterial drug residues may cause ecologic changes in the normal flora of the consumer. Small quantities of antibacterial drugs, as low as 2 ppm, have been shown to cause changes in antibiotic sensitivity patterns in gram-negative enteric bacteria (29). Population shifts may occur under the process of selection.
5. The effects that antibacterial drugs have on resident nonpathogenic flora and their involvement in the infectious transfer of antibacterial drug resistance to pathogens must be considered. Nonpathogenic *Escherichia coli* have the ability to acquire drug resistance and to pass the drug resistance to sensitive *Salmonella* pathogens without antibacterial drugs ever contacting the pathogenic *Salmonella*.
6. A significant problem is that pathogenic organisms may acquire resistance as a result of their exposure to antibacterial drug residues

in foodstuffs and then become a health problem by producing a disease in man that is refractory to the usual forms of treatment.

3. *Antibacterial Drug Residues in Livestock and Poultry*

In order to assess the antibacterial drug residue problems, two important questions must be considered:

1. Do slaughtered domestic animals contain drug residues?
2. If meat containing drug residues is cooked prior to consumption, will the heat destroy the drug residues and thus resolve the problem?

Let us consider the question of whether slaughtered domestic animals contain antibiotic drug residues? Our laboratory has developed methods to detect antibacterial drug residues and has surveyed several classes of livestock and poultry for the presence of drug residues and to determine if drug and feed additive withdrawal times were being followed. The prevalence of antibacterial drug residues in animals at time of slaughter was determined for more than 5000 animals.

Tissues, urine, or feces were collected from swine, sheep, veal calves, beef cattle, and poultry at slaughterhouses subjected to federal or state meat inspection. Samples were obtained from apparently normal animals sent to the abattoir for slaughter and subsequent use for human consumption.

A microbiologic *Bacillus subtilis* disc-assay method was used as a screening test to detect the presence of antibacterial substances (9). The disk-assay method for fluid milk was modified to test urine, feces, and tissues (10). A method involving electrophoresis of agar gel was modified and used to identify the antibacterial substances (11).

a. Swine

Eight groups of swine were tested in Illinois during each of the seasons. Twenty-seven percent of the 1381 slaughtered hogs showed positive results for the presence of antibacterial substances. Ten percent of the animals tested had positive penicillin residues (Table I).

Swine feces were collected from 118 animals showing a positive urine test. The feces were assayed to determine whether antibacterial drug residues were present because of drugs injected parenterally or attributed to medicated feed ingested orally. Animals that had positive urine tests also were found to have positive fecal tests. The large numbers of animals with positive urine and fecal tests, the lack of visible injection sites, and the metabolic and excretory patterns of the commonly used antibiotics indicated that most of the antibacterial residues were probably the result of oral drug administration rather than parenteral admin-

TABLE I

The Prevalence of Antimicrobial Substances in Swine

Group	Number tested	Positive urine	%
1	233	88	38
2	185	53	30
3	118	45	37
4	281	33	12
5	149	12	8
6	100	36	36
7	158	42	26
8	167	65	39
	1381	374	27 (av)

10% penicillinase positive

Comparison of feces and urine

Number tested	Positive urine	Positive feces
118	45/118 (37%)	34/118 (29%)

istration. A strong possibility exists that the drug withdrawal times are not being followed, the withdrawal times are inadequate, or the recommended dosages were exceeded.

b. Beef Cattle

Beef cattle showed the lowest prevalence of antibacterial residues of the domestic animals that were tested. In 5 groups of cattle tested, 9% of the 580 animals showed positive results. Two percent of the cattle with positive tests were found to have penicillin residues (Table II).

TABLE II

The Prevalence of Antimicrobial Substances in Beef Cattle

Group	Number tested	Positive urine	%
1	33	5	15
2	150	2	1
3	177	12	7
4	58	7	11
5	162	25	15
	580	51	9 (av)

2% penicillinase positive

TABLE III

Prevalence of Antimicrobial Substances in Veal Calves

Group	Number tested	Positive urine	%
1	216	18	8
2	337	78	23
3	82	18	22
4	70	6	9
5	51	1	2
6	32	11	34
	788	131	17 (av)
	7% penicillinase positive		

c. Veal Calves

Tests on the urine samples of veal calves that were collected at slaughter consisted of six groups. Seventeen percent of the samples had antibacterial residues; 7% of these were penicillinase positive (Table III).

d. Market Lambs

Four groups of market lambs were tested. Twenty-one percent (68 of 328 animals tested) contained antibacterial substances. Four percent of the animals tested positive for penicillin residues (Table IV).

e. Chickens

Four groups of chickens (798 laying hens and 128 broilers) were tested after their slaughter. Twenty percent of the chickens were shown to have had antibacterial substances present at the time of slaughter. Six percent of the positive tests indicated penicillin residues (Table V).

TABLE IV

The Prevalence of Antimicrobial Substances in Sheep

Group	Number tested	Positive urine	%
1	121	27	22
2	114	14	12
3	45	18	40
4	58	9	16
	328	68	21
	4% penicillinase positive		

TABLE V

The Prevalence of Antimicrobial Substances in Laying Hens

Group	Number tested	Positive feces	%
1	214	53	25
2	200	32	16
3	384	47	18
4	128	57	—
	926	189	20 (av)
	6% penicillinase positive		

4. *Identification of Antibiotic Drug Residues*

Electrophoresis of agar gel was used to identify the antibiotics: penicillin, dehydrostreptomycin, tylosin, neomycin, and members of the tetracycline group such as chlortetracycline or oxytetracycline. Other antibiotics and synthetic antibacterial drugs could have been included in the assay procedure, but only the more commonly used feed additive drugs were assayed. The assay of additional drugs such as erythromycin, nitrofuran derivatives, and others would have limited the information obtained on the more commonly used antibiotics.

The types of antibiotics found in domestic animal tissues and body fluids is presented in Table VI. The presence of antibiotics in the liver ranged from 16% for lamb livers to 5% for beef livers. The tetracycline antibiotics accounted for the largest number of identified antibiotics. The group classified as "others" consisted of antibiotics, sulfonamides, and others not identified.

Antibiotic residues in the kidney ranged from 7% for swine to 43% for beef. The high prevalence of dihydrostreptomycin residues in the bovine kidney probably occurred because the animals may have received an injection of dihydrostreptomycin with or without penicillin. Penicillin-dihydrostreptomycin preparations are easily available for lay or professional use and are frequently used to treat animals afflicted with respiratory diseases. Dihydrostreptomycin has the ability to be sequestered in tissues for long periods of time—30 days or longer. A high prevalence of dihydrostreptomycin residues indicates that this drug was administered parenterally because only a very small amount will be absorbed from the digestive tract.

Urine samples from numerous animals were subjected to electrophoresis to identify the various antibiotics. The highest prevalence of antibacterial residues occurred in a group of swine. Sixty-seven percent

TABLE VI

Identification of Antibacterial Residues

Tissues	Number tested	Positive, %	Penicillin	Dihydro-strepto-mycin	Tetra-cycline group	Tylosin	Others[a]
Liver							
Chicken	231	12	5	2	2	3	15
Chicken	132	11	0	3	1	0	10
Beef	198	5	0	5	0	0	5
Swine	242	13	1	1	19	0	11
Lamb	200	16	3	4	0	1	24
Kidney							
Chicken	132	8	0	4	2	0	5
Swine	101	7	1	0	0	0	6
Swine	182	15	10	2	8	0	8
Beef	199	43	13	64	7	0	1
Urine							
Beef	162	13	15	0	6	—	—
Swine	281	17	2	0	47	0	0
Swine	100	37	2	0	35	0	0
Swine	145	67	1	0	96	0	0
Swine	167	37	12	1	48	0	0
Calf	51	2	0	0	0	1	0
Calf	75	8	0	0	4	1	1
Calf	32	34	2	0	3	0	6
Lamb	58	16	0	0	9	0	0
Lamb	75	41	0	0	31	0	0
Feces							
Chicken	128	44	31	0	26	0	0

[a] Assay limited to five antibiotics.

of the swine had drug residues of which all except one were residues of the tetracycline antibiotics. The prevalence of antibiotics in urine ranged from 67% in 145 hogs to 2% in 51 veal calves. There was a higher prevalence of tetracycline residues than any other antibiotic. These compounds are readily absorbed from the gastrointestinal tract following oral administration, concentrated by the kidney, and excreted in the urine. Thus the common use of the tetracyclines as animal feed additives and the metabolic patterns supported the observations regarding the high prevalence of chlortetracycline and oxytetracycline in the urine.

The feces of 128 chickens were assayed; 44% contained either penicillin or a tetracycline antibiotic. Because it is impossible to obtain urine in the intact chicken, one can only speculate about the source of antibiotic in feed or by parenteral injection. Either the antibiotics were added to the feed, and the animals were fed medicated feed to the time of slaughter with no withdrawal period used; or the antibiotics were parenterally administered, metabolized, and excreted in the urine which was mixed with the feces in the cloaca. There is much variation in the prevalence of residues of the animals shipped to market. Some farmers use antibiotics wisely and follow the legal withdrawal times. Others do not, and many violators are not discovered or apprehended because of currently employed meat inspection procedures.

5. *Comparison of Residues in Meat and Milk*

There can be no doubt that antibacterial drug residues are present in meat and meat products. Some of the antibiotics are potent sensitizers involved in acute anaphylactic and anaphylactoid reactions and chronic allergic reactions. The prevalence of antibiotic residues in the meat supply for the United States is not known. Our results are limited to animals slaughtered in Illinois.

In our study the prevalence of antibiotic residues in tissues and body fluids of domestic animals exceeds that reported for milk in the United States during the 1950s before a successful milk monitoring program was initiated. The prevalence of antibiotic adulterated milk dropped from 11% to less than 0.5% with the initiation and continuance of a successful monitoring program. In 1964 in Britain, lacking a similar surveillance system, examination of 41,700 milk samples revealed that 11% were contaminated (12). It has been noted that urticaria and contact dermatitis disappeared in man if the intake of antibiotic adulterated milk or dairy products was discontinued.

In the United States the prevalence of antibiotic contamination of milk has been established in the past decade; however, the prevalence of antibiotic residues in meat and meat products is limited. The work conducted by our laboratory represents the largest number of animals tested in the United States (9). Dean *et al.* (13) tested 146 samples of wild and domestic meat and meat products for a period of 6 months and reported that 46% showed the presence of antibacterial substances. Of the positive tests, 32% were identified as penicillin. A vivid contrast was noted because meat from domestic animals had a much higher prevalence of residues than did meat from wild animals. Tests on reindeer, turtle, and buffalo meat were negative.

Some information regarding antibiotic residues in meat has been reported from Europe. In Denmark (14) 12% of the cattle, 58% of the calves, and 23% of the swine had antibiotic residues. Although the samples size was quite small, the results could be related to husbandry practices and drug usage. Another Danish antibiotic residue investigation of meat involved 1004 cattle, 3032 pigs, and 1893 calves (15). Antibiotic residues were reported for 77% of the calves and 1% of swine and cattle. In France (16) antibiotic residues were found in 4.1% of the meat samples assayed. They examined extracts of tissue and viscera of domestic animals.

When using antibiotics and other antibacterial drugs, we have a definite responsibility to use them most effectively and with an awareness that meat and animal products must be kept free of potential health risks to the consumer. It should be apparent that regulatory agencies would want to guarantee a meat supply as uncontaminated as the current milk supply in the United States. Currently, less than 0.5% of the milk is contaminated, because an effective monitoring and testing system has been established. In the United States the prevalence of antibiotic and antibacterial drug residues in meat and animal products is sufficient to constitute a health hazard to some consumers.

The differences in the prevalence of antibiotic residues in the various groups of livestock are probably related to the manner in which animals are raised and the legal uses of antibiotic drugs. Some countries restrict antibiotic drugs on a prescription basis; in others they are freely available to anyone wishing to use them. The latter case applies to the antibiotic drug situation for animal usage in the United States.

6. *Hazards of Antibacterial Drug Metabolites and Degradation Products*

A question frequently raised regarding drug residues is the effect of heat. Does cooking destroy residues in contaminated meat? Some think that pasteurization of milk or cooking of meat removes all harmful agents. Many pathogenic organisms will be destroyed, but there are several antibacterial drugs that will not be altered. Although some antibiotics such as penicillin may be partially degraded by heat, the degradation products and metabolites of the commonly used antibiotics may have a toxic impact equal to or more harmful than the original compound (e.g., penicillanic acid may have sensitizing properties as great as or greater than its parent compound, procaine penicillin G). Similarly, tetracycline metabolites and degradation products may have a greater hemolytic or hepatoxic effect. In addition, the temperatures attained during cooking have very little effect on the degradation of other antibiotics such as streptomycin and dihydrostreptomycin.

Temperatures incident to the process of cooking cannot be relied on to completely inactivate antibiotics and synthetic antibacterial drugs when they are present in meat or meat products because: (*1*) some drugs are very stable and resist heat; (*2*) drugs sensitive to heat may be only partially inactivated; and (*3*) degradation products and metabolites may be as important toxicologically as the original antibiotic or antibacterial drug.

B. Drug Resistance and Ecological Problems

The persistent and widespread use of antibacterial drugs in our environment for man, his animals, and crops emphasizes an urgent need to monitor ecologic changes.

Bacteria may become tolerant or resistant to drugs by various mechanisms. Some organisms become drug tolerant because hereditary material changes spontaneously through mutation. If the offspring from such bacteria are in a population exposed to antibacterial drugs, the number of drug-sensitive bacteria will greatly diminish; the antibacterial drug-resistant organisms will proliferate, resulting in a totally drug-resistant population while the antibacterial drugs are present in the environment. Furthermore, naturally resistant organisms in the environment are subsidized and facilitated to the point that they also greatly increase in number because of "selective" antibiotic pressures. This mechanism of resistance development is quite slow and specific for certain antibacterial drugs.

Antibacterial drug resistance may develop more rapidly via another mechanism: infectious or transferable drug resistance, which also enables the bacteria to become multiresistant to several drugs. Infectious or transferable drug resistance was discovered in Japan in 1959 (17). A resistant bacteria may have the ability to pass one or more resistance factors to other sensitive bacteria. In many instances the sensitive organism becomes multiresistant. The sensitive bacteria acquires the resistance without being exposed to antibiotics. The resistance factor can be transferred to different bacterial species of the gram-negative enteric bacteria. The multiresistant organisms that we observe today have probably acquired their resistance through the transferable drug resistance process. The speed at which this multiresistance has been acquired cannot be related to the older methods of resistance development by mutation and selection.

1. *Infectious or Transferable Drug Resistance*

Transferable drug resistance has been observed in gram-negative enteric bacteria. The gram-negative enteric bacteria are composed of pathogenic and nonpathogenic organisms. However, both groups can transfer or receive resistance factors, and it is not unusual for nonpathogenic organisms to transfer multiple drug resistance to pathogenic organisms of animals or man. If a resistant pathogen produces a disease, the treatment of a patient may be difficult if the pathogen is refractory to common forms of therapy.

The resistant factors are located in episomes. Episomes and chromosomes contain DNA and can replicate; however, the episome is not essential to the normally functioning cell. In order for the R factor to be transferred, a transfer factor must be present. Bacteria may contain only the transfer factor. This factor then may be passed to another recipient; if the recipient contains an R factor, then that bacteria will become drug resistant and have the ability to transfer the resistance to other sensitive organisms. Similarly, some organisms will have only the R factor; however, a cell has only the ability to transfer infectious drug resistance if it has both the R factor and the transfer factor.

Sensitive organisms may become multiple drug resistant *in vitro* within a matter of hours after contact with multiresistant organisms. The R factors can disseminate through a bacterial population more rapidly than the rate of growth of the population, indicating that episomal duplication is very rapid. *In vivo* transfer of R factors and transfer factors occur, but more information is needed on the rates of transfers and colonization in man and animals and from animals to man and vice versa.

The R factors have been found in organisms, *E. coli* preserved in 1942, prior to the widespread use of antibacterial drugs, but R factors and transfer factors probably do not augment survival abilities until the bacteria are subjected to antibiotic drug pressures. The widespread use of antibacterial drugs for nonmedical and medical uses in both man and animals provides an optimum environment for the dissemination of multiple drug resistance. A public health hazard must be considered because animals may serve as reservoirs for zoonotic diseases. Further, nonpathogenic organisms in animals may transfer multiple drug resistance to organisms pathogenic for man. Experimental data support the probability that organisms that are pathogenic to man acquired their resistance by contact with drug-resistant nonpathogenic organisms of animal origin (18). Nineteen of twenty strains of *E. coli* isolated from humans demonstrated transferable drug resistance. The drug resistance patterns were identical to 16 of 20 *E. coli* strains isolated from calves, 16 of 20 from pigs, and 18 of 20 from fowl.

Many gram-negative enteric bacteria have developed resistance to more than one antibacterial drug. The results of research conducted in England and Japan had been interpreted as alarming (19–21). The rate of resistance development is dependent on the existent antibacterial drug pressures. Small amounts of antibacterial drugs over a long period of time may have a greater ecologic environmental impact than the use of the same drugs at therapeutic concentration for short periods. Watanabe (19) described the resistance problems well when he stated:

> The public health threat posed by infectious drug resistance is measured by the range of bacterial hosts it affects, the number of drugs to which it imparts resistance and the prevalence of certain practices in medicine, agriculture and food processing that tend to favor its spread.
>
> In many parts of the world, antibiotics are routinely incorporated into livestock feed to attempt to promote growth and are also used to attempt to control animal diseases. Anderson and Datta (22) have clearly shown that the presence of antibiotics in livestock exerts a strong selective pressure in favored organisms, particularly *Salmonellae,* with resistance factors and plays an important role in the spread of infectious drug resistance.

Transferable drug resistance has been found in many parts of the world: Japan (23), Rumania (24), United States (25), South Africa (26), England (27), Germany (28). The R factors have been identified in pathogenic and nonpathogenic gram-negative enteric organisms isolated from man and animals. The R factors in *E. coli* of animal origin were observed in Great Britain by Anderson and Lewis (29) in association with an outbreak of *Salmonella typhimurium.* The *E. coli* and *S. typhimurium* organisms had identical resistance patterns to the antibacterial drugs. It has been suggested that a drug-sensitive pathogen can become resistant by contact with a resistant nonpathogenic organism because R factors have been identified in pathogenic and nonpathogenic *E. coli* isolated from both man and animals (18,30).

Antibacterial drugs not only insure the acquisition of R factors by gram-negative enteric organisms such as *S. typhimurium* (31), but also they facilitate the initiation of the infection by the pathogen because of the elimination or inhibition of drug-sensitive competitor organisms. Multiple drug resistance can be and is retained if the organism is exposed to any one drug included in its resistance spectrum.

The development of transferable drug resistance in man does not arise exclusively from the use of antibacterial drugs in livestock and poultry, because most of the drugs are used for both man and animals; however, the method of administration to man and to animals is quite different. For medical purposes the use of antibacterial drugs in man is essentially on an individual basis, but in livestock and poultry the animals are treated on a herd or flock basis, usually in feed or drinking water. In addition, many of the same antibiotics are used for nonmedical purposes—in feed additives for rate of gain. In the United States it is very difficult to obtain an animal feed that does not contain a drug or drugs. The drug is usually an antibiotic drug. Thus a majority, 77.8%, of the meat and eggs that we consume are derived from animals exposed to antibacterial drug pressures.

The nonmedical and prophylactic uses of antibacterial drugs favor the spread and maintenance of organisms that should be controlled (4). The situation in disease problems of livestock and poultry has been compared to hospital wards that have been exposed to the uncritical use of antibacterial drugs, promoting the emergence and spread and maintenance of drug-resistant strains of pathogens.

2. *Low-Level Usage and Drug Resistance*

When animal feed additives and drug resistance are considered, the following question is usually presented: Does low-level usage, such as 5–15 ppm, of antibacterial drugs cause the development of resistant bacteria or does this practice constitute a health hazard? Before the antibacterial drugs were used extensively for nonmedical animal production purposes, a health hazard was considered unlikely. However, in 1965 evidence was presented that the addition of antibacterial drugs to animal feeds in amounts as low as 2 ppm was found to produce transferable drug resistance (32). Animals fed antibiotics at the rate of 2–10 ppm had a greater prevalence of antibiotic multiresistant organisms that persisted for a longer period of time than did animals not fed antibiotics. Another study indicated that the administration of subtherapeutic amounts of antibiotics increased the number of animals from which multiresistant *Salmonella typhimurium* could be isolated (33). In our laboratory we have found that drug resistance can be produced more rapidly if organisms are exposed to small concentrations or subtherapeutic amounts of drugs rather than being exposed to therapeutic concentrations for a short time.

With the discovery of infectious or transferable drug resistance, the effect that antibacterial drugs have on nonpathogenic organisms also must be considered because drug-resistant nonpathogens can transfer their resistance to unrelated pathogens. For example, drug resistance may be transferred from a nonpathogenic *E. coli* to a pathogenic *Salmonella*.

It has been demonstrated that small amounts of antibacterial drugs have caused the development of drug-resistant organisms and amounts of 2 ppm or higher have had an ecologic effect on pathogenic and nonpathogenic bacteria.

3. *Organisms Resistant to Antibacterial Drugs*

The rates of development of bacterial resistance through mutation, selection, adaptation, and infectious drug resistance will determine the length of time that antibiotics, sulfonamides, and other chemotherapeutic drugs will be effective. Unfortunately, these drugs not only act on pathogenic bacteria, their intended targets, but also on nonpathogenic organisms. Furthermore, we must remember that organisms have the ability to adjust to new chemical insults that may result in new disease problems.

a. Animals

The problem of drug resistance in domestic animals has been the subject of two extensive reviews (21,37). Resistance problems have been recorded for many organisms, but only a few are considered in this study. *E. coli, Salmonella typhimurium,* and *Staphylococcus aureus* frequently have been involved in drug-resistant disease problems of animals and man.

Resistance development in the gram-negative organisms of the gastrointestinal tract has been a very severe problem. *E. coli,* isolated from diseased poultry, showed an increase in resistance to tetracycline antibiotics from 3.5% in 1957 to 63.2% in 1960, or an 18-fold increase during a period when the use of antibiotics was being expanded rapidly (35).

Smith and Crabb (36) investigated the effect of continuous administration of chlortetracycline in animal feeds. Pigs fed 4–30 ppm chloretetracycline in the animal feeds on 37 farms were found to have an *E. coli* population that was 88% resistant, whereas animals given antibiotic-free feed had 99% of their *E. coli* with a sensitivity ranging from "completely sensitive" to "mainly sensitive." The investigators encountered difficulties in locating farms where antibiotics were not included in the animal feed; many farmers were unaware that their feed contained the drugs.

Smith and Crabb (37) also surveyed the sensitivity patterns of *E. coli* in poultry and sheep. Organisms in poultry were found to be similar to those in swine. A predominantly tetracycline-resistant *E. coli* population was recorded for birds subjected to selective antibiotic feed pressures. A sensitive *E. coli* population was noted in birds not fed tetracycline antibiotics. Sheep had sensitive *E. coli* populations that coincided with drug usage. In England sheep are not fed antibacterial drugs.

Similar resistance patterns for *E. coli* have been reported for calves and poultry subjected to antibiotic pressures (21,38–40).

Edwards (41) demonstrated that the oral administration of streptomycin to calves at a dose of 300 mg for 1 week induced a rapid replacement of a predominantly sensitive *E. coli* flora to a resistant one.

In addition to determining antibacterial drug sensitivity patterns of *E. coli,* recent investigators have determined the prevalence of multiresistant organisms and R factors.

Walton (42) isolated multiresistant strains of *E. coli* from the feces of healthy pigs and calves that had consumed antibiotic feed; 99 of 135 multiresistant strains transferred their drug resistance to sensitive strains of *Salmonella typhimurium* and *E. coli.* The maintenance of multiresistant strains of *E. coli* was facilitated by the presence of antibacterial drugs in the feed.

Mitsuhashi *et al.* (43) isolated *E. coli* strains from 151 swine and 108 fowls and surveyed them for drug resistance and distribution of R factors. All the swine and 38% of the fowl excreted *E. coli* strains that were resistant to tetracycline, chloramphenicol, streptomycin, sulfanilamide, or certain combinations of these. Among 278 resistant cultures isolated from swine, 13% were resistant to only one antibiotic, whereas 87% were resistant to more than one antibiotic. Among the resistant strains, 40% carried R factors that were transferable by the usual conjugal process. The resistance patterns of these R factors included 36% singly resistant and 64% multiresistant. In 54 resistant cultures isolated from fowl, 24% were singly resistant and 76% multiresistant. Of the resistant strains from fowl, 22% carried R factors. The resistance patterns of the organisms containing R factors included 50% of the singly-resistant type and 50% of the multiresistant type.

Recently, we have studied resistance patterns and R factor prevalence in domestic animals and wildlife (44). The domestic animals studied were subjected to routine production and health practices which involved near maximal antibiotic pressures; for example, antibiotic drugs in the feed, drinking water, and, less frequently, antibiotics administered by injection. The wildlife had minimal exposures to antibiotic pressures. Although they were trapped in the wild, it is possible that a few of them, especially the rats, might have consumed animal feed, tissues, or feces containing antibiotic drugs. However, their exposure to antibacterial drugs would not approach that of domestic animals.

The higher percentage of *E. coli* isolates found in domestic animals than in wild animals might be related to antibiotic pressures, because antibacterial drugs seemingly favor the development of resistant bacteria through the process of selection while sensitive organisms decrease in number (4).

Multiresistant and R-factor-bearing *E. coli* isolated from domestic animals were more prevalent than sensitive non-R-factor-bearing *E. coli*

TABLE VII

Percentage of *E. coli* Resistant to Antibiotics

Antibiotic drug	Dairy cows	Dairy calves	Swine	Dogs	Wild animals
Oxytetracycline	56	94	91	72	11
Dihydrostreptomycin	69	94	82	88	8
Ampicillin	27	25	37	38	3
Neomycin	12	12	3	40	8
Chloramphenicol	0	0	0	43	3

(44). Dairy cattle *E. coli* isolates were 69% resistant to dihydrostreptomycin, 56% to oxytetracycline, 27% to ampicillin, and 12% to neomycin (Table VII).

Isolates of dairy calves were 94% resistant to oxytetracycline, 25% to ampicillin, and 12% to neomycin. Swine isolates were 91% resistant to oxytetracycline, 82% to dihydrostreptomycin, 37% to ampicillin, and 3% to neomycin. The prevalence of R factors in *E. coli* of domestic animals was as follows: 50–52% of the *E. coli* isolates from dairy cows, 36–37% from dairy calves, 62–63% from swine (Table VIII). The high prevalence of drug-resistant organism with R factors could be related to the prior exposure to antibiotic pressures occasioned by the use of medicated feeds.

A comparison of antibiotic sensitivity patterns in domestic animals and in wildlife showed several points of contrast. For example, oxytetracycline and chlortetracycline commonly and extensively are used for nonmedical animal production purposes in domestic animals. Both drugs are members of the tetracycline group and have relatively the

TABLE VIII

Prevalence of Resistance Transfer Factor in Isolated *E. coli*

Source of *E. coli*	% transferred RTF to sensitive *E. coli*	% transferred RTF to sensitive *Salmonella typhimurium*
Dairy cows	52	50
Dairy calves	67	36
Swine	63	62
Dogs	35	92
Wild animals	2	1

same antibacterial spectrum and patterns of cross-resistance. The high percentage of oxytetracycline-resistant *E. coli* found in domestic animals indicates that the therapeutic usefulness of oxytetracycline is limited, whereas most wildlife *E. coli* are still sensitive (44). Similarly, a significant percentage of domestic animal *E. coli* were found resistant to dihydrostreptomycin, a drug used extensively alone and in combination with penicillin and other antibiotics as animal feed additives and in medicinal animal health preparations. Approximately one-third of the domestic animals tested showed resistance to ampicillin. Ampicillin rarely is used in the United States to treat domestic animals; penicillin, however, has been used extensively as a feed additive, and it is possible that cross-resistance exists between ampicillin and penicillin.

An interesting observation was noted in the case of chloramphenicol. In the United States the use of chloramphenicol as a feed additive or in animal health medicaments for domestic livestock is not permitted. However, it can and is used for treating diseases of pets. The prevalence of chloramphenicol-resistant *E. coli* was found to be from 20 to 40 times greater in dogs than other animals. The increased prevalence was related to the extent of use. The prevalence of the resistance transfer factor was from 17 to 92 times greater in domestic animals exposed to high antibiotic pressures than in wild animals with minimal antibiotic pressures (44). Animals with minimal exposure to antibiotics have gram-negative organisms that are still very sensitive to antibiotics and other antibacterial drugs.

These observations on wild animals indicate that the extensive use of antibacterial drugs in domestic animals has facilitated the development of resistant gram-negative enteric organisms. The lack of an effective disease surveillance system in the United States makes it difficult to assess the extent that bacterial diseases have become refractory to commonly used antibiotic drugs. However, drug usage patterns by practicing veterinarians and changes in advertisement programs by drug manufacturers extolling the virtues of drug combinations (e.g., farmers are urged to use a combination of sulfamethazine, penicillin, and chlortetracycline instead of using chlortetracycline alone) would indicate that the development of resistant organisms is occurring at a rate necessitating changes in therapy and the amount of the therapeutic dose. Substantial evidence indicates that the drug sensitivity of animal pathogens has diminished greatly (4,21,34,38–40).

Antibiotic pressures exerted on domestic animals, at least in the United States, seem to be much greater than those exerted on people. Infections produced by resistant organisms are a greater problem today than 15 years ago.

b. Man and His Animals

The development of tetracycline resistance in *Salmonella typhimurium* was studied in man, cattle, and pigs (45). In 1959, *S. typhimurium* isolated from man, cattle, or pigs revealed that less than 16% of the strains were resistant to tetracyclines; however, the strains isolated in 1966 showed tetracycline resistance in 45–85% of the isolations. This increase in resistance was coincident with increased usage of antimicrobials.

The prevalence patterns of drug-resistant *Staphylococcus aureus* was investigated by Smith and Crabb (37). Strains were isolated from the nose and skin of animal caretakers. The animal caretakers consisted of three groups: (*1*) those who cared for swine that were fed feed containing tetracycline, (*2*) those caring for swine fed feed containing penicillin, and (*3*) those caring for swine received antibiotic-free feed. The caretakers of the pigs that were fed tetracycline were hosts to *Staphylococcus aureus* organisms that were 92.6% resistant to tetracycline. The caretakers of the pigs that were fed penicillin showed that 30.3% of the isolates were resistant to penicillin; whereas the caretakers of pigs that were fed no antibiotics had very few resistant organisms, 4.5% resistant to tetracyclines, and no organisms resistant to penicillin.

A similar study was reported in Poland in 1965 (46). Animal caretakers of swine that were fed oxytetracycline showed that 58% of the isolated *S. aureus* were resistant to oxytetracycline, whereas of caretakers of swine that were fed antibiotic-free feed, only 3.6% of *S. aureus* were resistant to oxytetracycline. It was demonstrated that bacteria common to man and animals can develop similar resistance patterns through the use of antimicrobial drugs in animal feeds. The zoonotic implications are considerable in this ecologic problem.

The frequency distribution of bacteriologic diseases in man prior to and after a substantial period of antibiotic use has been studied by D. H. Smith (47). Some diseases are still quite sensitive to antibiotic therapy (e.g., beta hemolytic *Streptococcal* infections). Today, they are less of an etiologic problem than they were in 1935. However, after the advent of antibiotic therapy, diseases caused by *Staphylococcus* and *Enterobacteria* were found to compose a larger percentage of bacterial diseases. *Staphylococcal* organisms accounted for 22% of the bacteremic diseases in 1935, whereas in 1957 they were the etiologic agents in 39% of the bacteremic patients. The *Enterobacteric* organisms causing diseases increased from 12% in 1935 to 34% in 1957.

The frequency distribution patterns of etiologic agents producing bacteremia and the percent of deaths in bacteremic patients were related.

The number of deaths in the Boston City Hospital produced by beta hemolytic *Streptococcal* organisms decreased from 23.4% in 1935 to 1.4% in 1957 (47). However, in diseases produced by organisms that have become resistant, the percent of deaths in bacteremic patients has increased. In 1935, *Staphylococcal* infections accounted for 18.6% of the deaths in bacteremic patients; whereas in 1957, the organism accounted for 40.1% of the deaths. Similarly, *Enterobacteria* infections were responsible for 9% of the bacteremic deaths in 1935, but in 1957 they increased to 47.7%.

The role of bacterial infections as a cause of death in a New York hospital and similar patterns were recorded (47). Organisms that had remained sensitive to antibacterials accounted for fewer deaths than those organisms such as *Staphylococcus* or *Enterobacteria* that develop resistance quite readily. A two to fourfold increase has occurred in the number of deaths produced by these organisms.

The prevalence of infective drug resistance in strains of *E. coli* isolated from diseased humans and domestic animals has been investigated (21). Most of the drug resistance possessed by strains of *E. coli* was the infectious type, and, in general, the prevalence of drug resistance in any species was directly related to the extent to which the drug had been used in that species.

The hazard of the widespread use of antibiotics for nonmedical and for indiscriminate medical purposes has been demonstrated by observations that resistant *E. coli* can transfer their resistance to pathogenic *Salmonella*. It has been observed that: (*1*) antibiotic therapy in hospitals has led to a greatly increased prevalence of antibiotic resistance among *Salmonella* isolated from patients and, (*2*) in England prophylactic and therapeutic use of antibiotics to control the salmonellosis in calves has led to an increased prevalence of resistance of *Salmonella* (4).

Furthermore, antibiotics have facilitated the spread of drug resistance in selected populations (4). The great ability of resistant *E. coli* isolates from domestic animals to pass resistance transfer factors to pathogenic *E. coli* and *Salmonella typhimurium* emphasizes the need for an urgent reevaluation of the nonmedical and indiscriminate medical uses of antibacterial agents.

V. SUMMARY AND FUTURE DEVELOPMENTS TO REDUCE ENVIRONMENTAL CONTAMINATION

The high prevalence of multiresistant R-factor-bearing *E. coli, Salmonella typhimurium,* and other gram-negative enteric organisms in our environment is related to man's use of antibacterial drugs for numerous purposes, nonmedical and medical; some are efficacious and some are not. The nonmedical and indiscriminate medical applications of antibacterial drugs have subjected most domestic animal populations in the United States to extensive antibiotic pressures. The exposures to antibacterial drugs has been extensive. Animal products that enter man's food chain have been derived from animals subjected to antibiotic pressures. These are pressures that have ideal prerequisites for creating ecologic problems—subtherapeutic quantities administered continuously for most of the lifespan.

The ecologic changes observed in domestic animals today is unique. No parallel situation exists for man. What would today's pediatrician use to treat bacterial diseases of children if 95–99% of them previously had been exposed to penicillin, dihydrostreptomycin, tetracyclines, sulfonamides? What would today's internist use to treat bacterial infections if 95–99% of the patients had undergone "selective antibiotic pressures" by the commonly used antibiotics and sulfonamides for most of their life? The physician would probably classify such a situation as untenable and hypothetical; yet this is the situation that confronts the veterinarian and the animal producer.

The effects on public health of widespread antibacterial drug usage are difficult to determine with the current "state" of the science. Many unanswered questions exist, and basic information is lacking, especially in the United States. Man can become seriously ill if he is infected with an organism such as *Salmonella typhimurium,* obtained from an animal source or on marketed meat or meat products. Other bacteria found in domestic animals have been implicated in diseases of man (e.g., enteropathic *E. coli*). Some public health investigators believe that domestic animals serve as reservoirs of pathogenic drug-resistant organisms or nonpathogenic organisms that have the ability to transfer drug resistance to organisms that are pathogenic for man (20). The epidemiologic studies relating drug-resistant *Salmonella typhimurium* in animals to the disease in man have been conducted in England. Similar studies have not been conducted in the United States. The possibility of drug-resistant animal *E. coli* passing resistance by infectious transfer to organisms pathogenic to man also has been proposed. The implications of feeding antibiotics to our livestock and poultry with the spread of animal organisms on meat and meat products during slaughter and meat processing and packaging is of gigantic proportions. For example, a housewife may unwrap a roast and place it on the kitchen counter prior to cooking. However, if the counter is not thoroughly disinfected, then R-factor-bearing organisms may come in contact with food (e.g., bread or rolls) that are consumed without being subjected to heat. The possibility of R-factor-bearing organisms from meat or meat products contaminating other food appears to exist.

In addition to the public health aspects of drug-resistant animal bacteria, we must appreciate the problem of treating animal diseases which are not direct public health hazards to man. The gram-negative bacterial infections of the gastrointestinal tract and of the reproductive and renal systems are exceedingly difficult to treat with the drugs being used as animal feed additives. The problem of animal diseases produced by drug-resistant organisms has been established as a serious problem.

The ingestion of antibacterial drugs, their metabolites, and degradation products through the meat portion of our food supply is a problem that is correctable with a satisfactory testing program. The prevalence of residues in meats exceeds that of fluid milk even before an adequate testing program was initiated. Certainly the testing of each animal at slaughter would be a cumbersome procedure; however, a random sample of 1 animal per group or lot, or 1 of 50 or 100 would help to monitor the problem. A realistic random sampling program, subsequent condemnation of the meat, and enforcement of the existing regulations would greatly minimize the residue problem. Until the producer is made aware of an adequate antibiotic testing program, the adherence to antibacterial drug withdrawal times probably will be limited.

Inconsistent results following the use of antibacterial drugs for nonmedical purposes in feeds draws attention to the need for studies pertaining to efficacy. If some nonmedical or medical uses are not efficacious, then no reason exists to continue their use, for the possibility of doing harm would greatly exceed any beneficial effect.

Restriction of antibacterial drug usage would contribute to reduction of the ecologic problems, improve the efficacy of therapeutic drugs, and help to produce meat protein more efficiently. If drugs were used only when necessary there would be significant financial savings to the animal producer. The widespread use of antibacterial drugs has been augmented by analogies to insurance programs; however, the benefits of insurance always appear to be the greatest when they are not needed.

Fortunately, the problems man has created in his environment with the injudicious uses of antibacterial drugs in animals can be greatly reduced. Certainly some interested groups will lose markets and income if drugs are restricted and nonefficacious uses are eliminated; however, the benefits of reduced public health problems and more efficient animal production will be appreciated by many.

References

1. T. D. Luckey, *Antibiotics: Their Chemistry and Non-Medical Uses,* H. S. Goldbert, Ed., Van Nostrand, London, 1959.
2. M. P. Plumlee, *Proc. Ill. State Vet. Med. Assoc.,* Chicago, Ill., Feb., 1963.
3. E. S. Anderson, *Brit. Med. J.,* **3,** 333 (1968).
4. E. S. Anderson, *Amer. Rev. Microbiol.,* **22,** 1506 (1968).
5. M. Swann, *Joint Committee on the Use of Antibiotics in Animal Husbandry and Veterinary Medicine,* Her Majesty's Stationary Office, London, November 1969.
6. M. H. Lepper, *Proc. Symp.: The Use of Drugs in Animal Feeds,* Publication 1679, National Academy of Sciences, Washington, D.C., p. 368, 1969.
7. H. Welch, *Antibiot. Ann.,* 1958–1959.
8. J. R. Agird, *Connecticut Med.,* **30,** 878 (1966).
9. W. G. Huber, M. B. Carlson, and M. H. Lepper, *J. Am. Vet. Med. Assoc.,* **154,** No. 12, 1590 (1969).
10. G. J. Silverman and F. Kosikowski, *J. Milk Food Technol.,* **15,** 120 (1952).
11. J. W. Lightbrown and P. de Rossi, *Analyst,* **90,** 89 (1965).
12. W. G. Huber, *J. Pure Appl. Chem.,* **20,** 337 (1970).
13. D. Dean, J. K. Bennett, and E. L. Breazeale, *Southwestern Med.,* **45,** No. 11, 352 (1964).
14. E. H. Kampelmacher, P. A. M. Guinée, and L. M. van Noorle Jansen, *Tijdschr. Diergeneesk,* **87,** 16 (1962).
15. M. Van Schothorst, *Tijdschr. Diergeneesk,* 579 (1965).
16. J. Pitre and P. Martinet, *Bull. Acad. Vet. France,* **36,** 175 (1963).
17. K. Ochiai, T. Yamanoka, K. Kimura, and O. Sawada, *Nippon Iji Shimpo,* **1861,** 34 (1959).
18. H. W. Smith and S. Halls, *Brit. Med. J.,* **I,** 266 (1966).
19. T. Watanabe, *Sci. Amer.,* **217,** 19 (1967).
20. E. S. Anderson, *Brit. Med. J.,* **2,** 1289 (1965).
21. H. W. Smith, *New Zealand Vet. J.,* **15,** 153 (1967).
22. E. S. Anderson and N. Datta, *Lancet,* 407 (1965).
23. S. Mitsuhashi, K. Harada, and H. Hashimoto, *Med. Biol. (Tokyo),* **55,** 157 (1960).
24. A. Sasarman, M. Surdeanu, and T. Horodniceanu, *Rumanian Nat. Congr. Med. Microbiol.,* 28 (1965).
25. T. C. Salzman and L. Klemm, *Antimicrobiol. Agents Chemotherap.,* **212** (1966).
26. I. J. Mare and J. N. Coetzee, *S. African Med. J.,* **39,** No. 23, 864 (1968).
27. N. Datta, *J. Hyg.,* **60,** 301 (1962).
28. G. Lebek, *Zentr. Bakteriol. Parasitenk. Abt. l. Orig.,* **189,** 213 (1963).
29. E. S. Anderson and M. J. Lewis, *Nature,* **215,** 89 (1967).
30. H. W. Smith, *J. Hug.,* **64,** 465 (1966).
31. E. S. Anderson and M. J. Lewis, *Nature,* **206,** 579 (1965).
32. P. A. M. Guinée, *Antonie van Leeuwenhoek J. Microbiol. Serol.,* **31,** 314 (1965).
33. E. S. Anderson, *Annalis Inst.,* **12,** 547 (1967).
34. A. Bandaranyake, *Ceylon Vet. J.,* **9,** 59 (1967).
35. W. J. Sojka and R. B. A. Carnaghan, *Res. Vet. Sci.,* **2,** 340 (1961).
36. H. W. Smith and W. E. Crabb, *Vet. Rev. Annot.,* **69,** 24 (1957).
37. H. W. Smith and W. E. Crabb, *J. Pathol. Bacteriol.,* **79,** 243 (1960).
38. N. A. Fish, *Can. Vet. J.,* **8,** No. 7, 152 (1967).

39. K. A. McKay and H. O. Branion, *Can. Vet. J.,* **1,** 144 (1960).
40. Anon, Hess, and Clark, *Research Digest,* **7,** No. 4, 1 (1969).
41. S. J. Edwards, *J. Comp. Pathol. Therap.,* **72,** 420 (1962).
42. J. R. Walton, *Lancet,* **1300** (December 10, 1966).
43. S. Mitsuhashi, H. Hashimoto, and K. Suzuki, *J. Bacteriol.,* **94,** No. 4, 1166 (1967).
44. W. G. Huber, D. Korica, T. P. Neal, P. R. Schnurrenberger, and R. J. Martin *Arch. Environ. Health,* **22,** 561 (1971).
45. E. H. Kampelmacher, *Proc. Symp.: The Use of Drugs in Animal Feeds,* Publication 1679, National Academy of Sciences, Washington, D.C., 318 (1966).
46. A. Pogorzelska and Z. Wencel, *Med. Doswaidczalna Mikrobiol., Warsaw,* **17,** 213 (1965).
47. D. H. Smith, *Proc. Symp.: The Use of Drugs in Animal Feeds,* Publication 1679, National Academy of Sciences Washington, D.C., 334 (1966).

Toxicity and Carcinogenicity of Aflatoxins

GERALD N. WOGAN
AND
RONALD C. SHANK
*Department of Nutrition and Food Science,
Massachusetts Institute of Technology,
Cambridge, Massachusetts*

I. INTRODUCTION

The discovery and development of knowledge about the aflatoxins and their occurrence as food contaminants have taken place during an era of increasing awareness of the importance of natural as well as man-made environmental contaminants. The remarkable intensity with which this problem has been investigated is reflected in the large number of publications on various aspects of the problem that have appeared since its emergence in 1960. This extensive literature has been reviewed numerous times during the intervening period, most recently in the form of a monograph dealing with all phases of the problem (1).

Several features of the aflatoxin problem have provided continued stimuli for its further exploration. The fact that some food-spoilage fungi were capable of producing toxic products, mycotoxins, has been recognized for many years (2). However, the importance of these agents as food contaminants was appreciated mainly by veterinarians, who frequently encountered outbreaks of poisoning in farm animals. The existence of aflatoxins was discovered as a consequence of mass outbreaks of poisoning in poultry, and it is possible that this may also have been relegated to the miscellaneous problems of agriculture, except for several features of the problem whose importance became obvious as the early stages of the story unfolded.

Important among these was the early recognition that the toxic agents were produced by common food-spoilage fungi, and the evidence of their wide geographic distribution by virtue of aflatoxin-contaminated commodities identified in practically every area of the world. Coupled with this information was the fact that the toxin-producing molds could produce their products on virtually all types of foodstuffs given adequate opportunity to grow. Perhaps the strongest impetus was provided by discoveries of the potency of the aflatoxins as poisons, and especially of their extraordinary potency as hepatocarcinogens in rats. The obvious potential implications of these various facets of the problem in relation to public health have stimulated extensive research into all aspects of the occurrence of the toxins, their toxicology and pharmacology, and means for prevention and control of their presence in foods for human consumption.

At this time, practically no direct information on human responses to aflatoxins exists, and it is unlikely that such direct information will be forthcoming. Nonetheless, increasing evidence can be found that human populations in some areas of the world are being exposed to aflatoxins and are therefore potentially at risk with respect to possible deleterious consequences of that exposure. In view of the lack of direct information on man, assessment of risk must be made on the basis of data from animals and other systems insofar as they can be extrapolated to the human. It is the purpose of this presentation to summarize the background of the evidence on the toxicology, pharmacology, and carcinogenicity of aflatoxins, and to attempt to evaluate their importance to man.

II. CHEMISTRY

A total of 12 structurally related compounds bearing the basic aflatoxin configuration have been identified to the present. For purposes of this discussion, it is convenient to group them into two series, according to their structural relationships to the two basic toxins originally identified, aflatoxins B_1 and G_1. Details of the structure elucidation and chemistry of many of these substances have been reviewed (3), and only the main features of this aspect of the problem are summarized here.

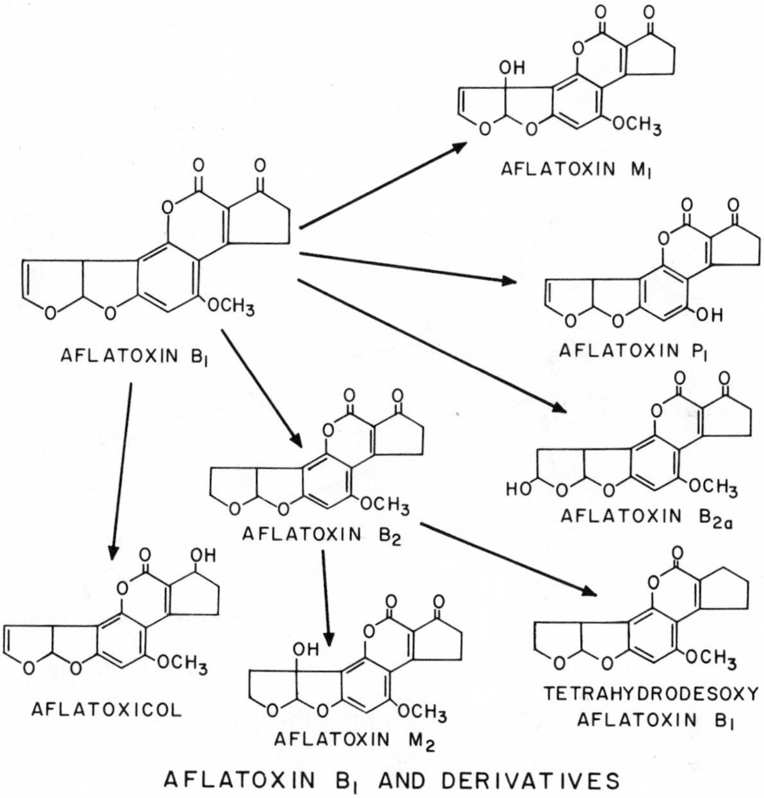

AFLATOXIN B_1 AND DERIVATIVES

Fig. 1

Figures 1 and 2 present the structures of presently identified aflatoxins and derivatives and suggest their relationships to the parent substances. Aflatoxins B_1 and G_1 were the principal components of the toxic culture extracts upon which the structure elucidation was carried out originally (4).

Figure 1 summarizes the relationships among aflatoxin B_1 and its derivatives. The parent substance, since its characterization in 1963, has proved to be the aflatoxin of primary interest. It is the major component of toxin mixtures produced by practically all aflatoxin-secreting fungi, and it is also the most potent member of the series with respect to its toxicity, carcinogenicity, and other biological effects. The derivatives shown in Figure 1 are arranged to indicate the various ways in which they are known or presumed to be formed.

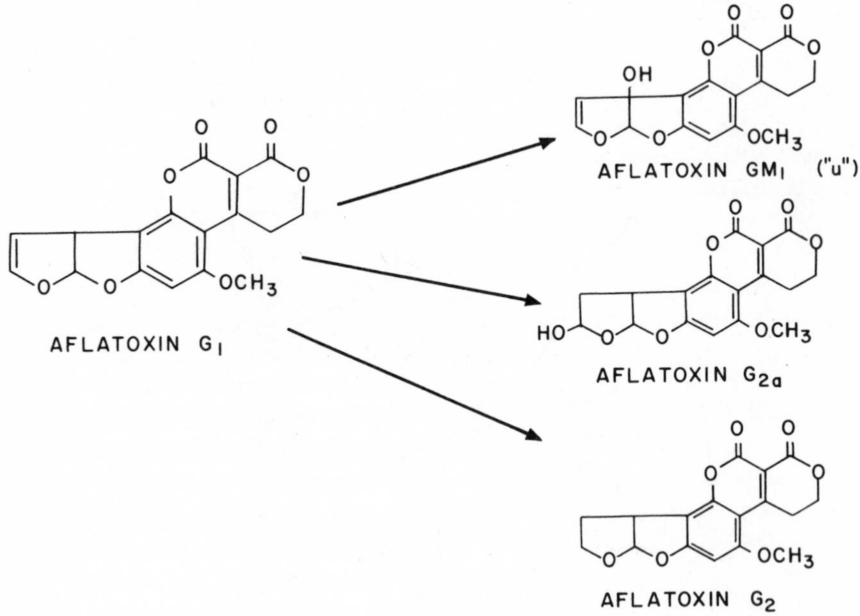

AFLATOXIN G_1

AFLATOXIN GM_1 ("u")

AFLATOXIN G_{2a}

AFLATOXIN G_2

AFLATOXIN G_1 AND DERIVATIVES

Fig. 2

Aflatoxins M_1 and M_2, products of metabolic hydroxylation of B_1 and B_2, were isolated originally from milk (hence, the designation M) of cows fed aflatoxins B_1 and B_2 (6,7). In addition, aflatoxin M_1 has been identified in the urine of many species of animals and in human urine following ingestion of aflatoxin B_1. Aflatoxins M_1 and M_2 are present as minor components of aflatoxin mixtures produced by some fungal strains.

Aflatoxin P_1, the product of O-demethylation of B_1, was recently isolated and identified from the urine of rhesus monkeys dosed with B_1 (8). In that species, it represents a major metabolic product, and its existence in other species is presumed but not yet established.

Aflatoxin B_{2a}, the hemiacetal of B_1, was isolated from liquid cultures of aflatoxin-producing fungi (9,10). This compound also forms readily through the acid-catalyzed addition of water to B_1. Its formation under these conditions has been applied as a confirmatory test for aflatoxin B_1 (11).

In addition to its connection to species differences in response, this information on domestic animals is relevant to the matter of human exposure because of the possible residual contamination of animal tissues or products used for human food. Several items of evidence are available regarding this point. It has been shown that eggs from hens fed aflatoxin-containing rations contained no detectable residue of toxin, as determined by biological assay (19). There is no evidence to suggest that aflatoxin is stored in tissues of cattle or pigs fed toxic rations over prolonged periods (20).

Therefore, it would appear that the only food products derived from farm animals fed aflatoxin-contaminated rations that might constitute a health hazard would be milk and milk products, because a small proportion of ingested aflatoxin B_1 is excreted in milk as aflatoxin M_1. This point emphasizes the necessity to avoid aflatoxin contamination of dairy cattle feed.

B. Toxicity to Laboratory Animals

The toxic effects of aflatoxins to many species of animals have been investigated under laboratory conditions (21). Aflatoxin B_1 has been most thoroughly studied, and a summary of its effects can serve as a basis for further information dealing with structure-activity relationships. With respect to its acute lethal potency in laboratory animals, single-dose LD_{50} values for aflatoxin B_1 have been established in several species. These values for the duckling, rat, hamster, guinea pig, rabbit, and dog are in the range of approximately 0.4–10.0 mg/kg body weight (22). As in the case of farm animals, the laboratory animals also tend to show considerable species differences in response. Thus while the rat is among the moderately susceptible species (LD_{50} at weanling age, 5.5 mg/kg), the mouse is much more resistant (LD_{50}, 60 mg/kg) (23).

Acute lethality of aflatoxin B_1 also has been compared among several species of fish, with interesting results. Halver (24) reported that the LD_{50} of the toxin in rainbow trout was in the order of 0.5 mg/kg body weight, while the coho salmon was 10–20 times more resistant, and the channel catfish was 30 times more resistant than the trout. Bauer *et al.* (25) obtained a similar value (0.81 mg/kg) for the LD_{50} of aflatoxin B_1 injected intraperitoneally in rainbow trout. Abedi and Scott (26) found that the presence of aflatoxin B_1 in water at a concentration of less than 1 μg/ml was lethal to zebra fish larvae, and suggested this system as a possible biological assay for detection of aflatoxin activity.

The evidence summarized here indicates that aflatoxin B_1 is lethal to a wide range of animal species from fish through birds and mammals. In all animals poisoned by the compounds, the liver appears to be the primary, if not the only, organ specifically affected. Details of the histopathological responses to aflatoxins in livers of experimental animals have been reviewed extensively by Newberne and Butler (27).

C. Toxicity to Primates

It is usually assumed that toxicity data from subhuman primates are of greater value than those from lower animals in predicting human response. Available information on responses of primates to aflatoxins is therefore of considerable import to this discussion.

Although the amount of information available on this point is still fragmentary, it is sufficient to indicate that monkeys are highly susceptible to poisoning by aflatoxins. In the earliest published study, Tulpule *et al.* (28) reported that young rhesus monkeys (1.5 to 2.0 kg body weight) died 14 to 28 days after they were fed a mixed aflatoxin preparation at doses of 0.5 to 1.0 mg per animal per day. Madhavan *et al.* (29) found that aflatoxin doses of 0.5 to 1.0 mg per animal per day were fatal to monkeys when administered over periods of 19 days or longer. The treatment induced extensive fibrosis and fatty infiltration of the livers. Subsequently, Madhavan *et al.* (30) investigated the effect of dietary protein level on this response and found that reduced protein intake increases the susceptibility to aflatoxin. At a dose level of 100 μg/day, animals on a protein-deficient diet died, while those receiving adequate protein intakes survived. Cuthbertson *et al.* (31) studied the effects of aflatoxin levels of 5 ppm in the diet on *Cynomologous* monkeys and found that this dose level rapidly induced liver cell damage and death. Lower levels permitted longer survival, but liver damage was induced in some animals.

More recently, Alpert *et al.* (32) dosed African monkeys with 0.01 to 1.0 mg per animal per day. Death occurred from 6 to 22 days after initiation of treatment, and a consistent pattern of liver toxicity similar to that occasionally observed in man occurred. Deo *et al.* (33) recently reported on an extensive study in rhesus monkeys dosed with a mixed aflatoxin preparation. Animals were treated with levels of 1 mg/kg daily; 0.25 mg/kg twice weekly; or 62 μg/kg once weekly. Animals on the highest dosage died within 3 weeks and showed extensive hemorrhagic necrosis of the liver. Animals at the two lower dosage levels survived for up to 2 years, but even the lowest level induced liver pathology. These effects were not influenced by dietary protein levels.

Taken together, these data indicate that primates are in the same range of susceptibility as other moderately sensitive species, with respect to the acute and subacute toxicity of aflatoxins.

D. Toxicity to *In Vitro* Systems

In addition to their potent toxicity in metazoan systems, aflatoxins also exhibit various biological activities to *in vitro* assay systems. Legator (34), in discussing the use of various responses as bioassay systems for aflatoxins, divides these responses into the categories, direct genetic and nongenetic damage.

Following an early report that extracts of aflatoxin-contaminated peanut meal were toxic to cultures of calf-kidney cells (35), many other studies have involved responses of *in vitro* cultures of mammalian cells to aflatoxins. Legator and Withrow (36) reported mitotic inhibition of cultured human embryonic lung cells, and later Dolimpio *et al.* (37) observed the same response in human leucocyte cultures. Gabliks *et al.* (38) demonstrated toxicity of aflatoxin B_1 to Chang liver cells, HeLa cells, and cultures of chick and duck embryo cells. Daniel (39) devised a precise dose-response relationship for aflatoxin toxicity using rat fibroblast cells.

Smith (40,41), using liver-slice preparations from rats or ducks, found that aflatoxins suppressed amino acid incorporation into liver proteins *in vitro*. Subsequently, Zuckerman *et al.* (42,43) demonstrated cytotoxicity of aflatoxin B_1 to cultures of human embryonic liver and also found that aflatoxins B_1, G_1, and G_2 inhibited (to varying degrees) the incorporation of precursors into RNA in these cells.

More recently, effects of aflatoxin B_1 in HeLa cells have been investigated extensively (44), with the conclusion that cytotoxicity to these cells is associated with inhibition of synthesis of ribosomal and heterodisperse RNA and inhibition also of protein synthesis. Savel *et al.* (45) recently have demonstrated that aflatoxin B_1 inhibited the stimulated uptake of tritiated thymidine by phytohemagglutinin in human lymphocyte cultures and decreased the stimulation usually induced by potent antigens.

In general, the aflatoxin dosages causing these various responses were in the range of 0.1 to 10 μg of compound/ml of medium. However, in some instances [e.g., Legator and Withrow (36) and Daniel (39)] activity was observed in the range of 0.01 to 0.1 μg/ml. Thus the toxins can be classified as highly potent agents in these systems.

Numerous studies have been carried out on microorganisms. Burmeister and Hesseltine (46) and Arai *et al.* (47) surveyed numerous organisms in a study of the antimicrobial activity of aflatoxins. Both surveys revealed that the toxins have weak antibiotic activity against a narrow range of microorganisms. However, a number of interesting findings with respect to aflatoxin effects in microbial systems have since been published. Wragg *et al.* (48) found that *E. coli* DNA polymerase activity was significantly lowered by aflatoxin B_1, while Legator *et al.* (49) showed that the toxin acted as a potent inducing agent with lysogenic bacteria. Lillehoj *et al.* (50,51) reported the production, by aflatoxin exposure, of filamentous forms of *Flavobacterium aurautiacum* and that exposure to aflatoxin greatly increased the vulnerability of this organism to rupture by sonication. Lillehoj and Ciegler (52) evaluated the toxicity of aflatoxin B_1 to *bacillus megaterinum.*

Recent publications deal with the induction of mutagenic effects by aflatoxins in microorganisms. Maher and Summers (53) reported a mutagenic action on transforming DNA of *Bacillus subtilis* as well as an inhibition of DNA template activity in a purely *in vitro* system. Ong (54) found aflatoxin B_1 to be mutagenic to *Neurospora crassa* and aflatoxin G_1 to have lesser potency in this regard.

The breadth of range of susceptibility of biological systems to aflatoxins is further illustrated by their actions in plant systems. Several of these responses are qualitatively similar to those caused in animal systems. Schoental and White (55) showed that exposure of seeds of watercress (*Lepidium sativum*) to aflatoxins inhibited germination or, at lower levels, prevented chlorophyll formation. A similar finding with respect to chlorophyll synthesis in maize leaves has recently been published (56). Lilly (57) described the production of chromosomal aberrations in the roots of *Vicia faba* seedlings upon exposure to mixed aflatoxin preparations. Black and Altschul (58) found that aflatoxin blocked the inducibility of α-amylase by gibberellic acid in cottonseeds, whereas Truelove *et al.* (59) reported that the uptake and incorporation of amino acids into proteins of cucumber cotyledon tissue is inhibited by aflatoxin B_1.

IV. METABOLISM

The manner in which aflatoxins are metabolized is of interest from several viewpoints. In particular, the metabolic fate of the compounds is important with respect to the biochemical mechanisms underlying their biochemical effects. The mode of action of the toxins must entail interactions of the parent substances, or derivatives of them, with cellular constituents. A knowledge of the metabolic products and the pathways by which they are produced is essential to an understanding of the molecular configuration responsible for biological activity. Such information is also indispensable for establishing the basis for various characteristics of the biological response, such as structure-activity relationships, tissue specificity, and species differences in response. Available information on aflatoxin metabolism has been derived through two kinds of experimental approaches, involving either studies on excretion of fluorescent metabolites, or studies using radioactive toxins.

Earliest investigations on fluorescent metabolites were those of Allcroft and Carnaghan (19), who showed that the milk of cattle fed aflatoxin-contaminated feed excreted a factor toxic to ducklings. De Iongh *et al.* (60) demonstrated that the toxic substance (the "milk toxin" ultimately to be identified as M_1) was produced from aflatoxin B_1 in both cattle and rats. Van der Linde *et al.* (61) showed that cows excreted about 1% of ingested aflatoxin through their milk.

Allcroft *et al.* (13) investigated the fluorescent metabolites excreted in sheep urine following aflatoxin dosing and proved ultimately that a compound (for which they suggested the designation aflatoxin M) identical to the "milk toxin" was excreted in sheep urine. Nabney *et al.* (62) studied the metabolism of aflatoxins B_1 and G_1 in sheep, and they found M_1 to be the principal metabolite excreted in milk, urine, and feces. Aflatoxin GM_1 ("U") was also identified as a minor derivative of G_1 in urine. In these studies, only about 8% of the administered aflatoxins were accounted for by excreted metabolites; the fate of the remaining material is unknown.

More quantitative information has been possible through the use of radioactive aflatoxin B_1. We (63) have studied the tissue distribution and excretion of the radioactivity from aflatoxin B_1-^{14}C after single doses to rats. The principal findings of these experiments are summarized.

During the 24 hours after a single dose of the compound, 70–80% of the dose is excreted; 20% is excreted via urine and the remainder through feces. The fecal contents appear to be derived mainly through biliary secretion. Liver contained more radioactivity than any other tissue throughout the course of the experiment, and it retained about 7% of the dose at the end of 24 hours. By use of two different forms of labeled toxin, it was concluded that ring cleavage does not seem to occur to any significant extent, but that O-demethylation represented a major degradative pathway.

In a comparative study of rats and mice, it was found that the principal difference between the two species was the larger total excretion (89 vs 80% of dose) by the mouse and the smaller retention of toxin in mouse liver (64). These differences are possibly related to the lower susceptibility of the mouse to toxic and carcinogenic effects of aflatoxins.

Recent evidence is of considerable relevance to the question of human exposure and response to aflatoxins. Campbell *et al.* (65) have reported that human subjects known to be ingesting aflatoxin-contaminated foods excrete aflatoxin M_1 in their urine. Although quantitative information was not obtained, the authors estimated that not more than 1 to 4% of ingested aflatoxin appeared in urine as M_1. Milk collected from mothers also consuming aflatoxin-contaminated diets did not, however, contain detectable M_1. These data suggest that man is similar to other species with respect to the ability to metabolize aflatoxin B_1 to M_1.

Dalezios *et al.* (8) have recently identified aflatoxin P_1 as the major urinary metabolite of B_1 in monkeys. This compound represented about 70% of the urinary aflatoxin metabolites (compared to about 1% in the form of M_1). If it can be established that man also produces this derivative, the screening of human populations for aflatoxin exposure should be greatly facilitated.

V. MODE OF ACTION

Many investigations have dealt with cellular biochemical alterations associated with toxicity responses to aflatoxins in susceptible biological systems. The objectives of these studies were elucidation of the mode of action of the compounds, and evidence has been sought for a sequence of biochemical events initiated by interaction of the toxins with cellular constituents that leads ultimately to cytologic manifestations of toxicity. It is not yet possible to define the precise sequence of biochemical events leading to morphologic manifestations of toxicity and/or carcinogenicity. However, general patterns of reactions have emerged that are thought to be associated with the toxicity response by virtue of the time-course and consistency of their occurrence and their potential importance in cellular metabolic phenomena.

Reactions involved concern alterations in nucleic acid and protein metabolism elicited by exposure to aflatoxins. Much of the available evidence has come from experiments dealing with alterations induced in liver tissue following *in vivo* or *in vitro* exposure to the toxins. Useful information has been derived from experiments on *in vitro* cell cultures and cell-free systems.

Administration of aflatoxin B_1 to rats is followed by rapid and marked inhibition of liver DNA and RNA synthesis, consequences of inhibition of DNA and RNA polymerases. Protein synthesis is also impaired, particularly under conditions in which the synthetic process is influenced strongly by altered RNA metabolism. The observed inhibition of nucleic acid polymerases is thought to be an indirect consequence of toxin-DNA interaction that results in modified template activity. Therefore, interaction with DNA is envisioned as the initiating event in the sequence of reactions. Most of the available evidence for such interactions has come from *in vitro* binding studies.

Several criteria have been applied in demonstrating binding, including alterations in the aflatoxin absorption spectrum that occur upon interaction with DNA. Sporn *et al.* (66) reported a shift in absorption maximum and hypochromism at 362 mμ upon binding of aflatoxin B_1 to calf-thymus DNA when the two compounds were equilibrated in phosphate buffer solutions. In addition, aflatoxin B_1 bound to native calf-thymus DNA and, to a lesser extent, to heat-denatured calf-thymus DNA and native *Escherichia coli* RNA. By the same criteria, aflatoxin B_1 did not bind to bovine serum albumin, calf-thymus histone, or enzymatically hydrolyzed calf-thymus DNA.

Clifford and Rees (67,68) demonstrated interactions between aflatoxins and DNA by measuring the spectral shift induced by the binding. By the same techniques, Clifford *et al.* (69) compared the interactions of aflatoxin B_1, G_1, and G_2 with DNA. Qualitatively, similar spectral shifts were induced by DNA-binding of the three compounds. However, the shifts were quantitatively different, being greatest with aflatoxins B_1, intermediate with G_1, and smallest with G_2. The extent of the toxicity and biochemical potency of the compounds were qualitatively related to the magnitude of spectral shift induced by DNA binding.

It is clear from these investigations that aflatoxins are capable of interacting with DNA under conditions in which the two compounds are brought into contact *in vitro*. Interactions with cellular constituents under *in vivo* conditions have been studied by Lijinsky *et al.* (70), using tritiated aflatoxins B_1 and G_1 to study interactions of the toxins with tissues of the rat. Maximal incorporation of both toxins into tissue RNA, DNA, and protein fractions was observed from 6 to 18 hr after injection. Radioactivity persisted in protein fractions for long periods after injection. However, it was not possible to correlate these results with tumor induction by the toxins.

Inhibition of DNA synthesis by aflatoxins is particularly evident in rat liver undergoing regeneration after subtotal hepatectomy. Frayssinet *et al.* (71) reported that aflatoxin B_1 inhibited net synthesis of liver DNA when administered to rats prior to or immediately after hepatectomy. Using similar techniques, de Recondo *et al.* (72,73) showed that administration of the toxin at 100 μg per animal inhibited the incorporation of thymidine–^3H into liver DNA. They further demonstrated that the enzymes responsible for DNA synthesis remained fully active under these conditions and concluded that the toxin exerted its inhibitory effects by interacting with DNA in such a way as to impair its ability to act as a primer for DNA synthesis.

Legator *et al.* (49) and Legator (74) reported that the toxin inhibited incorporation of thymidine–^3H into DNA of cultures of human embryonic lung cells at concentrations from 0.1 to 1.0 μg/ml in the culture medium. Zuckerman *et al.* (43) reported similar effects of aflatoxin B_1 in cultures of human embryonic liver cells, in which the compound had an LD_{50} value of 1 μg/ml.

Altered RNA metabolism as a result of aflatoxin treatment has also been demonstrated under a variety of experimental conditions. Changes of this type are among the earliest demonstrable effects in livers of rats dosed with the toxins. *In vivo* administration of the compound to rats or direct exposure of liver-slice preparations *in vitro* result in rapid and dramatic inhibition of precursor incorporation into RNA, particularly in the nucleus.

These effects were first observed in regenerating rat liver following partial hepatectomy (75). Subsequently, Clifford and Rees (67) reported than an LD_{50} dose of the toxin given to intact rats inhibited precursor incorporation into liver nuclear RNA within 3 hours after dosing, and Sporn *et al.* (66) reported similar findings.

This type of effect has been observed in rat liver-slice preparations exposed to the toxic compounds. Clifford and Rees (67) reported that orotic acid-^{14}C incorporation into total cellular RNA was strongly inhibited when rat liver slices were incubated *in vitro* in the presence of aflatoxin B_1. Under similar conditions, these investigators (68) found that aflatoxins G_1 and G_2 also inhibited RNA synthesis but with lower potency than B_1.

That the impaired RNA synthesis was attributable to inhibition of RNA polymerase activity was shown experimentally by Gelboin *et al.* (76). Activity of this enzyme was inhibited as early as 15 min after dosing, but the inhibition had nearly disappeared 12 hours after treatment. Subsequently, Pong and Wogan (77) and Friedman and Wogan (78) demonstrated that larger doses of the toxin produced inhibition of RNA polymerase activity that persisted for longer periods after dosing.

In view of the marked effects of the toxins on nuclear RNA synthesis, it might be anticipated that cytoplasmic RNA metabolism would be affected as well, and such findings have been reported (79). A single sublethal dose of aflatoxin B_1 to rats resulted in a decrease of approximately 50% in total liver RNA content over the 72 hours following treatment. Under similar conditions, marked alterations in polyribosome profiles take place. We have recently reported effects of aflatoxin B_1 on liver polyribosome profiles of rats treated with a single sublethal dose of the compound (80). This treatment resulted in large increases in monomer and dimer fractions and decreases in the polyribosome areas that were apparent within 3 hours after dosing. Polysomal disaggregation was maximal at 12 hours and completely reversed after 48 hours. The pattern of disaggregation and reaggregation after toxin treatment suggested that the observed effects were related to alterations in both RNA and protein metabolism.

Based on present knowledge of the mechanisms of gene transcription and translation, alterations in DNA-dependent RNA synthesis would be expected to result in changes in protein metabolism. The evidence previously summarized indicates that exposure of liver tissue to aflatoxin B_1 results in marked alterations in gene transcription as evidenced by impaired synthesis of nuclear RNA. Associated with this effect are significant and persistent losses of cytoplasmic RNA and polyribosomal disaggregation. Under these circumstances changes in protein metabolism, particularly impaired synthesis, would be anticipated, and such responses have indeed been demonstrated.

Systems that have proved to be very sensitive to aflatoxin blocking are inducible enzymes in rat liver. Clifford and Rees (67,68) showed that induction of rat liver tryptophan pyrrolase by cortisone was inhibited by prior treatment of the animals with aflatoxin. In an extensive investigation of this system, we concluded that the kinetics and other characteristics of the inhibition were consistent with the interpretation that inhibition was attributable to suppression of RNA synthesis by the toxin (81).

Inhibition by aflatoxin B_1 of inducibility of rat liver enzymes has also been demonstrated in a different experimental system. Pong and Wogan (82) showed that the induction of zoxazolamine hydroxylase (a microsomal drug-metabolizing enzyme of rat liver) by benzypyrene was completely blocked by simultaneous administration of aflatoxin B_1 and the inducer.

Thus the evidence summarized indicates that aflatoxin B_1 causes dramatic alterations in nucleic acid and protein synthesis in liver when administered in acute doses to rats. The observed inhibitions of DNA synthesis, RNA synthesis, and alterations in gene transcription appear soon after the compound has been administered. The time-course and other characteristics of these responses are consistent with the hypothesis that they are initiated as a result of interaction of the toxic material with DNA in such a way as to interfere with its transcription. Although binding to DNA by aflatoxin B_1 has been amply demonstrated under *in vitro* conditions, evidence that comparable interactions occur *in vivo* is much less extensive. It is not yet clear whether the interaction involves direct binding of the unaltered aflatoxin molecule or whether interaction of a type (e.g., alkylation) requiring metabolic conversion of aflatoxin B_1 is involved. The compound causes similar effects in widely diverse systems, which would seem to favor the hypothesis that it acts directly, without metabolic conversion. It is unlikely that all of these systems would possess the necessary capabilities for activation. However, the possibility of metabolic activation cannot yet be eliminated, particularly with respect to the long-term effects of the toxin.

Unknown also is what proportion of an administered dose of aflatoxin B_1 reaches the nuclei of liver cells and to what extent the material interacts with other nuclear constituents (e.g., histones) that might influence gene transcription. Whether the lesser biochemical effectiveness of the other aflatoxins (B_2, G_1, G_2) is attributable solely to their varying capabilities to interact with DNA remains to be determined.

Considerable further experimental data is required to relate the observed biochemical effects of aflatoxins directly to the subcellular and cellular events ultimately manifest in toxicity (necrosis) or in tumor induction. Current evidence indicates that the biochemical changes induced by single doses of aflatoxins are associated with the acute (toxicity) response to the compounds. Relation of these effects to the chronic processes leading to tumor formation is less certain. The importance of further investigation of the biochemical events associated with carcinoma induction by aflatoxins is obvious. In such studies the aflatoxins undoubtedly will provide useful model compounds, and their investigation may provide additional insights into the mechanisms underlying the carcinogenic process.

VI. CARCINOGENICITY

To date aflatoxins have been shown to have carcinogenic activity in three species: the rat, the trout, and the duck. Generally, the liver is the organ principally affected, in which the toxins induce malignant hepatocellular carcinomas. However, in a few instances significant incidences of tumors of organs other than the liver (especially the kidney) are apparently associated with aflatoxin dosing. Tumor distribution seems related to both the aflatoxin preparation and the animal strain studied.

Various types of aflatoxin preparations have been examined. Most of the earlier literature deals with effects of feeding diets containing ingredients (such as peanut meal) that were contaminated in nature, while more recent experiments have utilized purified preparations of mixed or individual toxins administered by feeding or dosed by other means.

A. Carcinogenicity in Rats

The earliest evidence of carcinogenic properties of aflatoxins was that of Lancaster (83), who found that the same peanut meal involved in the outbreaks of turkey poisoning in Britain also induced liver tumors when fed chronically to rats. These initial findings were confirmed and greatly extended in further experiments using the same peanut meal (21,84). Newberne (85) also studied liver tumor induction in rats by feeding peanut meal (originating in the United States) that was aflatoxin-contaminated. In the studies of Butler and Barnes (86) toxin levels were in the range of 0.1 to 5.0 ppm in the diet, which induced very high incidence (up to 100%) of tumors in both male and female rats when they were fed for varying periods of time. Newberne (85) described excellent correlation between dietary aflatoxin level and tumor incidence in rats in the range from 0.075 to 1 ppm.

In the experiments of Butler and Barnes (84) most of the tumors associated with aflatoxin ingestion appeared in the liver. However, elevated incidence of tumors of other sites was evident in treated but not controlled animals. These investigators called particular attention to the possibility that carcinoma of the glandular stomach, an uncommon tumor in rats, may have been associated with aflatoxin exposure (86), and in addition, they reported an elevated incidence of kidney tumors.

A recent report (87) indicates that rats fed a diet containing 80 ppb aflatoxin, derived from naturally contaminated peanuts, failed to develop liver tumors after long-term feeding. Because these data suggest lower sensitivity of rats to the carcinogenic effects of aflatoxin than is indicated by other work, the authors postulate the possible existence of protective factors in natural as compared to purified diets. However, other possible influences such as strain differences in susceptibility must be considered.

Carcinogenicity to rats of purified aflatoxin preparations has been demonstrated in numerous experiments of different types. Wogan and Newberne (88) studied the dose-response characteristics of the carcinogenic activity of pure aflatoxin B_1 administered orally to rats. When fed continuously from 60 to 80 weeks, the toxin induced very high liver cell carcinoma incidence in both male and female rats at dietary levels as low as 0.015 ppm (15 ppb). However, continuous feeding was not required for tumor formation, because a total dose of 400 μg per animal administered by intubation in 10 daily doses induced significant tumor incidence. This confirmed the earlier similar conclusion by Carnaghan (89), who showed that hepatomas were induced in rats surviving a single LD_{50} dose of aflatoxin B_1 given at weanling age.

As in earlier studies hepatocellular carcinomas were the principal tumors induced by aflatoxin B_1 in our experiments. A number of adenocarcinomas of the colon also were found in aflatoxin-treated animals (not the controls), but the incidence was too small to permit definite association with aflatoxin treatment. More recently, however, Epstein *et al.* (90) reported a very high incidence of renal epithelial neoplasias in rats fed pure aflatoxin B_1. In this case some animals also had hepatocellular carcinomas, but the kidney lesion was more prevalent. The only obvious explanation for these findings is a difference among rat strains with respect to site specificity of aflatoxin B_1 carcinogenesis.

Only a few systematic studies on purified aflatoxins other than B_1 have been reported. Butler *et al.* (91) compared the potency of aflatoxins B_1, B_2, and G_1. Liver tumors were induced in rats each given a total of 1 or 2 mg of aflatoxin B_1 by long-term feeding. Aflatoxin G_1 proved to be less potent as a liver carcinogen, but it induced a high incidence of kidney tumors, particularly when administered at high dose levels (6 mg per animal). No tumors were induced by aflatoxin B_2 at a dose of 1 mg per animal. Recently, we have reported results of a comparison of aflatoxins B_1, B_2, and G_1 as well as a series of structural analogues of B_1 (92). Our results can be summarized as follows: Aflatoxin G_1 was somewhat less active than B_1 in inducing liver tumors, but induced a high incidence of kidney tumors, in agreement with the findings of Butler *et al.* (21). Aflatoxin B_2 induced liver tumors, but only with a dose 150 times higher than that of B_1. None of the other structural analogues containing only portions of the aflatoxin molecule showed any signs of carcinogenic activity at doses 200 times higher than effective B_1 doses.

Numerous reports have dealt with factors that modify the carcinogenic response of rats to aflatoxins. Certain of these, especially those involving nutritional status, are particularly relevant to the problem of human response. Madhavan and Gopalan (93) found that superimposition of aflatoxin exposure on protein insufficiency sensitized rats to the hepatotoxic actions of the toxins, but appeared to reduce tumor incidence. Newberne and Wogan (94) found that exposure of protein-deficient rats to aflatoxin B_1 followed by a return to an adequate protein diet resulted in enhancement of tumor induction. These results suggested sensitization to carcinogenic action by protein malnutrition.

Newberne *et al.* (95,96) subjected rats to a number of liver insults before, during, and after exposure to aflatoxin. Among other results, a potentiating effect of choline-deficiency cirrhosis on the carcinogenic action of aflatoxin B_1 was observed, particularly when the animals were returned to a normal diet following induction of cirrhosis. Lee *et al.* (97) found evidence of slight potentiation of aflatoxin carcinogenesis in rats by simultaneous feeding of cyclopropenoid fatty acids, a synergism that is important in the response of trout to aflatoxins as discussed below.

In other studies, Newberne *et al.* (98) found no evidence of syncarcinogenesis between aflatoxins and urethane, but Newberne and Williams (99) found that simultaneous feeding of aflatoxin and diethylstilbestrol decreased the carcinogenic potency of aflatoxins in rats. Goodall and Butler (100) learned that hypophysectomy of rats resulted in nearly complete inhibition of aflatoxin-induction of liver tumors.

In addition to induction of tumors following oral administration, aflatoxins B_1 and G_1 also induce sarcomas at the injection site following repeated subcutaneous injection to rats (101,102).

B. Carcinogenicity in Rainbow Trout

Rainbow trout is a second species in which carcinogenic activity of aflatoxins has been documented extensively. Investigations in this rather unusual "experimental animal" were stimulated by large-scale epidemics of trout hepatoma in hatchery-reared trout and the subsequent search for the etiologic agent. Conclusive evidence indicates that aflatoxin-contaminated feed ingredients were responsible for the epidemic, as is shown by the history and development of the problem (24). In the course of this work and coinciding with the development of the general aflatoxin problem, much information was derived on the responsiveness of this species to aflatoxins. These data are summarized briefly.

The rainbow trout appears to be among the more sensitive species to the carcinogenic actions of aflatoxins. Halver (24) describes a dose-response curve for hepatoma induction by aflatoxin B_1 that indicates the minimum effective dose for 10% tumor induction (continuous feeding, 20 months duration) to be 0.1 μg/kg diet (i.e., 0.1 ppb). Extrapolation of the relationship to zero incidence under these conditions yields a dose level of 0.05 ppb. Sinnhuber *et al.* (103,104) found that aflatoxin B_1 induces liver tumors in rainbow trout at levels of 4 ppb. Sinnhuber *et al.* (103,104) found evidence of strong potentiation of aflatoxin carcinogenesis in trout by simultaneous feeding of cyclopropenoid fatty acids.

Aflatoxin G_1 is also hepatocarcinogenic to trout, but it has a lower potency than B_1 (24,103). Recently, Sinnhuber *et al.* (105) have also shown that M_1 is a potent carcinogen for rainbow trout when fed together with cyclopropenoid fatty acids. Ayres *et al.* (106) have compared the carcinogenicity of various aflatoxins studied individually in trout. They conclude that the intact B_1 or G_1 molecule is essential for full biological potency.

C. Carcinogenicity in Other Species

Aflatoxin-contaminated peanut meal is carcinogenic to ducks. Carnaghan (107) showed that continuous feeding of a diet containing 30 ppb of toxins induced liver tumors in 8 of 11 survivors after 14 months. Butler (21) reports that tumors have been induced in ferrets by feeding peanut meal containing aflatoxins for up to 37 months.

In contrast to these positive findings with respect to aflatoxin carcinogenesis, there are some interesting species differences in response. For example, aflatoxin levels far higher than those that induce liver tumors in rainbow trout are ineffective in producing tumors in brook trout or coho salmon (24). Sheep tolerate dietary levels that are acutely toxic to rats (14). Among the rodents, mice fail to show tumor induction when fed aflatoxin levels of 1 ppm to 150 ppm for up to 85 weeks (92), in contrast to rats that respond to 0.1 ppm.

The limited amount of information available at this time on the chronic effects of aflatoxins in primates has produced no findings of tumor induction. Cuthbertson *et al.* (31) found that *Cynomologous* monkeys survived up to 3 years on a diet containing 360 ppb of aflatoxins. None of the survivors showed evidence of toxicity or carcinogenicity. Furthermore, at a level of 1.8 ppm, which was toxic to most animals, the few survivors after 3 years showed no evidence of neoplastic or preneoplastic changes in any tissue. O'Gara (108) also found that monkeys treated with pure aflatoxin B_1 by oral or intraperitoneal dosing showed evidence of toxicity, but not carcinogenicity after 2 years or more of dosing. Therefore it appears that, while monkeys are very susceptible to the hepatotoxic actions of aflatoxins, they are relatively less responsive to the carcinogenic action, as compared with other sensitive species such as the rat and trout. This conclusion has obvious importance in assessing human responsiveness.

VII. SIGNIFICANCE FOR MAN

From this background information on various properties of aflatoxins, several obvious conclusions can be drawn that are relevant to an assessment of their possible importance as health hazards to man. It is clear that these compounds are potent, toxic agents to a wide spectrum of biological systems, and they are carcinogenic to some fish, birds, and mammals.

Despite the fact that significant differences in responsiveness are known to exist among mammalian species, it is reasonable to assume that man might respond to either acute or chronic effects of the toxins in the event that exposure took place through contamination of dietary components. Furthermore, it seems reasonable to assume that the character of human response might vary depending upon many factors, such as age, nutritional status, concurrent exposure to other toxic agents (e.g., herbal medicines), genetic factors, concurrent illness (e.g., viral hepatitis or parasitic infestation), as well as the dose and duration of exposure. Thus in the absence of direct evidence on human responsiveness to aflatoxin, and the improbability of such information becoming available, assessment of risk must be based on indirect evidence linking exposure to disease incidence. Evidence on this point is still fragmentary, but enough information has accumulated to permit a more thorough analysis of the problem than was hitherto possible.

It has become evident that there is much less risk of exposure to aflatoxins in technologically developed countries than in developing areas. The lower risk in developed areas is attributable to the combined effects of several factors that contribute to prevention of contamination. The use of such agricultural practices as rapid post-harvest drying of crops and controlled storage conditions tend to reduce mold damage in general and therefore reduce the risk of aflatoxin contaminations (109). For example, the apparent success of these measures in minimizing aflatoxin contamination is illustrated by results of surveys of cereal grains and soybeans in the United States for the presence of aflatoxins (110,111). Low levels (traces to 19 ppb) were found in 2 of the 531 wheat samples; 6 of the 533 sorghum samples; and 3 of the 304 oat samples. Positive findings were reported in 35 of the 1311 samples of corn and in 2 of the 866 samples of soybeans. In all cases, contamination occurred in the lowest grades of storage grain or oilseeds; they would not have been used for food manufacture.

Efforts of industry, regulatory, and governmental agencies have been combined effectively in devising selection and processing methods to minimize the risk of aflatoxin contamination of human foods in the United States and other countries (112).

In regions not equipped technologically to apply such practices, the risk of aflatoxin exposure is clearly much greater, as indicated by a number of recent findings. Since the discovery of aflatoxins, sporadic reports have appeared of the finding of contaminated samples of human dietary staples collected in various parts of the world (113,114). Although these findings suggested that the problem was widespread geographically, they provided little useful information on human exposure, because samples were generally collected in random fashion and it was unknown whether they would actually have been consumed. Recently, however, two studies have been completed that involved more thorough and systematic collection and analysis of samples and therefore provided more meaningful information on exposure.

In one of these studies, Alpert *et al.* (115) determined aflatoxin levels in 480 food samples collected in various parts of Uganda during 1966–1967. Among these samples, 29.6% contained detectable levels of aflatoxins, and 3.7% contained more than 1 ppm. The frequency of contamination and also levels of aflatoxins present tended to be highest in beans, maize, and sorghum, although most dietary staples were contaminated to some extent. The distribution of contaminated samples by district was particularly interesting, because significant differences in frequency and levels of toxins were observed among districts. Comparison of distribution of contamination with disease incidence suggested an association of the two parameters.

In a somewhat more extensive survey in Southeast Asia, we have examined foods and foodstuffs from the Philippines, Malaysia, Hong Kong, and Thailand for aflatoxin contamination (65,116).

Peanut samples collected in the Philippines, Malaysia, and Thailand were the most heavily contaminated. In Malaysia, 6% of 268 food samples contained aflatoxins but only at low levels; only 7 samples contained more than 10 μg aflatoxin B_1/kg (10 ppb). Only 22 of 878 food samples (3%) collected in Hong Kong over a 2-year period contained detectable levels of the aflatoxin (116). The concentrations were low (trace to 90 μg/kg aflatoxin B_1).

More complete studies made in Thailand indicate that the risk of exposure is greater in that area.

Over a 23-month period, 2180 samples of more than 170 varieties of foods and foodstuffs were collected from markets, mills, warehouses, distributors, farms, and homes throughout Thailand. A total of 204 (9%) of these samples were contaminated with aflatoxin (116). Peanuts and corn were the most frequently contaminated foodstuffs (49 and 35% of the samples contaminated, respectively), while 11% of millet and dried chili pepper samples contained aflatoxins. Total aflatoxin concentrations in what seemed superficially to be wholesome foods or foodstuffs destined for human consumption were as high as 772 ppb in dried fish, 966 ppb in dried chili peppers, 2.7 ppm in corn, and more than 12 ppm in peanuts. The mean aflatoxin B_1 concentration in 215 peanut samples examined was 426 ppb.

Such extensive contamination of market foods and foodstuffs warranted a dietary survey to measure more accurately the amounts of aflatoxins that were actually consumed. Based on results from the market study, three areas in Thailand representing suspected high, intermediate, and low levels of contamination were chosen for such a dietary survey (116). Three representative villages were selected in each area, and 16 families in each village, randomly chosen, constituted the survey population. A portion of the diet of each family was collected and assayed for aflatoxins for 3 separate 2-day intervals over a period of 1 year. Samples of as many prepared foods as were feasible were collected, and a sample of cooked rice was obtained from nearly every meal of each family. In most cases, 30–50% of the prepared foods in the diet were analyzed. In the Singburi area (suspected of high aflatoxin contamination), an average from 73 to 81 mg total aflatoxins per kilogram body weight on a family basis were consumed each day. In the second area, Ratburi, the average was 45–77 mg/kg body weight per day, and in Songkla (the area of lowest contamination) the average was 5–8 mg/kg body weight per day. Maximum toxin consumption in Singburi occurred during the rainy season and in Ratburi during the hot season. In a few instances it was possible to measure the amount of aflatoxins consumed in 1 day by individual members of survey families. The highest single day's consumption measured was for a 75-year-old woman who ingested 1072 mg total aflatoxins per kilogram body weight; the contaminated food in this case was cooked rice, probably leftover and resteamed several times.

The most heavily contaminated foods were a dish containing cabbage fried with pork and garlic (1.3 ppm total aflatoxins), sun-dried fish (0.8 ppm total aflatoxins), and a dish containing fresh shrimp fried with pork, garlic, and chili peppers (0.4 ppm total aflatoxins). While it would not be anticipated that these foods (none of which contained peanuts) would be among the most heavily contaminated dishes, the data are consistent with the results from the market study indicating significant contamination in garlic, dried chili peppers, and dried fish—all common ingredients in Thai foods.

Peanuts were uncommon to many prepared foods in Thailand. Indeed, peanuts and peanut products were eaten usually as snacks away from home, and it was not possible to determine directly and systematically aflatoxin intakes from peanut sources for this reason. However, several interesting observations were made. For example, children consumed peanuts more frequently and in larger quantity on a body weight basis than did adults. Children aged about 3 years, weighing 10–12 kg, were observed to eat up to 250 g of shelled boiled peanuts in 1 day during harvest season. The aflatoxin B_1 content of boiled peanuts has been as high as 6.5 ppm; thus a 10-kg child could consume as much as 163 μg aflatoxin B_1 per kilogram body weight in 1 day by eating 250 g of such highly contaminated peanuts.

Although the absolute exposures to the aflatoxins appear to be quantitatively small, the potency of these compounds as carcinogens in animals must be remembered to put these data into perspective. The highest values, in Singburi, based on the yearly average total aflatoxin consumption, amount to 20–30% of comparable intake values that induce nearly 100% tumor incidence in rats following continuous exposure (117). Because these are family averages, exposures to individual family members are undoubtedly higher.

It is evident from these data that exposure of human populations is highly variable. In assessing the possible health significance of this fact, it is necessary to determine whether any distribution pattern of diseases can be associated with the pattern of exposure. Because of the predilection of the liver as the target tissue for aflatoxins in animals, with respect to both toxic and carcinogenic actions, considerable attention has been given to the possibility that aflatoxins may be involved in the etiology of human liver cancer. This form of cancer occurs with a highly regional pattern, being especially high in Africa and Asia and low in Western countries. For this reason, the hypothesis that aflatoxins may be involved as causative agents has been put forth by a number of authors (118–121).

The amount of direct evidence bearing on this hypothesis is limited, the most direct being provided by the aforementioned studies in Africa and Asia. Alpert *et al.* (122, 123) studied the incidence of liver cancer in Uganda, with particular reference to the geographic pathology of case distribution in various regions and tribes of the country.

The data on incidence and its distribution combined with the pattern of aflatoxin levels in foods (described previously) showed that the frequency of aflatoxin contamination was particularly high in provinces with a high hepatoma incidence or where cultural and economic factors favored the ingestion of moldy foods. These data tend to support the hypothesis that the two parameters are associated.

In the Thailand study (previously described), at the same time as the diet survey for aflatoxin ingestion was being conducted, the incidence of primary liver cancer was also being measured in the high and low aflatoxin regions (116). In a population of 99,537 people (all ages) in the Ratburi area (high aflatoxin), six residents died with primary liver cancer within the 12-month study period; 4 of the cases were residents of one town with a population of 32,634. The incidence for the total Ratburi area was therefore 6.0 new cases per 100,000 people per year and 12.3 for the one town. In the Songkla (low aflatoxin) area, the incidence was 2.0 new cases per 100,000 people per year (base population 97,867). The average age of liver cancer cases in Thailand was in the latter half of the fifth decade (about 20 years younger than those seen in North American and European populations).

Results of the Thailand study also suggest an association between aflatoxin consumption and incidence of primary liver cancer. However, none of the studies conducted to date can be regarded as having established unequivocally that aflatoxins are causative agents in the induction of liver cancer.

Although the incidence of liver cancer in a given population can be measured with some accuracy, it is impossible to estimate aflatoxin intake during tumor induction because of the long lag time for tumor induction.

Whether today's diet reflects contamination levels of diets 10 to 30 years ago is uncertain. Attempts to associate aflatoxin exposure with an acute liver disease might therefore be a more fruitful approach to ascertaining effects of the toxin in man.

A number of reports offer circumstantial suggestions of acute toxicity involving aflatoxin exposure in humans. Ling *et al.* (124) describe an outbreak of toxicity, involving 26 individuals with 3 deaths, that took place in Taiwan in 1967. Although few definitive data are available, aflatoxin involvement was suggested on the basis that aflatoxin B_1 and an aflatoxin-producing mold were found in the rice sample alleged to have been eaten by the persons involved. Serck-Hanssen (125) describes fatal hepatitis in a 15-year-old Ugandan boy who showed histopathologic lesions of the liver said to be identical with those induced by aflatoxins in African monkeys. It was also alleged that the boy had consumed cassava containing 1.7 ppm aflatoxin B_1 and was suggested that an amount equal to a lethal dose in monkeys could have been consumed. Robinson (126) summarizes the circumstantial evidence suggesting that aflatoxins may be involved in the etiology of infantile cirrhosis in India.

More extensive data have come from studies in Thailand. There is in Northeastern Thailand a form of acute encephalopathy and fatty degeneration of the liver, kidneys, and heart prevalent in children aged 1–14 years (127). Apparently, several hundred children each year are affected by this disease that was fatal in about 80% of the cases identified in 1968. The disease is variously referred to as Udorn encephalopathy; encephalopathy and fatty degeneration of the viscera; or Reye's syndrome (128); it is characterized by a short prodrome, vomiting, convulsions, coma, and a rapidly fatal course. Autopsy reveals intense cerebral edema, and abnormal fatty accumulation within hepatocytes, renal tubular epithelium, and myocardial fibers. The incidence of the disease follows a seasonal pattern, reaching a maximum figure immediately following the rainy season.

Attempts to associate a virus with this disease in Thailand have so far been unsuccessful (129). No evidence has been obtained to suggest mushroom poisoning. However, in a few cases samples of cooked foods eaten by patients shortly before onset of illness were analyzed qualitatively for aflatoxin contamination (130). The samples contained not only aflatoxins but also toxigenic strains of *Aspergillus glaucus, Aspergillus ochraceous,* and *Aspergillus niger* as well as *Aspergillus flavus.*

Suspecting aflatoxin as a factor in the etiology of Reye's syndrome, crystalline aflatoxin B_1 was given orally to 2-year-old macaque monkeys at single doses up to 40.5 mg/kg body weight (131) in an attempt to reproduce the human disease. Doses near or in excess of the LD_{50} (7.8 mg/kg body weight) produced clinical and histopathological responses strikingly similar to those of Reye's syndrome in children. Significant levels of aflatoxin B_1 were detected in the bile at the time of death in two monkeys that received 13.5 mg aflatoxin B_1/kg body weight (150 μg B_1/ml bile, death 73 hours after administration; 125 μg B_1/ml bile, death after 148 hours); and in one monkey that received 40.5 mg B_1/kg body weight (163 μg B_1/ml bile, 80 hours after death). Aflatoxin B_1 was also found in brain, liver, kidney, and heart blood from animals that received these two high doses.

Similar analyses were then done on autopsy specimens from 22 cases of Reye's syndrome and 7 cases of children and adolescents resident in the same area, but who died from causes other than Reye's syndrome (130). Aflatoxin B_1 was found in brain, liver, kidney, stool, stomach, and intestinal contents, and in bile, blood, and urine; a compound suspected of being in aflatoxin B_2 was found in one specimen each of brain, urine, and stomach contents; a compound with chromatographic properties similar to aflatoxin M_1 was found in two urine specimens. In two cases the concentrations of aflatoxin B_1 in the liver were of the same magnitude as those in livers of monkeys dying 3 days after receiving a dose from 13.5 to 40.5 mg aflatoxins per kilogram body weight.

This information together with the data on liver cancer incidence is strongly suggestive of association between aflatoxin exposure and acute or chronic disease in man. However, unequivocal causal relationships obviously have not been established. Nonetheless, the data provide sufficient indication of possible risk to warrant continued investigation of public health hazards related to aflatoxins and other mycotoxins.

References

1. L. A. Goldblatt, Ed., *Aflatoxin—Scientific Background, Control, and Implications,* Academic Press, New York, 1969.
2. J. Forgacs and W. T. Carll, *Advan. Vet. Sci.,* **7**, 274 (1962).
3. G. Buchi and I. D. Rae, "The Structure and Chemistry of the Aflatoxins," in *Aflatoxin—Scientific Background, Control, and Implications,* L. A. Goldblatt, Ed., Academic Press, New York, 1969, pp. 55–76.

4. T. Asao, G. Buchi, M. M. Abdel-Kader, S. B. Chang, E. L. Wick, and G. N. Wogan, *J. Amer. Chem. Soc.,* **85,** 1706 (1963).
5. T. Asao, G. Buchi, M. M. Abdel-Kader, S. B. Chang, E. L. Wick, and G. N. Wogan, *J. Amer. Chem. Soc.,* **87,** 882 (1965).
6. C. W. Holzapfel, P. S. Steyn, and I. F. H. Purchase, *Tetrahedron Lett.,* 2799 (1966).
7. M. S. Masri, R. E. Lundin, J. R. Page, and V. C. Garcia, *Nature,* **215,** 753 (1967).
8. J. Dalezios, G. N. Wogan, and S. R. Weinreb, *Science,* (1971) (in press).
9. M. F. Dutton and J. G. Heathcote, *Biochem. J.,* **101,** 21P (1966).
10. M. F. Dutton and J. G. Heathcote, *Chem. Ind. (London),* 418 (1968).
11. P. J. Andrellos and G. R. Reid, *J. Assoc. Offic. Agr. Chemists,* **47,** 801 (1964).
12. R. W. Detroy and C. W. Hesseltine, *Can. J. Biochem. Physiol.,* **48,** 830 (1970).
13. R. Allcroft, H. Rogers, G. Lewis, B. Nabney, and P. E. Best, *Nature,* **209,** 154 (1966).
14. R. Allcroft, "Aflatoxicosis in Farm Animals," in *Aflatoxin—Scientific Background, Control, and Implications,* L. A. Goldblatt, Ed., Academic Press, New York, 1969, pp. 237–264.
15. M. R. Gumbmann, S. N. Williams, A. N. Booth, P. Vohra, R. A. Ernst, and M. Bethard, *Proc. Soc. Exp. Biol. Med.,* **134,** 683 (1970).
16. H. F. Hintz, A. N. Booth, A. F. Cucullu, H. K. Gardner, and H. Heitman, Jr., *Proc. Soc. Exp. Biol. Med.,* **124,** 266 (1967).
17. H. F. Hintz, H. Heitman, Jr., A. N. Booth, and W. E. Gagne, *Proc. Soc. Exp. Biol. Med.,* **126,** 146 (1967).
18. B. H. Armbrecht, W. T. Shalkopp, L. D. Rollins, A. E. Pohland, and L. Stoloff, *Nature,* **225,** 1062 (1970).
19. R. Allcroft and R. B. A. Carnaghan, *Vet. Record,* **75,** 259 (1963).
20. R. Allcroft and M. C. Lancaster (1969), (in press); *Aflatoxin—Scientific Background, Control, and Implications,* L. A. Goldblatt, Ed., Academic Press, New York, 1969, p. 262.
21. W. H. Butler, "Aflatoxicosis in Laboratory Animals," in *Aflatoxin—Scientific Background, Control, and Implications,* L. A. Goldblatt, Ed., Academic Press, New York, 1969, pp. 223–236.
22. G. N. Wogan, *Bacteriol. Rev.,* **30,** 460 (1966).
23. G. N. Wogan, M. Akao, and K. Kuroda, *Cancer Res.,* (1971) (in press).
24. J. E. Halver, "Aflatoxicosis and Trout Hepatoma," in *Aflatoxin—Scientific Background, Control, and Implications,* L. A. Goldblatt, Ed., Academic Press, New York, 1969, pp. 265–306.
25. D. H. Bauer, D. J. Lee, and R. O. Sinnhuber, *Toxicol. Appl. Pharmacol.,* **15,** 415 (1969).
26. Z. H. Abedi and P. M. Scott, *J. Assoc. Offic. Anal. Chemists,* **52,** 963 (1969).
27. P. M. Newberne and W. H. Butler, *Cancer Res.,* **29,** 236 (1969).
28. P. G. Tulpule, T. V. Madhavan, and C. Gopalan, *Lancet* **i,** 962 (1964).
29. T. V. Madhavan, P. G. Tulpule, and C. Gopalan, *Arch. Pathol.,* **79,** 466 (1965).
30. T. V. Madhavan, K. S. Rao, and P. G. Tulpule, *Indian J. Med. Res.,* **53,** 984 (1965).
31. W. F. J. Cuthbertson, A. C. Laursen, and D. A. H. Pratt, *Brit. J. Nutr.,* **21,** 893 (1967).
32. E. Alpert, A. Serck-Hanssen, and B. Rajagopolan, *Arch. Environ. Health.* **20,** 723 (1970).

33. M. G. Deo, Y. Dayal, and V. Ramalingaswami, *J. Pathol. Bacteriol.,* **101,** 47 (1970).
34. M. S. Legator, "Biological Assay for Aflatoxins," in *Aflatoxin—Scientific Background, Control, and Implications,* L. A. Goldblatt, Ed., Academic Press, New York, 1969, pp. 107–151.
35. S. Juhasz and E. Greczi, *Nature,* **203,** 861 (1964).
36. M. S. Legator and A. Withrow, *J. Assoc. Offic. Anal. Chemists,* **47,** 1007 (1964).
37. D. A. Dolimpio, M. Legator, and C. Jacobson, *Proc. Soc. Exp. Biol. Med.,* **127,** 559 (1968).
38. J. Gabliks, W. Schaeffer, L. Friedman, and G. N. Wogan, *J. Bacteriol.,* **90,** 720 (1965).
39. M. R. Daniel, *Brit. J. Exp. Pathol.,* **46,** 183 (1965).
40. R. H. Smith, *Biochem. J.,* **88,** 50P (1963).
41. R. H. Smith, *Biochem. J.,* **95,** 43P (1965).
42. A. J. Zuckerman, K. R. Rees, D. Inman, and V. Petts, *Nature,* **214,** 814 (1967).
43. A. J. Zuckerman, K. N. Tsiquaye, and F. Fulton, *Brit. J. Exp. Pathol.,* **48,** 20 (1967).
44. E. H. Harley, K. R. Rees, and A. Cohen, *Biochem. J.,* **114,** 289 (1969).
45. H. Savel, B. Forsyth, W. Schaeffer, and T. Cardella, *Proc. Soc. Exp. Biol. Med.,* **134,** 1112 (1970).
46. H. R. Burmeister and C. W. Hesseltine, *Appl. Microbiol.,* **14,** 403 (1966).
47. T. Arai, T. Ito, and Y. Koyama, *J. Bacteriol.,* **93,** 59 (1967).
48. J. B. Wragg, V. C. Ross, and M. S. Legator, *Proc. Soc. Exp. Biol. Med.,* **125,** 1052 (1967).
49. M. S. Legator, S. M. Zuffante, and A. R. Harp, *Nature,* **208,** 345 (1965).
50. E. B. Lillehoj and A. Ciegler, *J. Bacteriol.,* **94,** 787 (1967).
51. E. B. Lillehoj, A. Ciegler, and H. H. Hall, *J. Bacteriol.,* **93,** 464 (1967).
52. E. B. Lillehoj and A. Ciegler, *J. Gen. Appl. Microbiol.,* **54,** 185 (1968).
53. V. M. Maher and W. Summers, *Nature,* **225,** 68 (1970).
54. T. Ong, *Mutation Res.,* **9,** 615 (1970).
55. R. Schoental and A. F. White, *Nature,* **205,** 57 (1965).
56. I. Slowatizky, A. M. Mayer, and A. Poljakoff-Mayber, *Israel J. Botan.,* **18,** 31 (1969).
57. L. J. Lilly, *Nature,* **207,** 433 (1965).
58. H. S. Black and A. M. Altshul, *Biochem. Biophys. Res. Commun.,* **19,** 661 (1965).
59. B. Truelove, D. E. Davis, and O. C. Thompson, *Can. J. Botany,* **48,** 485 (1970).
60. H. de Iongh, R. O. Vles, and J. G. van Pelt, *Nature,* **202,** 466 (1964).
61. J. A. van der Linde, A. M. Frens, and G. J. van Esch, "Experiments with Cows Fed Groundnut Meal Containing Aflatoxin," in *Mycotoxins in Foodstuffs,* G. N. Wogan, Ed., M.I.T. Press, Cambridge, Mass., 1965, pp. 247–249.
62. J. Nabney, M. B. Burbage, R. Allcroft, and G. Lewis, *Food Cosmet. Toxicol.,* **5,** 11 (1967).
63. G. N. Wogan, G. S. Edwards, and R. C. Shank, *Cancer Res.,* **27,** 1729 (1967).
64. G. N. Wogan, "Metabolism and Biochemical Effects of Aflatoxins," in *Aflatoxin—Scientific Background, Control, and Implications,* L. A. Goldblatt, Ed., Academic Press, New York, 1969, pp. 152–186.
65. T. C. Campbell, J. P. Caedo, J. Bulatao-Jayme, L. Salamat, and R. W. Engel, *Nature,* **227,** 403 (1970).

66. M. B. Sporn, C. W. Dingman, H. L. Phelps, and G. N. Wogan, *Science,* **151,** 1539 (1966).
67. J. I. Clifford and K. R. Rees, *Nature,* **209,** 312 (1966).
68. J. I. Clifford and K. R. Rees, *Biochem. J.,* **102,** 67 (1967).
69. J. I. Clifford, K. R. Rees, and M. E. M. Stevens, *Biochem. J.,* **103,** 258 (1967).
70. W. Lijinsky, K. Y. Lee, and C. H. Gallagher, *Cancer Res.,* **30,** 2280 (1970).
71. C. Frayssinet, C. Lafarge, A. M. De Recondo, and E. Le Breton, *Compt. Rend.,* **259,** 2143 (1964).
72. A. M. De Recondo, C. Frayssinet, C. Lafarge, and E. Le Breton, *Compt. Rend.,* **261,** 1409 (1965).
73. A. M. De Recondo, C. Frayssinet, C. Lafarge, and E. Le Breton, *Biochim. Biophys. Acta,* **119,** 322 (1966).
74. M. S. Legator, *Bacteriol. Rev.,* **30,** 471 (1966).
75. C. Lafarge, C. Frayssinet, and A. M. De Recondo, *Bull. Soc. Chim. Biol.,* **47,** 1724 (1965).
76. H. V. Gelboin, J. S. Wortham, R. G. Wilson, M. A. Friedman, and G. N. Wogan, *Science,* **154,** 1205 (1966).
77. R. S. Pong and G. N. Wogan, *Cancer Res.,* **30,** 294 (1970).
78. M. A. Friedman and G. N. Wogan, *Life Sci.,* **9,** 741 (1970).
79. D. Svoboda, H. Grady, and J. Higginson, *Amer. J. Pathol.,* **49,** 1023 (1966).
80. R. S. Pong and G. N. Wogan, *Biochem. Pharmacol.,* **18,** 2357 (1969).
81. G. N. Wogan and M. A. Friedman, *Arch. Biochem. Biophys.,* **128,** 509 (1968).
82. R. S. Pong and G. N. Wogan, *Biochem. Pharmacol.,* **19,** 2808 (1970).
83. M. C. Lancaster, F. P. Jenkins, and J. McL. Philip, *Nature,* **192,** 1095 (1961).
84. W. H. Butler and J. M. Barnes, *Food Cosmet. Toxicol.,* **6,** 135 (1968).
85. P. M. Newberne, "Carcinogenicity of Aflatoxin-Contaminated Peanut Meals," in *Mycotoxins in Foodstuffs,* G. N. Wogan, Ed., M.I.T. Press, Cambridge, Mass., 1965, pp. 187–208.
86. W. H. Butler and J. M. Barnes, *Nature,* **209,** 90 (1966).
87. R. B. Alfin-Slater, L. Aftergood, H. J. Hernandez, E. Stern, and D. Melnick, *J. Amer. Oil Chemists Soc.,* **46,** 493 (1969).
88. G. N. Wogan and P. M. Newberne, *Cancer Res.,* **27,** 2370 (1967).
89. R. B. A. Carnaghan, *Brit. J. Cancer,* **21,** 811 (1967).
90. S. M. Epstein, B. Bartus, and E. Farber, *Cancer Res.,* **29,** 1045 (1969).
91. W. H. Butler, M. Greenblatt, and W. Lijinsky, *Cancer Res.,* **29,** 2206 (1969).
92. G. N. Wogan, G. S. Edwards, and P. M. Newberne, *Cancer Res.* (1971) (in press).
93. T. V. Madhavan and C. Gopalan, *Arch. Pathol.,* **85,** 133 (1968).
94. P. M. Newberne and G. N. Wogan, *Toxicol. Appl. Pharmacol.,* **11** (1968).
95. P. M. Newberne, D. H. Harrington, and G. N. Wogan, *Lab. Invest.,* **15,** 962 (1966).
96. P. M. Newberne, A. E. Rogers, and G. N. Wogan, *J. Nutr.,* **94,** 331 (1968).
97. D. J. Lee, J. H. Wales, and R. O. Sinnhuber, *J. Nat. Cancer Inst.,* **43,** 1037 (1969).
98. P. M. Newberne, C. E. Hunt, and G. N. Wogan, *Exp. Mol. Pathol.,* **6,** 285 (1967).
99. P. M. Newberne and G. Williams, *Arch. Environ. Health,* **19,** 489 (1969).
100. C. M. Goodall and W. H. Butler, *Int. J. Cancer,* **4,** 422 (1969).
101. F. Dickens and H. E. H. Jones, *Brit. J. Cancer,* **17,** 691 (1964).
102. F. Dickens, H. E. H. Jones, and H. B. Waynforth, *Brit. J. Cancer,* **20,** 134 (1966).

103. R. O. Sinnhuber, J. H. Wales, J. L. Ayres, R. H. Engebrecht, and D. L. Amend, *J. Nat. Cancer Inst.*, **41,** 711 (1967).
104. R. O. Sinnhuber, D. J. Lee, J. H. Wales, and J. L. Ayres, *J. Nat. Cancer Inst.,* **41,** 1293 (1968).
105. R. O. Sinnhuber, D. J. Lee, J. H. Wales, M. K. Landers, and A. C. Keyl, *Federation Proc.,* **29,** A568 (1970).
106. J. L. Ayres, D. J. Lee, J. H. Wales, and R. O. Sinnhuber, *J. Nat. Cancer Inst.* (1971) (in press).
107. R. B. A. Carnaghan, *Nature,* **208,** 308 (1965).
108. R. W. O'Gara, *Proc. Amer. Assoc. Cancer Res.,* **9,** 55 (1968).
109. C. Golumbic and M. M. Kulik, "Fungal Spoilage in Stored Crops and Its Control," in *Aflatoxin—Scientific Background, Control, and Implications,* L. A. Goldblatt, Ed., Academic Press, New York, 1969, pp. 307–333.
110. O. L. Shotwell, C. W. Hesseltine, H. R. Burmeister, W. F. Kwolek, G. M. Shannon, and H. H. Hall, *Cereal Chem.,* **46,** 466 (1969).
111. O. L. Shotwell, C. W. Hesseltine, H. R. Burmeister, W. F. Kwolek, G. M. Shannon, and H. H. Hall, *Cereal Chem.,* **46,** 454 (1969).
112. C. J. Kensler and D. J. Natoli, "Processing to Ensure Wholesome Products," in *Aflatoxin—Scientific Background, Control, and Implications,* L. A. Goldblatt, Ed., Academic Press, New York, 1969, pp. 334–359.
113. G. N. Wogan, *Federation Proc.,* **27,** 932 (1968).
114. J. M. Barnes, *J. Appl. Bacteriol.,* **33,** 285 (1970).
115. M. E. Alpert, M. S. R. Hutt, G. N. Wogan, and C. S. Davidson, *Cancer Res.* (1971) (in press).
116. R. C. Shank, G. N. Wogan, J. B. Gibson, and J. S. Gordon, *Food Cosmet. Toxicol.* (1971) (in press).
117. G. N. Wogan and P. M. Newberne, *Cancer Res.,* **27,** 2370 (1967).
118. H. F. Kraybill and M. B. Shimkin, *Advan. Cancer Res.,* **8,** 191 (1964).
119. A. G. Oettle, *S. African J. Med. Sci.,* **39,** 817 (1965).
120. M. E. Alpert and C. S. Davidson, *Amer. J. Med.,* **46,** 325 (1969).
121. H. F. Kraybill and R. E. Shapiro, "Implications of Fungal Toxicity to Human Health," in *Aflatoxin—Scientific Background, Control, and Implications,* L. A. Goldblatt, Ed., Academic Press, New York, 1969, pp. 401–441.
122. M. E. Alpert, M. S. R. Hutt, and C. S. Davidson, *Lancet,* **1265** (1968).
123. M. E. Alpert, M. S. R. Hutt, and C. S. Davidson, *Amer. J. Med.,* **46,** 794 (1969).
124. K.-H. Ling, J.-J. Wang, R. Wu, T.-C. Tung, C.-K. Lin, S.-S. Lin, and T.-M. Lin, *J. Formosan Med. Assoc.,* **66,** 517 (1967).
125. A. Serck-Hanssen, *Arch. Environ. Health,* **20,** 729 (1970).
126. P. Robinson, *Clin. Pediat.,* **6,** 57 (1967).
127. C. Bourgeois, N. Keschamras, D. S. Comer, S. Harikul, H. Evans, L. Olson, T. Smith, and M. R. Beck, *J. Med. Assoc. Thailand,* **52,** 553 (1969).
128. R. D. Reye, G. Morgan, and J. Baral, *Lancet,* **ii,** 749 (1963).
129. L. Olson, C. Bourgeois, N. Keschamras, S. Harikul, C. Sanyakorn, R. Grossman, and T. Smith, *Amer. J. Diseases Children* (1971) (in press).
130. R. C. Shank, C. H. Bourgeois, N. Keschamras, P. Chandavimol, *Food Cosmet. Toxicol.* (1971) (in press).
131. C. H. Bourgeois, R. C. Shank, R. A. Grossman, D. D. Johnson, W. L. Wooding, P. Chandavimol, and P. Tanticharoenyos (accepted by *Lab. Invest.,* tentative to revision).

INDEX